Healthcare Ethics on Film

M. Sara Rosenthal

Healthcare Ethics on Film

A Guide for Medical Educators

 Springer

M. Sara Rosenthal
Program for Bioethics, Departments of
Internal Medicine, Pediatrics, and
Behavioral Science
University of Kentucky
Lexington, KY, USA

Markey Cancer Center Oncology
Ethics Program
University of Kentucky
Lexington, KY, USA

ISBN 978-3-030-48817-8   ISBN 978-3-030-48818-5   (eBook)
https://doi.org/10.1007/978-3-030-48818-5

This Springer imprint is published by the registered company Springer Nature Switzerland AG
The registered company address is: Gewerbestrasse 11, 6330 Cham, Switzerland

To my maternal grandparents,
Jacob Lander, M.D. (1910–1989), who would
have especially loved the film, The Doctor;
and Clara Lander, Ph.D. (1916–1978),
a scholar on plague literature, who would
have especially loved the film, Contagion.

Preface

The Marriage of Healthcare Ethics and Film

Healthcare ethics issues generally deal with healthcare delivery and healthcare access as a subset of societal "goods." They may be entangled with wider social and political problems, such as discrimination and distributive justice issues; more organizational systems problems due to organizational decision making, leadership, and policies; or professionalism and other behaviors, which then interferes with the overall delivery of health care to patients. Healthcare ethics dilemmas and clinical ethics dilemmas are distinct. Clinical ethics dilemmas deal with more direct healthcare provider–patient issues, or "bedside" issues that comprise end-of-life care; decision-making capacity issues; weighing therapeutic benefits against risks and side effects; clinical trials and experimental therapies. These topics are covered in my previous work, Clinical Ethics On Film (2018). When we think about "healthcare ethics" dilemmas, on the other hand, we are dealing with broader systemic issues that involve patient populations and healthcare delivery, such as healthcare access, health disparities, and hospital organizational cultures affecting medical professionalism, professional virtues, values, and integrity, including the disclosure of medical error. There are a myriad of films that demonstrate the nuances and dilemmas in healthcare ethics; some notable films in this collection, such as *And the Band Played On* (1993), originated from journalists documenting healthcare crises and helping to blow the whistle on cover-ups and poor decision-making affecting large populations of patients in the interests of public health. Some, such as *Sicko* (2007) and *Dirty Dancing* (1987), are original screenplays that start a national dialogue: They challenge societal attitudes and legislation that restrict healthcare access for millions. For example, *Sicko* (2007), a documentary advocacy film, led to U.S. healthcare reform becoming an election issue in 2008 and led to the *Affordable Care Act* (2010). *Dirty Dancing* (1987) is a reproductive justice film disguised as a fun musical that forced renewed reflection over abortion access as the children of the "Roe" generation were beginning to take hard-won reproductive rights for granted. Still, other films such as *Miss Evers' Boys*

(1997), *You Don't Know Jack* (2010), and *Something the Lord Made* (2004), are history of medicine screenplays adapted from other works surrounding specific incidents or characters that shaped healthcare debates. *The Doctor* (1991) is based on an autobiography in which a physician-turned patient questions medical professionalism and organizational aspects of medical training. Finally, *Contagion* (2011) and *21 Grams* (2003) deal with challenges in scarce resource contexts, such as solid organs for transplant and scarcity during pandemics (see Afterword).

This book arose from years of work I had done in the area of bioethics pedagogy using film to teach medical students about healthcare ethics principles and dilemmas. With limited human resources in my small bioethics program, as well as limited classroom time available to medical students and residents for ethics teaching, I began a long project of creating several online teaching modules that used over 100 film clips in documentary-styled modules, each with an approximate running time of an hour. Ultimately, this work grew into a popular elective bioethics course called "Bioethics on Film" for our healthcare trainees, including medical students, nursing students, allied health students, and distance learning. The requirement to teach healthcare ethics at academic medical centers evolved as requirements from the national councils that grant accreditation to academic medical centers. The Liaison Committee for Medical Education (LCME), which is responsible for medical school accreditation, began to require bioethics teaching in 2005. In the area of graduate medical education, since 2007, the Accreditation Council of Graduate Medical Education (ACGME) has included a set of core competencies in its Common Program Requirements for medical residents, and the requirement to teach ethics has become more rigorous. Among these competencies is professionalism.

However, medical professionalism is far less about knowledge of ethical principles then it is about *organizational ethics*. Professionalism must also be modeled, which may be difficult in certain hospital cultures. For these reasons, the film, *The Doctor* (1991), based on the memoir of a physician who only learned about "professionalism" after becoming a patient himself, is much more of a healthcare ethics film than it is a "clinical ethics" film. The epiphany for this "doctor patient" was that medical education required reform, and his realization that medical students model their mentors. If we don't train doctors properly in how to treat their patients with empathy and compassion, healthcare delivery overall suffers.

Bioethicists come from a variety of core disciplines including philosophy, law, medicine, social science, social work, religious studies, and many more. My core graduate discipline is medical sociology, which informs my theoretical frameworks and my work as a bioethicist; it also informs the content in the films I've selected for this collection and my discussion of the film's social and historical location. My undergraduate disciplines—literature and film—inform the sections of text devoted to the film's production and director's intentions with telling a story.

Using film in the teaching of bioethics is certainly not new (Friedman 1995). Many of my colleagues have been using film in their bioethics teaching for over 20 years, and there are several interesting bioethics film courses peppering different academic centers. What is new is the creation of a "healthcare ethics" film collection that is distinct from a general "bioethics film" collection. This book also complements my previous work, Clinical Ethics on Film , which comprises the core clinical ethics films surrounding autonomy and beneficence, as well as more nuanced topics in clinical ethics. This book, in contrast, is a companion work that comprises the systemic and broader societal issues that create healthcare ethics dilemmas.

What defines the films selected here as "healthcare ethics" films? The criteria for film selection in the book are as follows: As a *central theme*, the film involves a healthcare ethics issue or dilemma, even if the film does not take place in a healthcare setting. Films with only occasional scenes or minor subplots involving health care are excluded. Next, the film is about wider healthcare issues relevant to populations of patients, rather than just one patient's experiences, which is distinct from the "clinical ethics" films in my previous work. Finally, the film is about healthcare delivery, affecting population healthcare access or delivery in some way, rather than wider bioethics social justice and policy issues seen in films such as *Gattaca* (1997), *Soylent Green* (1973), or *District 9* (2009).

Another consideration is whether a film is timeless or dated. For example, for many years, popular healthcare ethics films discussed among bioethicists were *The Hospital* (1971) and *The Verdict* (1980). *The Hospital* was a classic film about hospital error and organizational ethics that served as satirical commentary on dramatic changes to American healthcare delivery in the wake of Medicare and Medicaid. At the same time, *The Hospital* depicts a rape scene (that the woman ultimately "enjoys"), in which the protagonist (the Chief of Medicine) rapes an attractive daughter of a patient. With a brilliant script by Paddy Chayefsky, it was once hailed as the "quintessential" healthcare ethics film. Yet the rape scene, and the ensuing romance between the Chief of Medicine and the patient's daughter, is so disturbing in a current context, medical educators would spend more time tied up in threatened litigation by outraged students than in doing any teaching, no matter how many "handouts" about 1970s context he/she creates. One would have to show it as a "timepiece" of the unacceptable and sexist hospital culture of yesteryear, with trigger warnings, and risk managers present in the room. And so, readers who wonder why this obvious choice is excluded have my answer here.

Similarly, *The Verdict* (1980), once considered pedagogical "malpractice" for exclusion on the topic of medical error, contains a disturbing scene in which the healthcare lawyer litigating the medical error case (Paul Newman, who won an Academy Award for his role) punches the leading lady in the face (Charlotte Rampling), after she betrays him. Even if we try to argue that she "deserved it," this scene will create so many problems with outraged students over a misperceived condoning of violence against women, its shelf life as a teaching film has unfortunately expired. Such films are akin to showing *Gone with the Wind* (1939) as a reliable Civil War film.

Another important element in my selection is the quality of the film. *Is it a good film* regardless of the above criteria? Several films in this collection won critical acclaim and awards. I've divided this book into three healthcare ethics genres: Part I, "Films About Medical Harms," comprising *Miss Evers' Boys* (1997); *And the Band Played On* (1993); and *You Don't Know Jack* (2010), examine the films with respect to violations of the Principle of Non-Maleficence, the specific obligation not to cause intentional harm. Part II, "Films About Distributive Justice and Healthcare Access" comprising *Sicko* (2007); *Dirty Dancing* (1987); *21 Grams* (2003); and *Contagion* (2011), examine the films with respect to the Principle of Justice. This section of films looks at access to basic healthcare, reproductive justice, resource allocation frameworks, and rationing. Finally, Part III, "Films About Professionalism and Humanism", comprising *The Doctor* (1991) and *Something the Lord Made* (2004), examine the films from the perspective of professionalism and humanism in medicine, in which certain virtues are an important feature of the healthcare profession.

Ultimately, this book is a crossover of healthcare ethics and film. It is important to note, however, this is about American healthcare *ethics* and not film theory or film studies. This book should be viewed as interdisciplinary. It offers rich content for both health care and film pedagogy. You will find the expected discussions surrounding the backstory on each film in this collection, and the various historical accidents that led to the film's making. Perhaps more unique about this book is the analytical frameworks used to discuss the selected films: medical sociology, history of medicine, and healthcare ethics. This is how I teach these films to my students, and how the films can be best understood as healthcare *ethics* films. These films not only serve as commentaries on healthcare ethics issues, but are also timepieces that speak to a particular point in American medical and cultural history. Thus, each film in this collection is discussed within its social location and medical history context in addition to the specific clinical ethics themes expressed.

As a result, the healthcare themes of many of the films in this collection spoke differently to their original theatre audiences upon each film's release than to the audiences watching today. For example, a 1987 musical about a summer romance between a Jewish girl at the Catskills and her dance instructor—who gets caught up in a botched abortion disaster—spoke differently to a 1987 audience who thought abortion access fights were over, and were watching the film as an historical timepiece. Today, abortion access is greatly restricted, and thus, the film has renewed significance. Similarly, a 1993 audience watching *And the Band Played On* prior to the widespread access of AZT and AIDS drug cocktails, would have a different experience than today's healthcare trainee audience, which largely finds *And the Band Played On* to be a shocking piece of recent medical history about which they are first learning. I anticipate that for the healthcare reader, this book contains much more content about the film itself, rather than just using the film as a vehicle to debate bioethics issues (see Shapsay, S., *Bioethics at the Movies*, for example). Among the nine films discussed in this book through a dedicated chapter, most are based on works of nonfiction. As for the fictional films, such as *Contagion, 21 Grams*, and *Dirty Dancing*, they are inspired by wider healthcare debates. Friedman notes that for healthcare students:

[W]hile fictional films play an important role in my ethics classes, they function mainly to motivate students to examine their feelings about a particular issue. The artistry of such works—the narrative, visual, aural and dramatic elements that coalesce to make them such effective pedagogical tools—rarely enters our discussions, since the films remain a pathway to moral dilemmas rather then focus of critical analysis: the medium is certainly not the message.

However, regardless of whether films are based in fiction or nonfiction, how a director tells a story visually is part of the story, and healthcare students may still benefit from such discussions.

For the film scholar, this book has less content about the technical aspects of the film, but will medically locate the film both historically and sociologically, which I expect will broaden a film scholar's technical analysis of the film when engaged in such discussion with students.

Ultimately, this book should serve as a unique collection of healthcare ethics films for the myriad of disciplines attracted to it. Whether you are planning a course on healthcare films as a film studies scholar, planning a film course as a healthcare scholar, or like to buy film books as a film buff, my hope is to provide you with a robust collection for teaching.

Lexington, USA M. Sara Rosenthal

Reference

1. Friedman, L. (1995). See me, hear me: Using film in health-care classes. *Journal of Medical Humanities, 16,* 223–228.

Acknowledgements

I wish to thank Kimberly S. Browning, MPH, for her editorial assistance; the editorial and production teams at Springer for their professionalism and patience during the COVID-19 pandemic; and finally, Kenneth B. Ain, M.D., Professor, Department of Internal Medicine, University of Kentucky, and also my spouse and best friend, watched endless hours of films with me and provided helpful insights and resources surrounding this book's content.

Contents

Part I
Films About Medical Harms

Chapter 1
The Tuskegee Syphilis Study: *Miss Evers' Boys* (1997)

Every basic lecture in healthcare ethics, health disparities, bioethics, public health or research ethics must include some mention of the Tuskegee Syphilis Study—a study funded by the U.S. Public Health Service that ran from 1932 until 1972[1], when it was finally investigated, although the study did not formally close until March 1973 (Reverby 2009). The study followed over 400[2] African American males with syphilis but withheld the standard of care treatment, including penicillin when it became available in the late 1940s. This infamous unethical research protocol exploiting African American males with syphilis was, in many ways, like the disease it studied. It had an acute phase, a latency stage, and then a final stage where it entered the entire "body politic" of civil rights, racial disparities and healthcare delivery.

The first formal apology for the Tuskegee Syphilis Study (hereafter referred to as the Tuskegee study) by a sitting U.S. President did not occur until May 16, 1997—when President Bill Clinton apologized on behalf of the United States to the study subjects and their families. It was the film, *Miss Evers' Boys,* that ultimately coincided with the apology suggested by the 1996 Tuskegee Study Committee (see under Healthcare Ethics Issues). This 1997 film is based on the 1992 Pulitzer Prize-nominated play of the same name by physician and playwright, David Feldshuh. Feldshuh stated that the play was "suggested" by the first definitive scholarly book about the study, Bad Blood (Jones 1981). The film was released February 22, 1997,

---

[1]In virtually every writing on the Tuskegee study, its length is cited as 40 years, from 1932–1972. That is not exactly accurate. The study was "outed" in the press July 25, 1972, investigated by a federal government panel (chartered August 1972) and recommended to be shut down in October 1972, with its final report issued in April 1973. Congressional hearings were held in February and March of 1973 (a.k.a. the "Kennedy hearings"), and it was not until these hearings concluded that the study was formally closed down, and the men administered full treatment. In this chapter, I will continue to refer to the oft-quoted "1932–72" time frame as the 40-year period the study ran unimpeded without oversight.

[2]See under History of Medicine section for the trial enrollment details, as well as Reverby 2009, Appendix C for a full list of the subjects' names.

© Springer Nature Switzerland AG 2020
M. S. Rosenthal, *Healthcare Ethics on Film*,
https://doi.org/10.1007/978-3-030-48818-5_1

during the 21st Black History Month; Black History month was officially declared in 1976 by President Ford; it began as "Negro History Week" in 1926 (Zorthian 2016).

The film juxtaposes the Congressional hearings held in 1973 (presented in the film as "1972") with a 40-year timeline of the study itself. The film's central character is based on the actual public health nurse involved over the duration of the study— Eunice Rivers (in her later years, Eunice Rivers Laurie)—whose surname is changed to "Evers" in the film; the film is from her point of view. The film also disguises the names of other medical providers involved in the study through the years by creating composite fictional characters. Most importantly, the film presents a "composite protocol" based on the public record that included many conflicting reports, accounts and oral histories from multiple perspectives of those involved. Historians point out that there was no official or formal written protocol to review in the final analysis (Jones 1981, 1993; Washington 2007; Reverby 2009), but there were several papers published on the study's findings, while Nurse Rivers' involvement was mentioned in a 1953 paper (See Jones 1993, pp. 281–282 for a complete list of published papers on the Tuskegee study).

This chapter will discuss the origins of the film *Miss Evers' Boys* within its own time frame (1980s and 1990s), and then unpack the Tuskegee study protocol within the social location of its 40-year time span, exploring the roles of key study investigators, its African American collaborators and the study's whistleblowers. The chapter will next discuss this study within a history of medicine context that considers: (1) African American medical care and disturbing, yet prevailing misconceptions about African American physiology; (2) the treatment of syphilis and clinical equipoise (Freedman 1987) over optimal therapies; and (3) the clinical rationale for observational protocols in general—*many of which are still being done in multiple settings today*, including "observation-only" trials of various cancers. An important misconception of this topic is to solely cite this study as a singular example of an "observational protocol" without treatment. In fact, as I'll discuss in this chapter, it's important to stress to students that variations of questionable observational protocols are ongoing in many different disease contexts.

Finally, this chapter will explore the multiple healthcare ethics issues entangled in the Tuskegee study, comprising violations of the "do no harm" concepts embedded in the Principle of Non-Maleficence, as well as other research ethics considerations; organizational ethics considerations; clinical ethics considerations; and public health ethics considerations. When planning a curriculum that includes the Tuskegee study, screening the film *Miss Evers' Boys* helps to provide a context for how the U.S. Public Health Service (USPHS) prospective protocol called "The Tuskegee Study of Untreated Syphilis in [the] Male Negro", an observational study on the natural history of syphilis, became a flashpoint for civil rights violations, medical harms and abuses in medicine, and a framework from which to examine deep racial and health disparities and the social production of illness. Educators should also spend some time on explaining the evolving vernacular surrounding reference to an African American.

At different points in history, the term "Negro", "Black" and "African American" have been used, but there may be students who still find reference to the term "Negro" jarring. For clarity, the terms "Negro" or "Black" are used in this chapter only when citing other documents or organizations; otherwise, the term African American is used.

Origins of *Miss Evers' Boys*

The play and later, film, *Miss Evers' Boys* was inspired by the book, <u>Bad Blood</u> (Jones 1981), which was written by Southern historian and bioethicist James H. Jones, and was the first rigorous accounting of the Tuskegee study. The book's title referred to the vernacular used to deceive study subjects to enroll in the study. Subjects were told that enrollment in the study was for treatment of "bad blood"—a general term that meant everything from high blood pressure to arthritis; few study subjects understood they had syphilis, or that their ailments were related to syphilis specifically. Jones' book brought into focus that the Tuskegee study was a collaborative effort between white investigators and African American healthcare providers, who until that point, had been dismissed as either not responsible, not fully aware of the study, or "victims" when that was not the case. When Jones' book came out in 1981, it soon became a public health ethics classic, and began to be included in various healthcare-related curricula—particularly in the areas of public health and bioethics. David Feldshuh, the playwright of "Miss Evers' Boys" was a unique audience for Jones' book. He had initially studied acting before going to medical school. Feldshuh read <u>Bad Blood</u> when he was doing his residency in Emergency Medicine, and worked on the play throughout the 1980s, at the dawning of the AIDS epidemic that gripped medicine at the time. The play "Miss Evers' Boys" was published in 1989, winning The New American Play Award (Isenberg 1990) and began to debut in 1990 at various theaters (see further); it focuses on the experiences of the African American healthcare providers involved in the study as collaborators. Some scholars have criticized this work because it is a fictionalized and speculative account of the Tuskegee study with composite characters, and is not intended as a docudrama. The play and film have been described as an "interpretation" of the Tuskegee study. Yet the more time passes, the more "truth" there is to Feldshuh's interpretation. I argue that the standard critique of his play, and the film, *Miss Evers' Boys* is misplaced, and misses the merit, and point, of this work. The play (and film) was designed to address the public's burning question of "what were they thinking?" in the absence of any definitively satisfying historically recorded answers by the African American healthcare providers who served as collaborators. In definitive scholarly works on the study (Jones 1981, 1993; Washington 2007; Reverby 2009), "what were they thinking?" is discussed speculatively, too. The clear answer to that question was absent from history because the African American champions of the study from the Tuskegee Institute (Drs. Moton and Dibble) were deceased when the study was halted in 1972. Nurse Rivers Laurie (1899–1986) was 73 years old in 1972, and had been interviewed by a government-appointed panel at that time (see further), but the panel

Chair, Broadus Butler, decided to destroy her interview tapes presumably because they were deemed problematic (Washington 2007; Reverby 2009). (See under Social Location.) In the Spring of 1977, Rivers was interviewed privately by James Jones for his book (Jones 1993), but the transcripts were never made public. Rivers gave her last interview on October 10, 1977 to A. Lillian Thompson for The Black Women Oral History Project (Hill 1991). The Thompson interview transcripts have been used and analyzed by multiple scholars, such as Susan Smith (Smith 1996), but the entire oral history project, including the Lillian Thompson interview of Rivers, is now online (Harvard Library 2018). At the time of these interviews, Nurse Rivers Laurie was 77, had tired after about an hour, but her interviews with both Jones and Thompson vindicates much of what is portrayed in the play and film (see further under Social Location). Rivers was also interviewed in 1977 by Dr. Dibble's widow, Helen Dibble, who was a librarian, along with a Tuskegee University archivist, Daniel Williams (Reverby 2000).

Ultimately, since the July 25, 1972 Associated Press article broke the story (Heller 1972), scholars have spent almost 50 years trying to answer the "what were they thinking?" question to get to the "truth" of the Tuskegee study protocol. The play and later film, *Miss Evers' Boys*, offers a reasonable and important answer from the perspective of the African American healthcare providers who participated in the Tuskegee study. Current viewers who are looking back from a twenty-first century lens must remember that Moton, Dibble and Rivers were professionals living in the Jim Crow South, in a pre-Civil Rights era, when lynching was a commonplace activity championed by law enforcement at the time (see further under the Social Location section).

The Jones Book on the Tuskegee Syphilis Study (1977–1981)

The only other major scholarship on the Tuskegee study prior to Jones' book was a 1978 article by Allan Brandt, titled "Racism and Research: The Case of the Tuskegee Syphilis Experiment" (Brandt 1978). Brandt's article was published just four years after the government reached a settlement with the harmed survivors and families of the study (see further).

James H. Jones was the first medical historian to write a book on the Tuskegee study, and his interest in the topic was serendipitous. While looking at PHS archives in the late 1960s for a planned book he wanted to write about Alfred Kinsey's sex research, he came across some correspondence from the 1930s about the Tuskegee study, not really sure what he was reading, but made a mental note that it was interesting (Reverby 2009). Then, in 1972, when he heard about the study in the mainstream press while driving to Boston to start a postdoctoral position at Harvard university—a research position that was supposed to focus on his Kinsey book—he decided to write about the Tuskegee study instead (Reverby 2009). It's understandable that his interest and knowledge about Kinsey, whose research focused on the sexual habits of Americans, would make for an easy transition to writing about a syphilis study. After his Harvard position ended, Jones accepted a full-time position at the National Endowment for the Humanities; while there, he contacted attorney Fred Gray, who was amidst a class action suit on behalf of the study survivors, and they struck a "grand bargain". Jones had something Gray wanted, and Gray had something Jones wanted. Jones would give Gray copies of the 1930s PHS correspondence he had, while Gray would give him access to all the Centers for Disease Control (CDC) original files on the Tuskegee study he had obtained. Unbeknownst to Jones, Brandt also had the same PHS documents from the 1930s tracking the origins of the study, and used them for his 1978 article in the *Hastings Center Report*, including the materials from the congressional hearings just a few years prior (Brandt 1978). Brandt referenced the participation of African American professionals only in a footnote. Brandt's paper argued that the congressional hearings focused on informed consent, which was not the main transgression of the Tuskegee study. Brandt argued that the study's premise was grounded in racism. Jones' book picked up where Brandt left off, and would become the definitive work on the Tuskegee study until Susan Reverby's Examining Tuskegee (Reverby 2009).

Jones discovered that sorting through fact from fiction was difficult; there was a lot of "fiction" investigators added to the facts, and in some ways, the facts of the study are clearer in the fictional representation of the study in *Miss Evers' Boys*. When Jones first begins to research the Tuskegee study, it is still a fresh wound, but since he is successful in discovering and locating the critical PHS and CDC documents that uncovered how this observational study began, morphed, and ultimately continued for 40 years, the book became the definitive account.

Jones is researching the book in the late 1970s, during a time frame when the United States was in a reactionary mode from the traumas it endured in the 1960s, and still in recovery from Vietnam and Watergate. In fact, the Congressional hearings into the Tuskegee study overlapped with the Watergate investigation. The class-action lawsuit that had been filed by attorney Fred Gray (see further), on behalf of the Tuskegee study subjects, was still being negotiated when Jones began his research, but because of the PHS files Jones made available, Gray successfully negotiated a settlement in 1974 (see further).

Jones' groundbreaking book, Bad Blood: The Tuskegee Syphilis Experiment is released in June 1981; it coincides with the June 5, 1981 article in *Morbidity and Mortality Weekly Report MMWR)*, entitled "Pneumocystis pneumonia" (CDC 1981), the very first published medical article about HIV/AIDS. Meanwhile, the first review of Bad Blood appears in the *New York Times* on June 21, 1981, which opens with this text (Geiger 1981):

"Bad Blood" is the anatomy of a long nightmare—a particularly American episode in the treatment of black people. Some of the sentiments that inspired this horror, which ended a mere nine years ago, have an awful familiarity, and "Bad Blood" is as contemporary in its implications as yesterday's Medicaid rollbacks, today's food stamp cuts or tomorrow's definitions of the truly needy. "'Bad Blood" is more than mere history: As an authentic, exquisitely detailed case study of the consequences of race in American life, it should be read by everyone who worries about the racial meanings of governmental policy and social practice in the United States.

The *New York Times* review was written in the context of worrisome budget cuts announced by the new Reagan Administration, which would dramatically impact the CDC's response to AIDS (see Chap. 2). As Jones' book makes its way into a diverse and multidisciplinary readership, the AIDS epidemic really starts to bloom, and the African American community's response and engagement with this new deadly sexually transmitted virus is met with suspicion and resistance. Thus, the "legacy of Tuskegee" as it became labelled in the next century (Katz 2008) has been identified as a major cause of early resistance to AIDS education and research in the African American community because its members did not trust that they were being told the truth about AIDS. In 1993, Jones released an expanded edition of his book, with a new chapter on AIDS. Jones states in the preface to his 1993 expanded edition of Bad Blood the following: (Jones 1993: x):

> I have traced the Tuskegee Study's legacy in the age of AIDS. Briefly, I have shown why many blacks believe that AIDS is a form of racial genocide employed by whites to exterminate blacks, and I have examined how the Tuskegee Study has been used by blacks to support their conspiracy theory…I hope this new edition of Bad Blood will find its way into the hands of new readers who want to learn more about this peculiarly American tragedy, for the AIDS epidemic has added both a new chapter and a new urgency to the story. Today, as before, the Tuskegee Study has much to teach us about racism in the United States and the social warrant of medicine in people's lives.

When AIDS emerged in the African American community in the 1980s and 1990s based on the U.S. government's delayed actions to screen risk groups or protect the public blood supply (see Chap. 2), it was eventually understood to be a blood borne virus (truly "bad blood"). There were indeed uncomfortable echoes of the syphilis epidemic. But the legacy of the Tuskegee study made containment through research and education about HIV especially challenging in the African American community (see further under Healthcare Ethics Issues). HIV/AIDS was one reason why Jones' book on the Tuskegee study's story resonated in the 1980s and Feldshuh's play resonated in the early 1990s. In between these works on the Tuskegee study, a major bombshell book about AIDS would hit bookstores in 1987: And the Band Played On (see Chap. 2). In many ways, the story of AIDS was "déjà vu all over again." I discuss this work thoroughly in Chap. 2.

The Feldshuh Play (1972–1992)

Dr. David Feldshuh was born in 1944, and is the older brother of actress Tovah Feldshuh. David Feldshuh came of age during the 1960s, and initially studied drama and acting before entering medical school. He did his initial actor training at the London Academy of Music and Dramatic Art; he then joined the Guthrie Theatre, located in Minneapolis, Minnesota as an actor, then as an Associate Director. He was at the Guthrie theater for seven years, and then completed two terminal degrees: a Ph.D. in theatre, and then an M.D., followed by a residency in emergency medicine, which he still practices (Cornell University 2018). Feldshuh was 28 when three major news stories broke: June 17, 1972 was the first *Washington Post* story about Watergate, which would soon take the country by storm. The next month, on July 25, 1972, the *Associated Press* broke the first story about the Tuskegee study. Two months later, on September 5, 1972, 11 Israeli athletes at the 1972 Munich Olympics were attacked and later killed by a terrorist group known as Black September. It was the Munich story that most affected Feldshuh, who began to question whether medicine was a calling he should pursue at that time, in an effort to want to "do something more with his life". In fact, it was not until he read Bad Blood when he was doing his Emergency Medicine residency, that he began to think about the Tuskegee study. When Feldshuh saw the wall to wall news coverage of the Munich Massacre, "it made me wonder 'What am I doing here,' and I realized there were questions I wasn't really asking myself. I was relatively successful at a young age, so I knew the 'dis-ease' wasn't related to a lack of achievement in my career. I had to look elsewhere; I had to look within myself" (Isenberg 1990). Feldshuh next studied consciousness-raising techniques, such as Zen, Gestalt therapy and bioenergetics, and completed his Ph.D. in theater arts in 1975 with a doctoral dissertation entitled "Seven Consciousness Expanding Techniques and Their Relevance to Creativity in Actor Training," (Isenberg 1990). He next started medical school in 1976. Feldshuh was attracted to Emergency medicine because he was interested in something "more hands on and more concrete," (Isenberg 1990). He continued to support himself with acting and directing while training as an ER doctor. In fact, he saw parallels in the two careers: "In both cases, you are called upon to make decisions, frequently rapid decisions. . . . If you're not willing to make a decision and take responsibility for decisions, you shouldn't be directing and you shouldn't be doing emergency medicine. Neither are environments conducive to serene cogitation" (Isenberg 1990).

Feldshuh's introduction to the Tuskegee study was in 1981, when he first read a review of Jones' Bad Blood in a medical journal—it was likely the review that appeared in the December 4, 1981 issue of the *Journal of the American Medical Association*, or *JAMA* (Meyer 1981). "After he read it, he couldn't stop–he read medical reports and Congressional testimony and, as the years went by, did personal interviews as well" (Isenberg 1990). He began working on the play at that time, and completed a rough draft of "Miss Evers' Boys" by 1984, when he accepted an offer at Cornell University to become its Artistic Director for its Center for Theatre Arts. He continued to work on the play throughout the 1980s during his summers. "Miss Evers' Boys" went through 27 drafts, and a number of workshops and readings until it more formally debuted on the stage. It premiered November 17, 1989, at Center Stage, in Baltimore, and then began touring extensively in 1990. At the Baltimore premiere, during a scene depicting a lumbar puncture, one of the audience members fainted, and Feldshuh attended to the audience member as a physician. He stated: "It was an interesting kind of nexus of my life as a playwriting physician…I'm there as a playwright, looking at the play to see if it works. And I find myself practicing medicine. Does this mean it works? Or that it works too well?" (Isenberg 1990).

Regarding his motivation for writing "Miss Evers' Boys", Feldshuh stated this (Isenberg 1990):

> I was drawn to this subject because I recognized feelings in myself that concerned me…I
> asked myself a simple question: 'Would I have done what these physicians did had I been
> there?.' I don't know, but I think I was fearful enough that I wanted to find out exactly how
> it was they let themselves do it. I wanted to explore the process through which they allowed
> themselves to participate in something that was clearly in retrospect wrong.

Feldshuh continued to revise and rework "Miss Evers' Boys" throughout its touring, which took place at the beginning of the first Bush Administration, a period in which the Cold War was ending, the Berlin Wall was finally torn down, a new international order was reshaping; and the National Commission on AIDS met for the first time. The year 1989 was a good time to revisit "old wounds" and the right time for staging a play about the Tuskegee study. In essence, the play was a continual work-in-progress. The play was performed frequently to good reviews, and discussion of developing it into a film began as early as 1990 (Isenberg 1990), after it debuted at the major Los Angeles theatre, the Mark Taper Forum, July 8, 1990. With respect to interest in developing it as a film, the Mark Taper Forum producer stated to the *Los Angeles Times*: "There was some early interest around the time of the Baltimore production… We thought it was better to finish the journey of the play as a play. I'm sure that there will be some increased interest once the show is seen here (but) one tends to want to hold off" (Isenberg 1990). Reviews of the play as a fictionalized representation were positive. A July 20, 1990 review (Drake 1990) noted:

> Playwright Feldshuh, a theater man and medical doctor, has fictionalized [the Tuskegee
> study] but not the circumstances. He has reduced the 400 [men] to an emblematic four and
> the doctors involved down to two: a white Dr. John Douglas (Charles Lanyer) and a black
> Dr. Eugene Brodus (Bennet Guillory). The nurse who cares for these men is Eunice Evers
> (Starletta DuPois), a character modeled after real-life nurse Eunice Rivers, a black woman
> who participated in the study, though herself somewhat ambiguously misled into believing
> in its moral propriety.

The *New York Times* review of the play made the context of the play's debut during the height of AIDS clear: "In these days of AIDS, it's valuable to illuminate [the view that humans are subjects], particularly as it's applied to people whose lives, in an unspoken but implicit social contract, are deemed dispensable" (Winer 1989).

The Los Angeles production had cast Starletta DuPois as Rivers, who had studied nursing at the Harlem Hospital School of Nursing before going into acting and read Jones' Bad Blood. Feldshuh took the opportunity with Dupois to expand the character and role of Rivers because of the actress' passion and intense interest in the character (Isenberg 1990). Feldshuh stated at the time that he wanted to focus on the presumed moral distress of Rivers, or "doubts she might have had but was unable to express…Here is a woman who exists as a black in a white world, female in a male world, nurse in a doctor's world. She loves her patients and has at the same time great allegiance to her profession. Now what is she going to do when the profession demands that she treat her patients from a distance and not up close? A doctor tells her, 'You've got to step back and look from a distance,' and she says 'I can't.' (Isenberg 1990). Feldshuh describes his creation of Miss Evers as a character who "seems to be making the right decisions at each point but the overall journey is clearly down the wrong road. She is trapped in a vortex of small decisions, and it ultimately not only pulls her down but finally pulls her apart. It separates her from herself" (Isenberg 1990). Regarding the treatment of syphilis by the Tuskegee study healthcare providers, Feldshuh notes that syphilis "was also called 'the great imitator' because of how diversely it presented itself. Syphilis could have no effects, a mild or a devastating effect. It was a complex, subtle disease and that made moral thinking about it also complex and subtle. You could point to somebody and say 'It's not so bad. We're not treating him and he's fine'" (Isenberg 1990).

The Feldshuh play was nominated and was a finalist for the Pulitzer Prize in 1992, and it inspired several scholarly and artistic projects about the Tuskegee study, including a 1993 documentary he co-produced entitled *Susceptible to Kindness* (Cornell University 1993), which included a Study Guide developed by an African American bioethicist and legal scholar, Larry Palmer. *Susceptible to Kindness* included original interviews with survivors, and filmed scenes from the August 22–31 1991 production of "Miss Evers' Boys" performed at the Illusion Theater, at Cornell University's Center for the Theatre Arts. This documentary won the International Health and Medical Film Festival award in 1994. Feldshuh also later received a distinguished service award from the National Center for Bioethics at Tuskegee University.

On the 20th anniversary of the closing of the Tuskegee study, Feldshuh's play was well-known, and began to inspire mainstream news documentaries. In 1992, ABC's *Primetime Live* news magazine did a segment about the Tuskegee study, interviewing one of the surviving physician investigators, Sidney Olansky (see further) as well as a surviving subject, Charles Pollard. Olansky came off as unapologetic (Reverby 2009), and makes a case for why "Miss Evers' Boys" was best as a fictionalized account. Feldshuh's play no doubt inspired PBS' *NOVA* series to produce its own documentary on the Tuskegee study, which aired in 1993, titled *Deadly Deception*. The public's renewed focus on the Tuskegee study in this time frame also helped to derail the 1995 Senate confirmation of African American physician, Henry W. Foster Jr., (Lewis 1995; Reverby 2009) to the position of Surgeon General because he worked at the Tuskegee Institute in the final years of the study; there was an accusation that he knew of the study's existence and may have been complicit with it, or did not speak out.

The Play's Staging and Stages

Feldshuh's play has been praised for its organization and staging details, which makes it an easy play to produce. Aside from Miss Evers, four study subjects (one character to symbolize each of the 100 men in the study to represent the oft-quoted 400); one white U.S. Public Health Service physician as a composite character for several through the years, and one African American physician who oversees the healthcare delivery for the study subjects at the Tuskegee hospital, as the composite for Dr. Eugene Dibble, Medical Director of the teaching hospital at The Tuskegee Institute (see further). The play is set in the Possom Hollow Schoolhouse, juxtaposed with "testimony areas" in which a 1972 Senate subcommittee investigation is taking place, with a composite character probably for Senator Ted Kennedy. (In reality, the Kennedy hearings, discussed further, occurred in 1973). Nurse Evers is pulled in and out of action to give her testimony to the audience. (Theatrically, some of this technique echoes the interesting staging of Arthur Miller's "Death of a Salesman" in which Willie Loman is in and out of different realities and contexts.). Act One takes place in 1932, where we meet the four study subjects who enroll and are initially treated with mercury and arsenic until funding runs out, and the study rationale for observation without treatment is dramatized. Act Two, takes place in 1946, when penicillin is debated as a treatment but not offered. There is an Epilogue in which Miss Evers raises questions about what has occurred. One of the most central criticisms of "Miss Evers' Boys" is the fact that it is has "factual errors" due to its *fictionalized* characters. Fred Gray, the attorney for the Tuskegee study subjects and families, for example, who spoke out against the film version (see below), did not like how the study subjects were portrayed. The surviving study subjects had all seen the play (they were filmed seeing it in the PBS documentary), and while pointing out differences in their experiences in private (Reverby 2009), they did not disapprove of the play and saw its merit. Gray did not like that the characters were part of a local dance/entertainment troupe in the play, when this was not reality. (Portrayal of African Americans as performers is frequently considered stereotyped representations.) Feldshuh stated that he created the dancer characters so that he could, as a playwright, visually show the effects of an "invisible" disease (syphilis); when one of the characters begins to have difficulty dancing in the play after a long latency period, it is a sign that tertiary stage syphilis has begun, and is affecting him.

Again, it is imperative that any teaching of *Miss Evers' Boys* emphasize that it is a work of fiction based on a real study, but that its characters are fictionalized. This is not unlike the fictional representations of a study in the film, *Awakenings*, for example, in which "Dr. Sayers" is a fictionalized Oliver Sacks (See Clinical Ethics on Film, Chap. 10). It's important to emphasize that there is a difference between reality and art. A critical factor in Feldshuh's work is its historical timing. When Feldshuh was creating this work, it was still not distant enough to "name names"; artistic interpretations of the truth are sometimes the best way to tell the story when wounds are too fresh, and historical figures or their immediate descendants or family members are still alive. This was evident by some of the damning interviews with the still-living physicians in both the PBS and ABC documentaries. For example, at one of the play's performances, a white woman confronted Feldshuh and stated: "How dare you portray white doctors like this?" (Reverby 2009). Several directors were confronted by angry African American patrons or actors demanding that the play should show how the government "gave syphilis" to the study subjects, when this was not the case at all (Reverby 2009). On the flip side, one of Tuskegee's mayors in the 1990s "stated that staging the play in the city 'helped to absolve a sense of shame' and he made sure that survivors came to see the production" (Reverby 2009). Indeed, a fictionalized representation of the Tuskegee study was the most prudent pathway to show it throughout the 1990s.

The Color Purple Problem

There is an "elephant-in-the-room" issue with "Miss Evers' Boys" as a play and then film (see further), which I call The Color Purple Problem. "Miss Evers' Boys" was not created by an African American writer, and so its "truths through fiction" genre was not as readily accepted as other works of fiction inspired by socio-political truths, such as Alice Walker's The Color Purple, which was published in 1982, and won the 1983 Pulitzer Prize for fiction for which Walker was the first African American woman to win. (Her book was indeed controversial for its portrayal of African American males.) But then her novel was made into a film by Steven Spielberg in 1985, which also launched the careers of major African American stars, Whoopi Goldberg, Oprah Winfrey, and Danny Glover. Spielberg's film was nominated for 11 academy awards but won nothing due to an interesting controversy over race—his. There was a perception that a white Jewish man should not have been allowed to take ownership of any part of this story—that it belonged to the African American community. In many ways, the film, *The Color Purple* mirrored the critiques that befell the film, *Miss Evers' Boys* (see further). There was a perception with this work, too, that remains: white Jewish, privileged males have no business telling stories about the African American experience and therefore, any putative "truths" about the African American experience should not be trusted or valued from this "whitewashed" messenger. Walker, in fact, was involved in the film adaption, and her friend, Quincy Jones, recommended she choose Spielberg as the director for her book's film, but she regretted the choice, and hated Spielberg's filmed adaptation of her book. Her screenplay was edited/rewritten by a white Danish male writer (Holt 1996), who she felt changed the story and its truths. She made her critique very public around a decade later—just as "Miss Evers' Boys" was being made into a film (see further), too. According to Walker, "Steven's version of 'The Color Purple' would not deserve the name. And so I created an alternative title for his film" (Holt 1996). Remarkably, Walker was so profoundly disturbed that Spielberg made his film, she wrote another book, The Same River Twice (1996), entirely about her difficult experience with the film's making, as it brought so much focus to her novel and to her. The Same River Twice was published during principal production for *Miss Evers' Boys*, and shines a light on the problematic relationship between Hollywood and the African American experience.

Miss Evers' Boys: *The HBO Film (1996–1997)*

The play "Miss Evers' Boys" was adapted into a film by HBO, which was the first pay television network to make quality films using the same production standards as feature films. By 1997, HBO had already aired several quality films, including *And the Band Played On* (1993).

The Home Box Office (HBO) network debuted the same year the Tuskegee study was exposed in the press: 1972. At the time, it was just in a very small market of Allentown, PA., and not many Americans knew of its existence. HBO started to expand through the 1970s into the Manhattan market; designed as a subscription cable channel, it was primarily known for showing feature films and sports events, and first adopted a 24/7 schedule by December 1981. It began to produce its own feature films in 1983, and by the 1990s had completely transformed "made for television" films into high quality films that were comparable to feature films.

The most important fact to get across about the film, *Miss Evers' Boys*, is that it is not an original screenplay about the Tuskegee study, but adapted from Feldshuh's play. Thus, it is the same fictionalized representation of the Tuskegee study as the play, never intended to be a docudrama or documentary about the study. Many academic scholars miss this, and then criticized the film for being "not accurate". Harriet Washington, in her book, <u>Medical Apartheid</u>, merely refers to it briefly as "HBO's irresponsible film" (Washington 2007). Thus, those new to the film must understand that it was adapted from a long-running, Pulitzer prize-nominated play intended to dramatize the "moral truths" about the Tuskegee study through mostly composite characters, and addressed questions about the study no scholar had actually yet uncovered: what were the African American healthcare providers thinking? We must also remember that "Miss Evers' Boys" debuted on-stage as another government deception had started to become a public health scandal surrounding knowingly exposing the general public to HIV through the blood supply (see Chap. 2). When the HBO film was released to a new generation of audiences through film, it was often treated as an original work and its stage origins were often forgotten. For example, the *New York Times* review of the film mentioned it was based on the fictional play but then stated this (Marriott 1997):

> Yet while "Miss Evers' Boys" attempts to illuminate its story with the glare of authenticity, questions of factual fidelity always arise—as with Oliver Stone's "Nixon," or Ron Howard's "Apollo 13"—when pages of history books are rewritten into pages of a director's script. This is no less true for the cinematic retelling of the Tuskegee study, a sensitive subject that is still creating reverberations, including a willingness among some blacks to consider maladies like AIDS and crack addiction part of a genocidal conspiracy against them by whites.

The Producers

Miss Evers' Boys was filmed in Georgia, and produced by HBO New York, in association with Anasazi Productions, a small production company launched by married actors, Ted Danson and Mary Steenburgen, who met on the film set of *Pontiac Moon* (1993), a box office failure, but clearly a "hit" for their personal lives. When they met professionally, Danson and Steenburgen had both recently ended long marriages, fell in love and married in 1995; they are still married. At the time, Danson was most known for his role on the long-running television sitcom, *Cheers*, and Steenburgen had a prolific film career, starring in high profile films such as *Philadelphia* (1993), discussed in Chap. 2. Steenbergen was also the narrator in *The Long Walk Home* (1990), in which Whoopi Goldberg plays an African American maid who is actively boycotting public transit in the famous 1955 Montgomery Bus Boycott—a case initiated by Rosa Parks who was represented by attorney Fred Gray. Danson and Steenburgen each had interesting personal relationships prior to the making of *Miss Evers' Boys* that are relevant. Danson, who had been very active politically surrounding social and environmental justice issues, had a highly public romance prior to Steenburgen with Whoopi Goldberg in 1992–1993. As a high profile mixed race couple, they each became flashpoints for the changing attitudes surrounding race in the United States, which had reached a boiling point in Los Angeles then due to the Rodney King beating, a precursor to the 1992 L.A. riots. However, in one terrible decision the power couple made, Danson appeared at a 1993 Friar's Club Roast for Goldberg in "blackface" and ate a watermelon. Roger Ebert opined at the time: "[The audience] cringed in disbelief during the opening monologue by … Ted Danson who appeared in blackface and used the [N-word] more than a dozen times during a series of jokes that drew smaller and smaller laughs, until finally the audience was groaning." (Conner 2019). Goldberg tried to defend it as a piece of social commentary through satire and performance art: "Let's get these words all out in the open. It took a whole lot of courage to come out in blackface in front of 3000 people. I don't care if you didn't like it. I did." (Conner 2019). Several African American artists and politicians were deeply offended and Danson and Goldberg ultimately issued a public apology (Levitt 1993), and then ended their relationship in November 1993. (Goldberg went on to solo-host the 1994 Academy Awards, being the first African American woman to do so).

Meanwhile, Steenburgen, born in Arkansas, Bill Clinton's home state, had a close personal friendship with Hillary Clinton, who was then the First Lady of the United States. Steenburgen met the Clintons in Little Rock when her father, a retired railroad worker had gone to hear Bill Clinton speak when he was a local politician. Clinton mentioned her in his speech as an example of local talent. Steenburgen recalled in an interview: "[Bill] goes, 'And there's a young woman in our community and she's become an actor and it just shows you the kind of talent that's out in Little Rock and North Little Rock and out in Arkansas…And my dad said [when he went up to him Bill Clinton after the speech], 'Well, I'm Maurice Steenburgen and if you're gonna talk about my daughter that way, I think you oughta meet her…"And so my so-called Hollywood connection with Bill and Hillary Clinton came through my freight train conductor dad. And so we've been friends ever since." Steenburgen had noted that Hillary helped to inspire her to become politically active. (Gebreyes 2016). In this time frame, Danson's fresh wound with racial politics from his Goldberg relationship, and Steenburgen's involvement with the President and First Lady—at a time surrounding the first Presidential Apology for the Tuskegee study, undoubtedly made the *Miss Evers' Boys* film project meaningful for them personally and politically.

The President of Anasazi Productions, one of Danson's previous drama teachers, Robert Bendetti, was billed as the Executive Producer of *Miss Evers' Boys*, and had a very similar acting and drama career as Feldshuh himself. Bendetti also authored several books on acting and drama. The writer hired to adapt the Feldshuh play to a screenplay was Walter Bernstein, who had been blacklisted during the McCarthy era, and who had written the screenplay for *The Front* (1976) about the blacklist.

Bernstein created some departures from the play: mainly, the romance between Miss Evers' and the character, Caleb is more of an emphasis, while a scene where the nurse accompanies the men to one of their local performances shows her dancing in a provocative manner, something Fred Gray and some Tuskegee study survivors found disrespectful and insulting to her memory when the film debuted (Reverby 2009).

Joseph Sargent (see further), a veteran of made for television films on other networks, was selected as director. The film was adapted with the understanding that Feldshuh's play was a fictionalized representation, but still very sensitive, potentially explosive subject matter. HBO, which had already produced a "whistleblowing" film about a public health disaster—*And the Band Played On* (1993)—was a natural host for *Miss Evers' Boys*. Benedetti recognized that this material was not particularly well-suited for commercial network television, and stated (King 1997):

> Once in a while the networks will do as a special event something worthwhile, but generally if I get a piece of material like this as producer, I immediately think cable… I was very happy to say that [HBO] was insistent that we not reduce the moral ambivalence [of the play]….The lead characters are somewhat ambivalent….You don't know whether what they have done is right or not. Usually for television, you like to have fairly simple moral issues. You like to have good guys and bad guys. This is a situation where there were no good guys and no bad guys. The real villain is institutionalized racism, rather than any particular person… [And] I think the movie is fundamentally accurate. There are certainly no misrepresentations in it (Marriott 1997).

Laurence Fishburne, who plays Caleb Humphries in the film, was also Executive Producer. *Miss Evers' Boys* was the second HBO film Fishburne had done in which the word "Tuskegee" figured prominently, as he had starred in *The Tuskegee Airmen* (1995), based on the first African-American combat pilots' unit in the United States Army Air Corps. In fact, these pilots had trained in Moton Field in Tuskegee Alabama—so named after the very Robert Moton who was the second president of the Tuskegee Institute, and been involved with the Tuskegee study in its early years. Fishburne, by 1997, brought enormous star power to the film. He had attended the High School of the Performing Arts, and was only 14 when cast in Francis Ford Coppola's Vietnam tour-de-force, *Apocalypse Now* (1979); he spent both his 15th and 16th birthdays making that film. "[One interviewer] quoted Fishburne as saying that shooting *Apocalypse* was 'the most formative event' of his life. He had a chance to observe several luminaries of American film acting—Marlon Brando, Robert Duvall, Martin Sheen, and others—and to consult them for advice. Coppola taught Fishburne that acting 'could be taken seriously, as art, with potential for educating, entertaining and touching people.'" (Encyclopedia.com 2018). In 1983, Fishburne played a part in the PBS drama *For Us the Living*, based on the story of civil rights figure, Medgar Evers. Fishburne stated that "this is a gig where I had to put myself up and pay my own transportation, but to be involved with Roscoe Lee Browne, Howard Rollins, Dick Anthony Williams, Irene Cara. Well, that was my ancestors saying to me, 'OK, here's some work we can do (Encyclopedia.com 2018)".

He also had a minor role in *The Color Purple*, discussed earlier. As an adult, Fishburne's major breakthrough came when he starred in a culturally ground-breaking film, *Boyz n the Hood* (1991), which was about the toxic conditions for African Americans living in South Central Los Angeles. Fishburne played Cuba Gooding Jr.'s father, even though he was only six years his senior. Fishburne had a good understanding of the differences between film and stage; he had starred in a long-running August Wilson play, "Two Trains Running," for which he won a 1992 Tony Award and several other awards. He also played Othello in a film version of the Shakespeare play. Fishburne was also cast as the abusive Ike Turner opposite Angela Bassett in *What's Love Got to Do with It?* (1993). Fishburne said of *Miss Evers' Boys*: "For a Black woman in the '30s to be involved with this sort of government experiment and to be privy to the sort of information she knew, it would be risky for her to go out on limb and say, 'This is wrong.' She would have to be thinking to herself, 'Who would believe me?... My character, Caleb, is sort of the conscience of the movie because his point of view is very clear from the beginning. He answers some of the questions Miss Evers should be answering, but can't, or the ones that she doesn't really want to answer" (Jet 1997).

The other producers on the project were Derek Kavanagh, who worked on *Dances with Wolves* (1990) and Kip Konwiser, who had worked on several Oliver Stone projects before *Miss Evers' Boys*.

Director and the Star

Joseph Sargent had worked on a number of diverse projects. Of note, one month after the Tuskegee study was publicized and shut down in 1972, a prescient film was released that Sargent directed called, *The Man*, about the first African American president played by James Earl Jones. The premise of the film is that Jones (who plays Douglass Dilman), as the President pro tempore of the U.S. Senate, succeeds when the President and Speaker of the House are killed in a building collapse at a Summit in West Germany; but since the Vice President is suffering from a terminal condition and refuses to assume the office, Dilman is sworn in as President (Canby 1972). The new President Dilman faces many of the same things Barack Obama would confront 36 years later; in the film, he especially experiences opposition to a minority rights bill he champions. Sargent would also direct *Something the Lord Made* (2004), discussed in Chap. 9.

In interviews, Joseph Sargent said he was drawn to the project because (King 1997):

> [I]t's not just a piece about a racial decision by a white government bureaucracy. It also reveals the unwitting collaboration of several black doctors and nurses, and that's what gives dimension to this piece. It gives it a lot more substance…Basically, everyone had good intentions. Miss Evers is making the best of a terrible situation, and the inner struggle that produces makes the thing so dramatic….

He later shared (Kagan 2004):

> [Miss Evers' Boys] tended to put me into an angry frame of mind when I took this on, and I had to get away from that. I didn't want to be in a position of commenting, politically, on

the subject matter, because the subject matter takes care of it...Sometimes you are dealing with content that makes you so uncomfortable that you have to transcend your emotional feelings and get into a neutral position so you are able to stay objective, so you don't tip the boat too far. And that was the case here.

Alfre Woodard, who played the starring role of Eunice Evers, had distinguished herself as a major stage and screen character actress by the mid-1990s. She did her training in drama and acting at Boston University and began her career in stage plays. In 1976 she moved to Los Angeles. She later said, "When I came to L.A. people told me there were no film roles for black actors...I'm not a fool. I know that. But I was always confident that I knew my craft." (Dougherty 1987). Woodard first received accolades for her television roles. She won her first Emmy award in 1984 for an extraordinary performance that still resonates in the "Black Lives Matter" era: she played the mother of a young boy on *Hill Street Blues* who was accidentally shot and killed by police. She next won an Emmy for her role as a woman dying of leukemia on *L.A. Law*, and she was then cast as an obstetrician and gynecologist in 1986, in the medical drama, *St. Elsewhere*, where her love interest is played by Denzel Washington; she was nominated for more Emmy awards with those roles. When Woodard migrated to film in the 1990s, she rose to prominence in several notable performances such as Winnie Mandela in *Mandela* (1987), also an HBO production; *Grand Canyon* (1991)—one of the "toxic L.A." genre films of the period; and *Passion Fish* (1992), a nursing ethics film where she plays a nurse in recovery from addiction caring for an alcoholic paraplegic played by Mary McDonnell (who was also in *Grand Canyon*). Woodard won an Independent Spirit Award for Best Supporting Actress award for *Passion Fish*. She later said this about the role of a nurse (Jet 1997): "Doctors come in and are basically technical with their patients, but the nurses often form relationships with them because they're the ones who have to soothe fears. They are the ones mopping your brow when you're feverish in the middle of the night. They are the ones helping you to bathe when your body breaks down and betrays you." (Jet 1997).

Throughout the 1990s, Woodard's range in roles was limitless and prolific—she played a judge; she starred in "chick flicks"; she starred in comedies; she starred in sci-fi roles. By the time she was cast as Miss Evers, she brought many dimensions to the role.

Woodard had a living memory of the Tuskegee study hearings, as she had been demonstrating against the Vietnam War at Boston University at the time (Mills 1997). She recalled: "I was sitting on the trolley tracks on Commonwealth Avenue in Boston and the president of the university unleashed busloads of cops on us and anybody else they could grab and hit. In the midst of all that, learning about the hearings, I said, 'Of course!' What happened in Tuskegee didn't surprise me then, out of a fired-up youthful defiance. And it doesn't surprise me now, out of my more informed historical perspective" (Mills 1997).

Woodard did not initially warm to the role of Miss Evers because of her distaste for playing the "mind-set of a person who could be duped this way and then, in turn, sort of carry the banner [of the study]. The reason I couldn't not do it was, I was going to be playing this person opposite Laurence, and you can't walk away from a character who is so complex. The exercise is to stay out of the way and don't bring your opinions into it and really find out what this person was thinking." (King 1997).

With respect to her thoughts on the character and the real Nurse Rivers, she stated the following (Mills 1997; King 1997):

> Miss Evers was smart. She was educated. How could she have gone along with it?" ...I understand that she was between a rock and a hard place, and I understand that she was told that the study was good for humanity, but I draw the line at people's lives. I'm a hard-liner that way. Yes, I blame her. I'd never care to spend even tea-time with her....I do remember the hearings...I do believe [Rivers] was a small cog in a big rolling machine...Miss Evers wasn't the culprit. She wasn't the instigator or the sustainer of the injustice. But she was part of it, and everyone has to take responsibility for their part in a derailment. Even if her intentions were good and even if she did a lot of good, she kept a lot more good from being done.....When things get out of control, as the Public Health Service did with this, people have a way of putting on blinders. And that doesn't surprise me. No, as a person of African descent, as an American with Native American bloodlines, that doesn't surprise me.

Although in the play/film, Feldshuh demonstrates that Miss Evers has moral distress over her role, Woodard never believed that, and stated, that she did not think Nurse Rivers felt guilty at all (King):

> I think that's why she was able to do it all of those years....[To her] these field hands, who have never been given the time of day, even by black people around them, that were a class above them, these guys were getting vitamins and tonics and things that even black people with jobs didn't get. They were getting this specialized attention. She's a smart woman and she had to have a smart logical reason [to do what she did].

In fact, when transcripts of actual interviews with Rivers were finally published, that is exactly what she said (see further under Social Location).

Reception and Criticism

Miss Evers' Boys "color purple problem" (see earlier) was more pronounced with the film than the play, but the film project was championed by all of the African American artists involved—particularly since Fishburne had been one of the Executive Producers. For example, Obba Babatunde, who played Willie Johnson, stated: "After reading the script and then hearing about the actual case, I chased this project...I wanted to be one of the people to help to tell the story and, in a sense, pay homage to the men who sacrificed their lives. I believe the time is long overdue for this part of history to be told," (Jet 1997).

Notes Reverby (2009):

> Before the federal apology in 1997, the play/film appears to have functioned both to put the Study into wider cultural circulation and to provide a form of healing...Feldshuh's focus on the nurse pushes viewers into what Holocaust survivor/essayist Primo Levi calls the 'grey zone.' With a focus on those caught in the middle, 'grey zones' provide seemingly more dramatic tensions [and for the viewer] 'to reflect on the larger question of moral culpability and engage in an exercise of historical empathy.'

Two months after the film aired on HBO, the attorney representing the men, Fred Gray, called a press conference to correct inaccuracies in the film—again, not fully appreciating that the film was based on the play, *a work of fiction*. Gray took particular offense to scenes depicting Miss Evers as more seductive and dancing with the men. However, Nurse Rivers in interviews had actually discussed that she had enjoyed banter with the men, and had occasionally joined in on risqué jokes with them, and when she drove them around. She said: "So when the want to talk and get in the ditch, they'd tell me, 'Nurse Rivers, we're all men today!...Oh we had a good time. We had a good time. Really and truly" (Reverby 2009).

Gray stated in his press conference: "Miss Rivers was always professional and courteous to them. She did not accompany them to nightclubs. They did not dance, play music and entertain people...The entire depiction of them as dancers is a great misrepresentation" (Gray 1998).

After the official Presidential Apology in May 1997 (see further), things got worse in 1999, when a panel discussion at a bioethics conference at the Tuskegee University featured David Feldshuh and Fred Gray. They got into a defensive exchange with Gray criticizing Feldshuh's play/film, Feldshuh defending it as an artist, and Gray making clear that it had misrepresented the true experiences of the study subjects and what occurred.

For these reasons, when screening or assigning *Miss Evers' Boys* as part of a healthcare ethics/public health ethics curriculum, it's important that it be accompanied by other readings and/or lectures on the facts of the Tuskegee study, covering the content I suggest in the History of Medicine section.

Synopsis

Miss Evers' Boys is based on The Tuskegee study, and tracks it from its origins of initially offering a treatment protocol to continuing the study as an observation-only trial of untreated late syphilis when funding for treatment dries up. The story is told from the point of view of Nurse Evers (Alfre Woodard), a fictionalized Eunice Rivers. In fact, pictures of Eunice Rivers and Woodard's Evers are identical. There is a composite character for the white PHS doctors, "Dr. Douglas," played by Craig Sheffer, who initially helps to conceive and champion the observation trial (probably a composite for O.C. Wenger, active from 1932–1950, Austin V. Deibert, active 1936–40; and Sidney Olansky, active in the 1950s). A scene with an older white physician authorizing/planning the study was likely a composite for the two Venereal Disease (VD) Division Chiefs at the PHS who thought up the observation trial (Taliaferro Clark, active 1932–33 and Raymond Vondehlehr, active 1932–1940s). In reality, there were a series of PHS physicians and VD Division Chiefs who championed and continued the study through the years, but they essentially functioned as "composites" in reality, too—always approving, and never halting, the protocol.

Dr. Douglas approaches the African American physician in charge of the medical facility at The Tuskegee Institute, "Dr. Eugene Brodus" to help the PHS carry out the study, and Brodus agrees to collaborate. Brodus is played by Joe Morton, and is a fictional character for Dr. Eugene Dibble (see further). It's possible Feldshuh named the character "Brodus" as a nod to Dr. Broadus Butler, one of the original Tuskegee airmen, the President of Dillard University (a historically black college/university), who later chaired the ad hoc advisory panel on the Tuskegee study in 1972.

The film cuts back and forth between the story of the study and a Congressional hearing in which an older Nurse Evers is providing her testimony. (In reality, she never testified, and thus, has her say in the film.) The Senate Chairman asking questions, played by E.G. Marshall is a composite character for Senator Ted Kennedy, who held such hearings in 1973 (see further). Among the composite study characters, Caleb Humphries (Laurence Fishburne) eventually—like many men in the study—gets treatment with penicillin when he joins the army, while Willie Johnson (Obba Babtunde)—like many in the study—suffers deteriorating health through the years until he dies because he never receives penicillin. The film also covers the issue of clinical equipoise over the use of penicillin treatment for late stage syphilis, particularly in scenes where Nurse Evers and Dr. Brodus argue about whether it would be beneficial or not.

Geography and Demography Are Destiny: The Social Location of the Tuskegee Study

To really understand the origins of the Tuskegee study and its socio-political context, it's critical to understand the demographics of the population that was studied in Macon County, Alabama—in the rural deep South in Depression America, where living conditions for African Americans at the time were "appalling" (Jones 1993).

The political conditions of apartheid and segregation permitted and enabled a completely racist environment in which African Americans did not have civil rights or civil liberties under the "Jim Crow" laws—state and local laws that enforced segregation in all public facilities. Jim Crow laws were enacted in the late nineteenth century and essentially remained in place until the *Civil Rights Act* passed July 2, 1964, a few months before Rivers retired from the study (Jones 1993; Smith 1996; Reverby 2009), both Drs. Moton and Dibble had died, and the *Archives of Internal Medicine* had published a paper on the 30th year of data from the Tuskegee study (Rockwell et al. 1964). "Jim Crow" was a minstrel character created by the "father of American minstrelsy," Thomas D. Rice (1808–1860), a pre-Civil War white traveling actor, singer and dancer who was popular circa the 1830s. He would do his "Jim Crow" character in black face, and in 1828 created a catchy song called "Jump Jim Crow" after this character. This character became the worst stereotype of an African American sharecropper that defined how whites thought about African Americans; by 1838, the term "Jim Crow" became a derogatory term for African Americans. (Padgett 2018). The segregationist laws that local and state governments began to enact became known as the "Jim Crow" laws after this well-known character—shorthand for laws in former slave states of the South that sanctioned both apartheid and racial terrorism targeting the African American population (Ferris State University 2018). Jim Crow laws were challenged, but upheld by the U. S. courts under the farce "separate but equal" (actually written as "equal but separate") in the 1896 *Plessy v. Ferguson* decision. Separate but equal meant that there were inferior facilities for African Americans (or none at all), whereby almost all social goods and economic advantages—such as education, employment, housing, bank lending, voting, hiring practices, and the bare necessities for health and wellbeing (healthcare access; healthy food; or clean running water) were frequently absent, inferior, or blocked (as in voting—in which literacy tests or poll taxes kept African Americans from voting). Moreover, there were various unspoken laws that African Americans were to abide. For example, African Americans who did not "know their place" could be lynched and killed. Lynchings were not even fully acknowledged as a risk of daily African American life—especially in Alabama—until 2018, with the opening of the National Memorial for Peace and Justice in Montgomery, Alabama. The *New York Times* described the new museum as being "dedicated to the victims of American white supremacy. [The museum] demands a reckoning with one of the nation's least recognized atrocities: the lynching of thousands of black people in a decades-long campaign of racist terror" (Robertson 2018).

Lying in complete juxtaposition to these terrible living conditions of the rural population, was the noted geographical location of the study hospital: the center of African American academic, intellectual and medical training—the Tuskegee Institute, founded by Booker T. Washington on July 4, 1881, as one of the first historically black colleges and universities (HBCU) and centers of higher education and learning in the South. Dr. Robert Russa Moton (1867–1940) became the next President of Tuskegee University in 1915, and was in that role until 1935, when he resigned due to his declining health. Moton was 29 years old when the *Plessy v. Ferguson* decision upheld segregation as the law of the land, which also led to the establishment of land grants for black colleges and universities in states where there were also land grants for white colleges and universities that restricted African Americans from attending. Moton died 14 years before the *Brown v. Board of Education* decision in 1954 (see further), which declared racial segregation of children in public schools was unconstitutional, and the legal doctrine of "separate but equal" as unconstitutional; he died 24 years before the *Civil Rights Act* was passed, ending the Jim Crow era, and 25 years before the *Voting Rights Act* was passed in 1965, when African Americans could finally vote in the South without being harassed or lynched. During Moton's lifetime in the Plessy era, *any* study that could potentially improve *anything* for African Americans living under their present conditions was viewed as a "good thing".

Living Conditions in Macon County, Alabama

The 1930 Census documented that 82% of the population in Macon County, Alabama—27,000 residents—was African American. The percentage was the same in 1970 (Jones 1993) when half the residents were living below the poverty line, and a third did not have indoor plumbing. In 1930, conditions were even worse. Notes Jones (1993):

> The typical dwelling was a tumble-down shack with a dirt floor, no screens, little furniture, a few rags for bedding, and a [toilet] only when underbrush was not nearby. Drinking water came from an uncovered, shallow well, often totally unprotected from direct surface drainage…the people who lived in this rural slum ate a pellagrous diet [leading to niacin deficiencies]..Salt pork, hominy grits, cornbread, and molasses formed the standard fare…while red meat, fresh vegetables and fruit, or milk seldom appeared on their tables. As a result, chronic malnutrition and a host of diet-related illnesses were serious health problems.

Clearly the health problems in this region were of particular concern. "In the early twentieth century, many rural African Americans lived in unhealthy surroundings and faced a range of health problems including malaria, typhoid fever, hookworm disease, pellagra, and venereal disease, along with malnutrition and high infant and maternal mortality rates" (Smith 1996).

Notes from Uva M. Hester, a Tuskegee Institute graduate in nursing, and the first black public health nurse in this region stated that "she was appalled by the flies, the dirt and the small rooms in the cabins she visited" (Smith 1996). Her diary noted the following from a June 1920 visit to a patient (Smith 1996):

> I visited a young woman who had been bedridden with tuberculosis for more than a year. There are two openings on her chest and one in the side from which pus constantly streams. In addition, there is a bedsore on the lower part of the back as large as one's hand. There were no sheets on her bed…The sores had only a patch of cloth plastered over them. No effort was made to protect the patient from the flies that swarmed around her.

The schools in Macon County were inferior (they did not improve after *Brown v. Board of Education*) and most residents were illiterate despite the proximity to the Tuskegee Institute. There was virtually no access at all to medical care, except for the Veteran's Association hospital, which had a segregated unit of African American healthcare professionals. In 1906, the Tuskegee Institute successfully obtained funding to create The Movable School project to service the community. This comprised of traveling public health nurses and educators to teach rural residents about basic hygiene, and to also try to teach residents how to farm, so they could transition from tenants to landowners. This was a "racial uplift" project that reflected the worldview of Booker T. Washington, and ultimately became the worldview and philosophical context of most African American educators and professionals at the Tuskegee Institute founded by Washington; such professionals were known as "race men/race women" (Smith 1996; Reverby 2009). Nurse Rivers joined The Movable School project in Tuskegee in 1923, a year after she had graduated from the Tuskegee Institute School of Nursing, and traveled to rural families teaching public health hygiene such as teeth brushing; safe birthing procedures; social hygiene; household hygiene; first aid care, etc. (Smith 1996; Reverby 2009). The Movable School project lasted until 1944, through the formative years of the Tuskegee study.

The John A. Andrew Memorial Hospital was founded on the campus of the Tuskegee Institute, and it primarily serviced the staff and students of the Tuskegee Institute. There were 16 private physicians (one who was African American) in Macon County during the 1930s who would treat African American patients for a fee; that essentially made seeing a doctor inaccessible for this population, while the sole African American physician was overwhelmed.

Remarkably, a commissioned sociological study published as a book in 1934, entitled Shadow of the Plantation by Charles Johnson, an African American professor of sociology and later, President of Fisk University, tracked living conditions during this time frame in Macon County. This book was based on qualitative interviews with roughly 600 families in Macon County in 1932, and was sponsored by the Rosenwald Fund (see under History of Medicine). Johnson was previously noted for his 1931 book, The Negro in American Civilization (Johnson 1931). As one scholar notes: Shadow of the Plantation "took on a racial myth, the conception of the easy-going plantation life and the happy Negro, and replaced the myth with the objective truth: Macon County was a twentieth century form of feudalism based on cotton cultivation" (Encyclopedia.com 2019). Ultimately, qualitative analysis of Johnson's interviews at this time paints a picture of underdeveloped infrastructure, a population living in impoverished conditions with a myriad of chronic health problems; deaths from several preventable diseases for that time frame; and very poor medical literacy. Johnson conducted several interviews surrounding health and community interactions with doctors involved in treating syphilis through a Rosenwald funded study (see under History of Medicine). He observed that none of the interviewees adequately understood or were informed about what they suffered from, and what the treatments they were given were designed to do. Johnson also served as an Advisor to the Hoover and Roosevelt administrations regarding rural issues, and his grandson, Jeh Johnson, would go on to serve as Director of Homeland Security in the Obama Administration.

The Generation Gaps: The View Between Plessy *and* Brown

Miss Evers' Boys squarely confronts the role that the African American healthcare providers played as either enablers, victims or champions of the Tuskegee study, creating ethically problematic assessments of their historical legacies. Hence, what must be stressed when teaching this film is that there were three distinct generations within the African American experience relevant to the Tuskegee study.

For example, there is an enormous chasm in the ethical, legal and social experiences that defined the world views of sociologist Charles Johnson, author of Shadow of the Plantation (Johnson 1934), who was born in the "Plessy Generation" (born

1893 and died 1956) and his grandson, Jeh Johnson, who was born in the "post-Brown Generation" in 1957—three years after the landmark Supreme court decision *Brown v. Board of Education* that overturned the "separate but equal" legal doctrine (see further). Essentially, the adult worlds of Charles and Jeh Johnson—grandfather and grandson—were as starkly different as being born into slavery or freedom and there was a critical generation in-between: "The Moses Generation". The Moses Generation would become the leaders and participants of the Civil Rights movement, who would eventually define the pre- and post-Civil Rights era. The Moses Generation roughly comprised African Americans born between 1925 and 1945, who were between 20 and 40 years-old when major Civil Rights legislation was passed. For example, Malcolm X was born in 1925; Martin Luther King Jr. was born in 1929, and Jesse Jackson, a close "disciple" of King's, was born in 1945. Those who were too young to participate in the Civil Rights movement, like Jeh Johnson and Barack Obama, were the direct beneficiaries of those struggles.

This means there are different African American generational lenses from which to judge the Tuskegee study, and the answer to how morally complicit Moton, Dibble or Rivers were in "aiding and abetting" what is now viewed as an unethical, racist protocol depends on this generational lens.

Moton, as mentioned above, was born two years after the Civil War ended, was 29 years old when *Plessy v. Ferguson* was decided, and died a year before the attack on Pearl Harbor. Eugene Dibble, involved in the Tuskegee study for most of its duration, was three years-old when *Plessy v. Ferguson* was decided, and died at age 75 on June 1, 1968—two months after Martin Luther King was assassinated, and less than a week before Robert F. Kennedy would be assassinated. That year marked one of the most demoralizing and disturbing years in twentieth century American history.

Eunice Rivers was born in 1899, three years after *Plessy v. Ferguson* was decided, when there was not even a *theoretical* constitutional right to vote because she was female. Although Rivers was 22 when the 19th Amendment was passed in 1921, it was meaningless for African American women in the South (Staples 2018).

Rivers was 33 years-old when the Tuskegee study began; she didn't marry until 1952—when she was 53—and she retired at age 66 in 1965, when the *Voting Rights Act* was passed. Rivers spent most of her life living under the apartheid conditions of the Jim Crow South, but also in a pre-women's rights era. By 1972, when the Tuskegee study was made public by the press, she was 73 years-old. Most African Americans under 65 at that time were horrified and offended by what they saw as an overt racist protocol, but perhaps not that surprised given their lived experiences. But those born in the Plessy Generation, which included Dibble and Rivers, saw things quite differently. As for the impact of the 1954 *Brown* decision, it would be well into the 1970s before anything changed in the deep South because there was no firm deadline for desegregation provided in the *Brown* decision (see further). Even the American Medical Association (AMA) would not desegregate until 1968 (see further). I discuss the professional roles of Dibble and Rivers in the Tuskegee study in the History of Medicine section further on.

Rivers Runs Deep

Eunice Rivers—dubbed "You Nice" in *Miss Evers' Boys*—was born a "Negro woman" in the Plessy era and essentially reached about the highest status possible within the racial and gender boundaries that existed for her. Rivers was encouraged by her parents to get an education to avoid working in the fields. Because of the *Plessy* decision, states had to comply with "separate but equal" by ensuring there were some places African Americans could go for post-secondary training because there needed to be African American professionals to take care of "their own". Since African Americans were barred from white establishments, they could not be taught by white teachers; they could not go to white hospitals; they could not be treated by white nurses, and so forth. Thus, the Plessy system encouraged all black post-secondary institutions, and the "Black College/University" system began, which truly offered excellent post-secondary and graduate educations. Eunice Rivers thrived in that system, and graduated with her nursing degree at age 22 in 1921, just in time for the 19th Amendment that gave mostly white women the right to vote. Working in a respected profession, with a lot of autonomy and a living wage, Eunice Rivers, who remained single until her 50s and childfree (potentially recognizing how trapped women became when they had children), was living a much more productive, and financially independent life than most women were at that time, regardless of race.

Eunice Rivers was a product of the African American professional worldview known as "racial uplift" and "racial betterment". As a proponent of this philosophy, she was known as a "race woman" (Reverby 2009; Smith 1996). This was a philosophy of "progress for the race" informed by a segregated existence, which emphasized community improvement through education, research and training. This was the "long game"—perseverance with the goal of creating a strong professional middleclass that would be able to self-govern within the confines of "separate but equal". Terms like "credit to his/her race" were used frequently as a "compliment" in this time frame. For example, actress Hattie McDaniel, also born in the Plessy Generation in 1895, and essentially a contemporary of Rivers, was the first African American to win an Academy Award for Best Supporting Actress for her role as "Mammie" in *Gone With the Wind* (1939). In her acceptance speech on February 29, 1940 she stated: "I sincerely hope I shall always be a credit to my race and to the motion picture industry." (Academy Awards 2019). We all cringe now when we see this speech, but it is an important reminder of the historical context of the Plessy era. Even in 1972, when the Tuskegee study was made public, white audiences did not view the portrayal of African Americans in *Gone With the Wind* (1939) as racist, and even accepted it as a reasonable portrayal of the Civil War. The next African American woman to win an Oscar after McDaniel would be Whoopi Goldberg in 1991 for *Ghost* (1990), just around the time she began dating Ted Danson, who produced *Miss Evers' Boys*.

Rivers' career was much more focused and successful than most women could hope for at that time. At 22, she had a job in The Moveable School Project, in which she helped hundreds of families as a public health nurse and educator, including educating midwives and registering births and deaths for the state. She saw terrible conditions, and saw her role as truly contributing to the "racial uplift/racial betterment" movement. Cutbacks at the start of the Depression led to her being laid off, and she picked up a night supervisor position at the Tuskegee Institute hospital. She did that job for about 10 months, considered leaving to work in New York, when she was tapped to be the part-time "scientific assistant" for the 1932 Tuskegee study (Reverby 2009). (I discuss that role more under History of Medicine.) Undoubtedly, Rivers considered this role to be very worthwhile given the medical viewpoints surrounding race, medicine and syphilis at the time (see Under History of Medicine). She also stated in interviews that she was noted to be politically skilled at handling (presumably sexist and racist) "white doctors" and that O.C. Wenger could "yell all he wants. I don't even hear him…He'll be the one dying of high blood pressure, not me" (Reverby 2009).

When Rivers agreed to take the part-time position, it initially paid $1000 annually, plus $600 a year for gas money and transportation expenses (Jones 1993), which in 1932 was equivalent to $22,000. By 1933, the transportation stipend was reduced to $200. That was for a *part-time* role. She continued to work for the health department's maternity service; worked as school nurse; and taught in the Tuskegee Institute's School of Nursing. All of her jobs likely added up to at least the equivalent of an annual salary of $35,000 in today's numbers. These were extraordinary earnings for a single African American woman in this time frame, when during the Depression, more than 24% of the country suffered from unemployment, with African Americans shouldering an even larger percentage of the burden. According to statistics published by the Department of Labor (Olenin and Corcoran 1942) the average white male factory worker in 1932 earned $17.86 per week at a full-time job for $857.00 annually, which was considered a living wage intended to support an entire family. In 1932, you could buy a decent house for about $500 in most areas (the very nicest homes in the country were around $4000). In Alabama specifically, "personal annual income fell from an already low $311.00 in 1929 to a $194.00 in 1935" (Downs 2015). In 1932, Nurse Rivers was very successful, exceeding the income of even most white males.

Rivers was interviewed only once by the ad hoc advisory panel in 1973 but she never gave Congressional testimony (Reverby 2009; Jones 1993; Washington 2007), which is why Feldshuh provided her with a fictionalized forum in which to "testify" to the audience. Remarks Reverby (2009): "Her only defense of her role publicly available is in three oral history interviews done in the late 1970s, and in her deposition from the class action law suit filed by Fred Gray…The real Nurse Rivers did not leave us a lot to understand if she was conflicted" (Reverby 2009). Yet I would venture to dispute this claim when teaching about Rivers. All one needs to do is look at the historical context and lived experiences in the Plessy era to understand Rivers and her peers.

African Americans in Medicine and the Tuskegee Institute 1932–1972

There was a thriving African American medical community in 1932, however it was indeed segregated, and physicians were barred from becoming members of the AMA until it desegregated in 1968. As a result of segregation, African Americans formed their own medical association known as the National Medical Association (NMA), which was instrumental in helping to get civil rights legislation enacted, and in helping to desegregate healthcare for African Americans. The NMA continues to focus on improving health disparities. Here is how the NMA describes its origins and early agenda (NMA 2018).

> Under the backdrop of racial exclusivity, membership in America's professional organizations, including the American Medical Association (AMA), was restricted to whites only. The AMA determined medical policy for the country and played an influential role in broadening the expertise of physicians. When a group of black doctors sought membership into the AMA, they were repeatedly denied admission. Subsequently, the NMA was created for black doctors and health professionals who found it necessary to establish their own medical societies and hospitals....
>
> The discriminatory policies of the nation at the time the NMA was founded manifested countless examples of the inadequacies of a segregated health care system. A priority item on the first NMA agenda was how to eliminate disparities in health and attain professional medical care for all people.
>
> Racism in medicine created and perpetuated poor health outcomes for black and other minority populations. In the South, hospital accommodations were frequently substandard. If blacks were admitted to general hospitals at all, they were relegated to all-black wards. In some instances, white nurses were prohibited from caring for black patients. Conditions in the North were also inequitable. It is reported that as late as 1912, only 19 of New York City's 29 hospitals would admit black patients, and only three gave black physicians the right to tend to their patients or perform operations.

With respect to medical training for African Americans, some physicians had trained at white institutions in the North, which accepted the occasional African American applicant, but most trained at a number of black colleges and universities that grew as a result of segregation, reinforced by the *Plessy* decision. This is discussed more in Chap. 8.

The Tuskegee Institute and its medical training were thus anchored and established in the post-*Plessy* context. The Tuskegee Institute started with a nursing school and expanded to medical school training in 1902. Eugene Dibble co-authored a history of the Tuskegee Institute's medical school training in 1961, and pointed out that the expansion of the medical school was also intended to "provide hospital facilities in which qualified Negro physicians had full privileges to treat their patients" (Dibble et al. 1961). Dibble's article also noted:

> John A. Kenney who completed his residency at Howard University, became resident physician and superintendent of the Tuskegee Hospital and Nursing school in 1902. In 1921, Dr. Kenney organized and conducted the first postgraduate course in medicine and surgery for

Negroes in the South. Dr. Eugene H. Dibble, Jr. was Dr. Kenney's first assistant and aided in making the course a success.

Ultimately, the Tuskegee Institute and John. A. Andrew Memorial hospital developed into a robust academic medical center and teaching hospital for African American students.

Dibble received his medical degree in 1919 from Howard University Medical School, and interned at the hospital affiliated with Howard University—Freedmen's Hospital in Washington, D.C. He then completed his surgical residency at the John A. Andrew Memorial Hospital in Tuskegee, Alabama in 1923, where he served as medical director from 1925–36. Dibble next did a stint at the Veterans Administration (VA) hospital in Tuskegee as its medical director and manager (1936–46), and then went back to serving as medical director at the John A. Andrew Memorial Hospital from 1946–1965. Dibble was essentially a continuous presence throughout most of the Tuskegee study years; he was forced to retire due to a cancer diagnosis and died three years later. Dibble also served as a member of the Board of Trustees at Meharry Medical College in Nashville, Tennessee, and was on the Editorial Board of the *Journal of the National Medical Association*—the same journal that published his 1961 article about his academic home institution (JNMA 1962; Reverby 2009). In that paper (Dibble et al. 1961), he even made a brief proud reference to the Tuskegee study:

> Included in the long list of cooperative projects between the John A. Andrew Hospital and State and Federal health agencies is the U.S. Public Health Service study of syphilis in the Negro male in Macon County which began in the fall of 1932. The hospital cooperated in this plan to give each of 600 patients a complete physical examination, including chest x-rays.

Dibble's specific role in the Tuskegee study is discussed further (see under History of Medicine), but it was he who convinced the President/Principal of the Tuskegee Institute, Robert Moton, one of the most respected African American university presidents, to participate in the USPHS syphilis study (Jones 1993; Reverby 2009). Moton's role in the study was limited, and essentially amounted to approving the study and providing university resources for it. Moton retired three years after the Tuskegee study began (in 1935), but there was no reason to believe that Moton would not have felt the same way as Dibble about the research project and have a similar world view regarding racial uplift projects in a segregated society based on his own words and writings. Moton was a keynote speaker at the opening of the Lincoln Memorial on May 30, 1922, invited to "speak for his race" and also agreed to having his speech edited/censored by the organizers. Of course, given the Plessy era, he was also not allowed to sit with the other white speakers in the audience (NPS 2019; National Museum of American History 2019). On that notable day, long planned by the Lincoln Memorial Commission (chaired by former President Taft), and the American Commission of Fine Arts (whose Chair had perished on the Titanic), Abraham Lincoln's 78 year-old son, Robert Todd Lincoln, was in the audience when Moton spoke. His speech was also broadcast on the radio (Furman 2012). As one historian observes (Fairclough 1997):

Robert Russa Moton, viewed the Lincoln Memorial as a moral symbol of the African-American fight against discrimination. He intended to deliver a passionate plea for racial justice. The speech Moton actually read out, however, contained no language of militant protest: at the insistence of the Lincoln Memorial Commission, he substituted soothing bromides reminiscent of Booker T. Washington's Atlanta Exposition address of 1895... The affair also throws an interesting light on Moton, himself, now a largely forgotten figure, but then, by virtue of his position as president of Tuskegee Institute—Booker T. Washington's successor—a man widely regarded by whites as the preeminent spokesman for Black America.

Moton's uncensored speech, included the following in his closing remarks (Fairclough 1997):

[S]o long as any group does not enjoy every right and every privilege that belongs to every American citizen without regard to race, creed or color, that task for which the immortal Lincoln gave the last full measure of devotion-that task is still unfinished...More than sixty years ago [Lincoln] said in prophetic warning: 'This nation cannot endure half slave and half free: it will become all one thing or all the other. With equal truth it can be said today: no more can the nation endure half privileged and half repressed; **half educated and half uneducated**; half protected and half unprotected; half prosperous and half in poverty; **half in health and half in sickness**; half content and half in discontent; yes, half free and half in bondage...This memorial which we erect in token of our veneration is but a hollow mockery, a symbol of hypocrisy, unless we together can make real in our national life, in every state and in every section, the things for which he died.... A government which can venture abroad to put an end to injustice and mob-violence in another country can surely find a way to put an end to these same evils within our own borders....honor. Twelve million black men and women in this country are proud of their American citizenship, but they are determined that it shall mean for them no less than any other group, the largest enjoyment of opportunity and the fullest blessings of freedom. Let us strive on to finish the work which he so nobly began, to make America the symbol for equal justice and equal opportunity for all.

And here is what Moton actually said that day (Fairclough 1997):

Here we are engaged, consciously or unconsciously, in the great problem of determining how different races cannot only live together in peace but cooperate in working out a higher and better civilization than has yet been achieved. At the extremes the white and black races face each other. Here in America these two races are charged...with the responsibility of showing to the world how individuals, as well as races, may differ most widely in color and inheritance, and at the same time make themselves helpful and even indispensable to each other's progress and prosperity. This is especially true in the South where the black man is found in greatest numbers. And there today are found black men and white men who are working together in the spirit of Abraham Lincoln to establish in fact, what his death established in principle.... In the name of Lincoln, twelve million black Americans pledge to the nation their continued loyalty and their unreserved cooperation in every effort to realize in deeds...

Moton's May 30, 1922 remarks about race, though cryptic and filled with subtext, essentially paraphrases the "racial uplift" philosophy that was espoused by his predecessor, Booker T. Washington, and demonstrates how he resolved to work within the Plessy system with his white peers. But it also informs how he likely saw the proposed USPHS syphilis study that Dibble brought forward for his approval a decade later. Undoubtedly, Moton, Dibble, Rivers and their peers at the Tuskegee Institute, regarded the USPHS syphilis study as a way of "showing to the world how individuals, as well as races, may differ most widely in color and inheritance, and at the same time make themselves helpful and even indispensable to each other's progress and prosperity" (Moton 1922; Fairclough 1997). Moton also wrote in 1929 (Moton 1929):

> Here met the three elements—the North, the South, and the Negro—the three elements that must be taken into account in any genuinely satisfactory adjustment of race relations... Up to this time the Negro had usually been the problem and not regarded as an element worthy of serious consideration, so far as any first-hand contribution was concerned that he could make toward the solution of any large social question....and a cooperation vitally necessary in the promotion of any successful work for the permanent betterment of the Negro race in our country....

> No greater or more serious responsibility was ever placed upon the Negro than is left us here at Tuskeegee. The importance of the work and the gravity of the duty that has been assigned the principal, the officers, and the teachers in forwarding this work cannot be overestimated. But along with the responsibility and difficulties we have a rare opportunity, one almost to be envied-an opportunity to help in the solution of a great problem, the human problem of race, not merely changing the mode of life and the ideals of a race but of almost equal importance, changing the ideas of other races regarding that race.

Ultimately, Moton's many works (Moton 1913; Moton 1916; Moton 1920; Moton 1922; Moton 1929) reveal his perspective as a proponent of the racial uplift philosophy.

Dibble as a "Race Man"

As discussed in earlier sections, Dibble's world view regarding race, like Moton's and Rivers', was situated in "racial uplift/racial betterment" framework in a segregated society he thought would never change. Dibble was motivated to participate in research endeavors that would provide scientific stature to his institution, his peers and mentees. In his correspondence surrounding the Tuskegee study, he echoed Moton above and saw it as an "opportunity" to bring focus and attention to African American health needs in a time when the population was completely ignored (Reverby 2009; Jones 1993). In some ways, the African American medical community at that time mirrored the professionalization goals of their Jewish peers; Jewish physicians, too, needed to establish exclusively Jewish hospitals so they could train their own doctors, as there were severe quota restrictions in traditional academic medical centers discussed more in Chap. 9.

Dibble's aim to bring his institution into national focus with important research was best exemplified by the Tuskegee Institute's HeLa cell research in the 1950s. In 1951, an African American patient, Henrietta Lacks, was treated at Johns Hopkins University hospital for ovarian cancer, and died. Her tumor cells were propagated in a laboratory by George Gey, and became the first immortal cell line, known as HeLa cells. The Tuskegee Institute would become well-known the following year for its research with that cell line. In 1952–3, "a staff of six black scientists and technicians built a factory at Tuskegee unlike any seen before." HeLa cells were "squirted" into one test tube after another, and "the Tuskegee team mixed thousands of liters of [George Gey's] culture medium each week, using slats minerals, and serum they collected from the many students, soldiers, and cotton farmers who responded to ads in the local paper seeking blood in exchange for money" (Skloot 2010). The science journalist, Rebecca Skloot, author of The Immortal Life of Henrietta Lacks, noted this:

> Eventually the Tuskegee staff grew to thirty-five scientists and technicians, who produced twenty thousand tubes of HeLa—about 6 trillion cells—every week. It was the first-ever cell production factory and it started with a single vial of HeLa that Gey had sent [to a researcher] in their first shipping experiment, not long after Henrietta's death. With those cells, scientists helped prove the Salk vaccine effective. Soon the *New York Times* would run pictures of black women hunched over microscopes examining cells, black hands holding vials of HeLa and this headline: 'Unit at Tuskegee helps Polio fight. Corps of Negro Scientists has key role in evaluating of Dr. Salk's vaccine.' Black scientists and technicians, many of them women, used cells from a black woman to help save the lives of millions of Americans, most of them white. And they did so on the same campus—and at the very same time—that state officials were conducting the infamous Tuskegee syphilis studies. (Skloot 2010: 96–97).

In 1964, while the Tuskegee study was in its last stages, an egregious trial at a Jewish hospital in Brooklyn would be "outed" that involved injecting live HeLa cancer cells into unsuspecting elderly patients (Beecher 1966; AHRP 2014). It's important to note that the Tuskegee study had plenty of company in research ethics infamy (see further under Healthcare Ethics Issues), partially due to a vacuum of research ethics awareness, training, and appreciation of vulnerable populations.

The Impact of Brown v. Board of Education

"Inferior school facilities for children and widespread illiteracy went hand in hand with poverty in Macon County, despite the presence of the Tuskegee Institute. Year in and year out Alabama ranked at or near the bottom nationally in the amount of money spent per pupil on education, and Macon County was not a leader in black education within the state" (Jones 1993). In 1932, 23 out of 1000 whites were illiterate in Macon County; in the African American population 227 out of 1000 were illiterate (Jones 1993). See Chap. 9 for more on this.

As a result of the *Brown* decision, states were required to dismantle dual systems of higher education, and white institutions were ordered to accept African American students and desegregate. However, in the South, desegregation was very slow due to specific wording in the *Brown* decision that was seen as a loophole for delaying desegregation: "with all deliberate speed" (Harvey et al. 2004; *Brown v. Board of Education* 1954). The effect of the *Brown* decision on higher education and black colleges and universities such as the Tuskegee Institute was a slow erosion of resources that one can see in Dibble's 1961 paper on the goals for his medical institute, which had conducted a needs assessment published in the aftermath of *Brown*, which he called the "Tuskegee Institute Study" (having nothing to do with the syphilis study). He wrote (bolded emphasis mine) the following as a request for funding which clearly demonstrates Dibble's desire to be recognized as a leading academic medical center (Dibble 1961):

> As part of the comprehensive Tuskegee Institute Study, a committee representing the medical and administrative staff of John A. Andrew Hospital has compiled, studied and analyzed data involving the hospital's operation over the past 15 years. Some of the country's leading hospital and medical consultants and other qualified individuals in the health field who are familiar with John A. Andrew's program agree with this committee's report **that more adequate physical facilities and equipment are needed to provide up-to-date methods of treatment and for continued high quality medical care...**

> [Our goals for the future are]: To engage in research; to provide post-graduate training for physicians...to train nurses; to train doctors... [to serve our] medically indigent pregnant women and new mothers in the hospital...[and to] provide facilities, and personnel dedicated to the treatment of many medically indigent people in this area...To study, **in cooperation with County, State, and Federal health agencies and other health organizations,** the incidence of disease and high morbidity and mortality rates in this area...**To organize clinics in rural areas to demonstrate what can be done in cooperation with existing public health agencies...**

Years after the *Brown* decision, black colleges/universities were still segregated but suffered budget cuts, lack of adequate libraries, research equipment, and so forth (Coaxum 2018; Brownstein 2014). The Tuskegee Institute was no exception, and Dibble's 1961 article makes clear that he was concerned about his shrinking resources. Ultimately, the Tuskegee Institute would become a shell of its former self by the 1970s and closed in 1987 as the last black hospital in Alabama, reopening in January 1999 as part of the Apology "package" as the National Center for Bioethics in Research and Healthcare in January 1999.

The Tuskegee Institute in the Civil Rights Era

What would be marked historically as the "civil rights era" began with the *Brown* decision, and its social impact in Alabama was enormous, as that state would became a focal point of non-violent protests led by the Moses Generation. For example, on December 1, 1955, 42 year-old Rosa Parks, born in 1913 (when Dibble was 20), became the modern "Plessy" by refusing to sit at the back of the bus, just as Plessy had refused to sit in a different train compartment. The Montgomery Bus Boycott (1955) followed, as the Moses Generation began to lead its people to the Promised Land. Moton's speech at the Lincoln Memorial occurred seven years before Martin Luther King Jr. was born, who would, of course, go on to deliver his "I Have a Dream" speech 41 years later at the very same memorial, culminating into the *Civil Rights Act* in 1964, followed by the *Voting Rights Act* of 1965, as well as landmark healthcare reform legislation (see further). By then, Rivers had retired; Dibble was struggling with cancer and not only retired, but was at the end of his life. It seems plausible that neither of them had particular faith that "The Times they are A-Changin'" because they had never known anything except a segregated world. It's likely they both felt "nothing was really changing" when they witnessed the Martin Luther King assassination. But it would be another document that would ultimately have much more impact on Dibble's and Rivers' professional legacies, and the fate of the Tuskegee Institute, published the same year as the *Civil Rights Act*; that document would be the World Medical Association's Declaration of Helsinki (1964) (see under Healthcare Ethics Issues). After Dibble's and Rivers' retirements, their roles would continue with others in their places. The Tuskegee Institute, a decade after the *Brown* decision, was now an HBCU by this period in history, but had not really left the Plessy era in terms of its mission. What would make the Tuskegee Institute into a truly historic—if not, anachronistic—institute was actually the dawn of a new era in medicine that came of age with Civil Rights: the post-Holocaust era and birth of modern medical ethics.

The Tuskegee Study Meets the Post-Holocaust Era

In the immediate aftermath of the Holocaust and the Nuremberg trials in the late 1940s, medical professionals were probably much more interested in penicillin than the Nuremberg Code (see under History of Medicine). Real post-Holocaust analysis and scholarship did not really emerge until the early 1960s, but Anne Frank: Diary of a Young Girl diary had been first published in the United States in 1952, with a Foreword by Eleanor Roosevelt (Anne Frank.org 2018). During the Civil Rights era, American Jews were particularly active and devoted to the cause of civil rights because it so deeply resonated with them. Many coming of age in the 1960s had either been the children of Holocaust survivors, or had relatives who perished. They viewed what was happening in the Jim Crow South as a comparable example to what had recently occurred in Nazi Germany.

Standing in remarkable juxtaposition to the *Civil Rights Act* of 1964 and The Declaration of Helsinki (1964), reaffirming the need for informed consent, was an unassuming journal article published in December of that same year, entitled "The Tuskegee Study of Untreated Syphilis" reviewing a 30-year study of "Negro males" who were followed, but not treated for syphilis (Rockwell et al. 1964). The paper was rather bland, and was written in a manner that was typical for medical journals. To Jewish medical professionals it started to raise alarm bells. One physician, Irwin Schatz, born the year the Tuskegee study began in 1932, was just four years out of medical school at the time (Roberts and Irwin 2015), and was prompted to write to the journal authors in early 1965 in disbelief that the study had been done. Schatz never received a reply, but archives show that co-author Anne Yobs indeed read, and disregarded, his letter (Reverby 2009). Another Jewish healthcare professional, Peter Buxton, who came across the study around 1966 when he was working for the CDC, would begin to review all of the documents about the long-running Tuskegee study and ultimately would become the main whistleblower who would bring it to its end (Jones 1993; Reverby 2009). I discuss the "whistleblowing phase" of the Tuskegee study further on.

The History of Medicine Context

When teaching *Miss Evers' Boys*, it's important, of course, to cover the history of the Tuskegee Syphilis Study, which actually comprised two studies over the course of its 40-year run. The first study was in fact, a treatment protocol sponsored by the Rosenwald Fund; it was the *second* study that was the infamous observational protocol, which became known by its 30th Anniversary by its journal article title: "The Tuskegee Study of Untreated Syphilis" which begins like this (Rockwell et al. 1964):

> The year 1963 marks the 30th year of the long-term evaluation of the effect of untreated syphilis in the male Negro conducted by the Venereal Disease Branch, Communicable Disease Center, United States Public Health Service. This paper summarizes the information obtained in this study—well known as the "Tuskegee Study"—from earlier publications, reviews the status of the original study group, and reports the clinical and laboratory findings on those remaining participants who were examined in the 1963 evaluation.

However, none of the Tuskegee study's history will make any sense to your students without the accompanying context of the history of African American healthcare, and what actually counted as "medical knowledge" published in prestigious medical journals in a prevailing false narrative about medicine and race throughout the nineteenth and early twentieth centuries. The Tuskegee study cannot be understood, either, without explaining the history of treating syphilis, which includes clinical equipoise over the use of penicillin in later stages of syphilis. Finally, to put this into a more current perspective, it's important to discuss current trends surrounding observational protocols in other diseases, such as prostate cancer, cervical cancer and thyroid cancer, in which patients have been/are being observed, but not treated with the standard of care.

History of African American Healthcare

First, access to healthcare in the general American population has always been a problem for the poor or underclass, for whom seeing a doctor was considered to be a luxury. There really was no such thing as regular primary care for most Americans until the early twentieth century, and most did not see a doctor if they didn't have an acute problem. In the antebellum period, slave owners would have doctors come and treat their slaves as a matter of "household maintenance" since they were considered property. Sick slaves were not good for business. "Apart from humanitarian considerations, the economic value of slaves made their health a matter of solicitous concern" (Jones 1993).

After the Civil War, when the African American medical community began to grow, so did awareness of health disparities in urban areas. But Macon County, Alabama serves as a microcosm for what healthcare access really looked like for the poor, rural, African American population; the health of its people was surveyed and documented in 1934 (see earlier), and the results were alarming.

In the Plessy era, the majority of African Americans typically had no access to white physicians or other healthcare providers either due to segregation or cost. Most "went from cradle to grave deprived of proper medical care. 'I ain't had a Dr. but once in my life and that was 'bout 15 years ago' an elderly black resident confessed in 1932" (Jones 1993). Cost prevented them from seeing private practice white physicians who would take care of any patient regardless of race so long as they got paid. People with incomes less than a dollar day had absolutely no way to pay for healthcare. There was also consensus that chronic poor health in the African American population affected white Americans, too, as their economy depended upon a healthy "servant class" or cheap labor class. This is why there was essentially a white medical-establishment consensus that black medical schools and black nursing schools within the black college/university system were critical.

As discussed earlier, the conditions in Macon County, Alabama were what we might call "third world" today, and healthcare access was essentially a pipe dream, as even basic living essentials were scarce. This is why the Tuskegee Institute's Movable School project (see earlier) included public health/hygiene house calls, as well as maternity care, as most babies were born at home. By the later nineteenth century, there was a burgeoning African American medical community, which is why the National Medical Association was established in the first place. These medical professionals, from the start, indeed serviced their communities and the patients they could, but there were limited human resources for the need. Many physicians also sought to improve access to healthcare through community/public health research projects.

African American veterans' care was a particular issue. At the Tuskegee Institute, one major development was for Robert Russa Moton to make a major push to have an all-black VA Hospital on its premises, to serve returning World War I African American veterans he had personally visited in Europe at the request of President Woodrow Wilson. Of note, African American soldiers were believed to have higher incidences of syphilis than white soldiers, but there were never any studies that actually determined incidence. However, they were still allowed to serve when white soldiers with syphilis were not (Reverby 2009). Moton even describes his fight for a black VA hospital in his 1921 autobiography.

Thus, when the Tuskegee study began in 1932, although access to healthcare for the majority of rural African Americans living in Macon County was minimal (with the exception of the traveling public health nurses in the Movable School project), there was at least recognition by the white medical establishment that this community was in poor health, and health disparities were a real problem.

Medical Misconceptions About Race

Layered onto the healthcare access problem were absurd and false theories about physiological, psychological, and medical differences between black and white races.

These were not sensationalized articles the masses read from non-credible sources; these were published as medical facts in high quality peer-reviewed medical journals, and the false theories about race were read, and accepted, by physicians of all colors and religions. In fact, many of these theories helped to support a flourishing eugenics movement. It's critical to note that Eugene Dibble was not immune to many of the then-accepted scientific facts about race—one reason he was motivated to participate in the Tuskegee study.

For example, there was a pervasive view that African Americans were intellectually, mentally and physiologically inferior (Jones 1993); did not feel pain the same way as whites (Washington 2007); had uncontrollable, violent or larger sexual appetites (Jones 1993); and did not experience diseases (e.g. malaria) the same as in white patients, and so may not require the same treatments (Jones 1993; Washington 2007; Reverby 2009). At the same time, prominent physicians also considered them more prone to disease and "contamination" as the "weakest members of society" who were "self-destructive" (Jones 1993); many viewed syphilis as a result of immoral sexual indulgence (Jones 1993). Social realities helped to falsely validate some of these beliefs, which is why the leaders of the Tuskegee Institute at the time strongly believed that the education of African Americans was tied to the health of African Americans—one of the key missions of the Institute's racial uplift framework. Indeed, illiteracy and low education gave the appearance of lower intellect, which translated for the white medical profession into fixed beliefs about African American patients' inability to consent to, or comply with, treatment, justifying deception in "the patient's best interests".

A revealing book that illustrated the psychological effects of institutionalized white supremacy was authored by the Tuskegee Institute's president, Robert Russa Moton in 1929. It exposed an obsequious tone that tells us a lot about Moton's modus operandi: working cooperatively within the white establishment. The book was called: What the Negro Thinks (1929). Its Foreword, which is cringe-worthy today, starts:

> This volume aims to place on record some facts concerning a phase of the Negro problem of which, up to this time little has been known outside of the race; that is, what the Negro, himself, thinks of the experiences to which he is subjected because of his race and colour. The subject has lately excited a growing interest, especially among those who would approach the problem of the Negro's presence in America with sympathy and understanding…

Of note, Moton's 1921 autobiography, Finding a Way Out, actually discussed how well he and his family members were treated by their white masters, and that the abuse of slaves was really more at the hands of the white overseers. Clearly, Moton, who was born in 1867, was of an entirely different era. His agreement to cooperate with the Tuskegee study is not at all a mystery, given his writings and worldview.

The second myth of not feeling pain was reinforced by African Americans not "seeking help" for many acute and chronic health problems, which translated for the medical profession as thinking about African Americans as more robust research subjects. The third myth about hypersexuality was reinforced by the legal system perpetuating false rape charges (see below), which translated into beliefs that African

Americans were "syphilis soaked" (Jones 1993) and a threat to white civilization. But the wildest racial misconception of all—which "underwrote" the Tuskegee syphilis study was that the white race was actually superior, scientifically justifying that using African Americans in experiments was more ethical than using actual "human subjects" (meaning, *white* human subjects). These sentiments of using an underclass that was seen as "subhuman" justified the medical experiments on Jews during the Holocaust, too.

Interracial Sex and Rape

Within the white medical establishment, there was recognition that interracial couples existed, even if they were closeted, which certainly meant they could infect one another with diseases. When such relationships were "outed" in the South between a white woman and African American male, a common theme was for the African American male to be either lynched or put on trial for rape, and then convicted with little evidence. Or, they would be falsely accused when the perpetrator was white. In fact, a major trial in the news in 1931 Alabama, involved nine young African American males, known as the "Scottsboro Boys", who were arrested on false charges for raping two white women (Reverby 2009). These consequences helped to deter interracial intercourse. In fact, this was such a common American experience and narrative that when Harper Lee based her Pulitzer prize winning novel, To Kill a Mockingbird (1960) on her own childhood observations, it resonated deeply in the civil rights era with many whites who had moral distress over these mockeries of the legal system. Lee's novel is set in 1936 Alabama, and it is her father's moral integrity for justice that served as a role model for many white lawyers litigating civil rights cases who read the book growing up.

Actually, it was much more common for white males to either gangrape, or individually rape an African American female with no consequences, upholding a long tradition of white slave owners abusing their slaves. But syphilis experts of that time were less concerned with legal justice and more concerned with implications for public health. If a white abuser became infected, he could presumably infect his white wife, and even unborn child through congenital syphilis, although the latter was not well understood at the time (Jones 1993). There was an unsubstantiated medical theory that syphilis in the white population was a potentially more serious disease than syphilis in the African American population (Jones 1993; Reverby 2009).

Within African American culture sexual abuse was also a big problem, and the painful tale of Alice Walker's rural 1930s Georgia character, Celie, in The Color Purple (1982) "outed" the secret lives and suffering of African American females within their own family units.

All of this context is necessary to truly understand the origins of the Tuskegee syphilis study. Reverby (2009) states that "historians and members of the public… who cannot comprehend the context of racism and violence that shaped what happened," are analyzing this trial in a vacuum.

The Pathogenesis and Treatment for Syphilis

Syphilis first appeared in Naples, rumored to have been brought back to Europe by Columbus' team of men, leading to the "syphilization of Europe" (Reverby 2009). No one understood the disease's stages until the early twentieth century (Reverby 2009), and in 1906, Paul von Wassermann developed "a complement fixation serum antibody test for syphilis" (Ambrose 2016), known as the Wassermann reaction test. This new diagnostic test could detect antibodies to syphilis based on a reaction when blood or spinal fluid was introduced to another chemical; the more severe the reaction, the more progressive the disease. However, the Wassermann reaction test produced false positives and negatives and there was long debate over its specificity. It was the gold standard however well into the 1970s.

There was *no good treatment for syphilis in 1932*, and no cure. Treatment was based on mercurial, arsenical and bismuth-based therapies (referred to as the "heavy metal" treatment), but it was not a cure, had many side-effects, and not considered very effective in later stages. Fever therapies were also used for neurosyphilis, which involved infecting the patient with malaria or a bacteria to induce a fever (Ambrose 2016; Reverby 2009).

One of the least understood aspects of the Tuskegee study surrounds who was being followed and denied treatment for syphilis. There are four stages of syphilis: primary, secondary, a latency period, and a tertiary stage (a.k.a. late-stage syphilis). The Tuskegee study followed, and did not seek to treat, men who had entered their latency phase, or who had late-stage syphilis; the study was never designed to enroll men with primary or secondary syphilis. The study did not enroll any women, or treat any women at-risk even though some physicians at that time saw more women with syphilis than men (Jones 1993). Finally, the Tuskegee study did not infect anyone with syphilis, which was a rumor historians have spent decades trying to dispel (Reverby 2009). That said, by not treating syphilis in the men they followed, opportunities for new infections occurred in women, which lent weight to that rumor.

The primary stage of syphilis lasts from 10–60 days, starting from the time of infection. During this "first incubation period" the chancre appears at the point of contact, such as the genitals. This is the primary lesion of syphilis. The chancre heals without the need for treatment and will leave a scar but no other symptoms. The second phase begins at this point, with a rash that looks like the measles or chicken pox. The bones and joints can become painful and there may be complications with the circulatory system. There may be fever, indigestion, patchy hair loss, headaches, and a range of diffuse complaints. Sometimes there are moist open sores with spirochetes that are infectious. After these symptoms, prior to tertiary syphilis, there is a latency period that can last from several weeks up to 30 years; it was men in this phase who were being followed in the Tuskegee study, although the sloppy science and methods did not always follow men in later stages, and some had much earlier stages. In the latency phase, syphilis symptoms mostly disappear outwardly, but the spirochetes begin to do a lot of internal damage to the central nervous system (CNS). Many people with syphilis can live almost full lives and "coexist" with the disease, but more often, the signs of tertiary syphilis contribute to shortened life span. Tertiary syphilis affects the functioning of bones, liver, cardiovascular system, and the brain with neurosyphilis.

Due to the complexity of syphilis, a separate medical specialty of "syphilology" had developed by the early twentieth century, which essentially disappeared by the time penicillin became the standard of care. In addition to the heavy metals therapy available, there were also a number of ineffective home remedies and patented medicine designed to purify "bad blood" (Reverby 2009) often exploiting the poor. In the African American community, the term syphilis was never used to describe the disease; the term "bad blood" was used instead. But that term was a non-specific way to describe a number of health problems (Jones 1993; Reverby 2009).

From a public health standpoint, in the 1920s, there was interest in reducing contagion of syphilis, which had long been ignored because it was considered a disease associated with "immoral behaviors"—not unlike the AIDS discourse in the 1980s (see Chap. 2). A 1941 textbook noted that the heavy metals therapy regimen was the standard of care, and could remove contagion after "at least 20 injections each of an arsphenanimine and a heavy metal" (Reverby 2009). That was the standard regimen suggested for patients who were untreated and had syphilis for more than five years. But "more than anything else, syphilologists wanted to understand the disease's natural history to determine how much treatment was really needed" (Reverby 2009).

Legitimate Questions About 1932-Era Treatment

When the Tuskegee syphilis study began, it was before the penicillin era, and treatment for syphilis involved only the heavy metal treatment. "In the 1930s, one treatment schedule for [syphilis] included three intramuscular injections sequentially over a six- week period of arsphenamine, a bismuth salt, and a mercury compound." (Ambrose 2016).

In 1908 Paul Ehrlich was awarded the Nobel Prize for his discovery of an organic arsenical therapy for syphilis, which was later developed into the drug arsphenamine, known as Salvarsan, or the "magic bullet" (Frith 2012) and in 1912, neoarsphenamine, or Neo-salvarsan Arsenic had many toxic side-effects, was hard to administer, and required a lot of intramuscular injections over long periods of time. By 1921, bismuth was found to be effective, too, and made arsenic more effective when they were combined. Ultimately, "Arsenic, mainly arsphenamine, neoarsphenamine, acetarsone and mapharside, in combination with bismuth or mercury, then became the mainstay of treatment for syphilis until the advent of penicillin in 1943" (Frith 2012). In 1986, one author noted "the heavy-metal cure often caused thousands of deaths each year" (Frith 2012).

Ultimately, the delivery system for the heavy metals therapy generally involved salves and intramuscular injections, which is what is shown in *Miss Evers' Boys*. Because the heavy metals treatment was thought to be more beneficial in the early phases, and could have lethal side-effects, there were legitimate medical and clinical ethics questions about whether the current standard of care was beneficent beyond the latency period. But race interfered with the science, and polluted any legitimate study design. The 1932 study sought to address two medical research questions about syphilis, which the Tuskegee Institute's leadership also thought was worthy, and beneficial to their community: (a) Does the pathogenesis of syphilis manifest as the same disease in African American males as in Caucasian males? (b) Is treatment necessary once it reaches the latency phase, given that the side-effects of the 1932 heavy metals therapy can be worse than the disease?

Clearly, the first question was based on racist medicine frameworks. But this was a question the African American medical community wanted answered, too, because if the answer was "the disease is equally bad or worse" then health disparities in syphilis treatment could be improved. The second question was legitimate until the penicillin era emerged (see further).

It's imperative to impress upon students that the scientific rationale for a study of late-phase syphilis was based on *genuine clinical equipoise over treatment beyond latency*, as there were no good treatments at that time. This rationale became harder to defend in the penicillin era (see further). "Describing the dangers of the 1930s treatment regimes, [Nurse Rivers] claimed they were "really worse than the disease if it was not early syphilis...If syphilis was not active the treatment was worse than the disease' " (Reverby 2009).

Notes Jones (1993):

[A]s one CDC officer put it, the drugs offered 'more potential harm for the patient than potential benefit' " ...PHS officials argued that these facts suggested that the experiment had not been conceived in a moral vacuum for if the state of the medical art in the early 1930s had nothing better than dangerous and less than totally effective treatment to offer, then it followed that, in balance, little harm was done by leaving the men untreated...Apologists for the Tuskegee Study contended that it was at best problematic whether the syphilitic subjects could have been helped by the treatment that was available when the study began.

Notes Reverby:

By the time the study began in 1932, concern over treatment, debates over racial and gender differences, and the problematic accuracy of the [diagnostic] blood tests filled medical journals and texts. It was becoming clear that not everyone died from the [syphilis] or even became seriously sickened by it. From a public health perspective, as Surgeon General Thomas Parran argued in 1938: 'with one or two doses of arsphenamine, we can render the patient promptly non-infectious [but] not cured.' ...Discrediting the efficacy of mercury and salvarsan helped blunt the issue of withholding treatment during the early years, but public health officials had a great deal more difficulty explaining why penicillin was denied in the 1940s.

The first Tuskegee study subjects were enrolled when Herbert Hoover was President, while the country slid into the worst economic Depression in its history. The study design is historically located in Hoover's America—pre-Nuremberg Code: there was no informed consent; there was deception; there was coercion; there was bias. But ethical abuses notwithstanding, later data analysis revealed highly flawed methodology that became an insult to scientific integrity because the data became so badly polluted (see further).

The Oslo Study

The Oslo study, discussed in the film, had been the only study on the effects of untreated syphilis in the literature prior to the Tuskegee study. In this retrospective study, by Bruusgaard (1929), the investigators "had reviewed the medical records of nearly 2000 untreated syphilitic patients who had been examined at an Oslo clinic between 1891 and 1910. A follow up had been published in 1929 and that was the state of published medical experimentation on the subject before the Tuskegee Study began" (Jones 1993). This study was referenced in the 1973 Ad Hoc Panel Report like this:

> The Oslo study was a classic retrospective study involving the analysis of 473 patients at three to forty years after infection. For the first time, as a result of the Oslo study, clinical data were available to suggest the probability of spontaneous cure, continued latency, or serious or fatal outcome. Of the 473 patients included in the Oslo study, 309 were living and examined and 164 were deceased. Among the 473 patients, 27.7% were clinically free from symptoms and Wassermann negative.

In 1955, Gjestland, then chief of staff at the same Oslo clinic, revisited this data and published the findings (Gjestand 1955); he also advised the PHS study investigators in 1952 (see further).

Rivers was aware of the Oslo study, too, and relayed in interviews that she thought of the Tuskegee study as a comparative study in the African American male, having made the assumption that there could be racial differences (Reverby 2009).

Ultimately, the Oslo study became the "touchstone" for the Tuskegee study design, but there was an enormous difference: the Oslo study was retrospective, having assessed untreated patients in a time frame where there was no good treatment. The Tuskegee study design that became notorious (see further) was a prospective study that actively harmed its subjects. Ethically, Oslo and Tuskegee were very far apart.

The First Tuskegee Treatment *Study: The Rosenwald Fund Demonstration Project (1929–1932)*

The entire saga of the Tuskegee syphilis study essentially started, and ended, because of Jewish humanitarian efforts. It was Jewish philanthropist, Julius Rosenwald who got the whole ball rolling. Julius Rosenwald was born in 1862 during the Lincoln Administration. As a major player in the garment industry and other investments, he eventually became co-owner of the huge Sears, Roebuck and Company, touted as the "Amazon" of its day. Rosenwald's biographers describe him as having been passionate about human rights causes, and he became a philanthropist who also established the current donor model of "matching funds". Through discussions with his friend, Paul Sachs, co-founder of Goldman Sachs, and his own rabbi, Rosenwald was bitten by the philanthropy bug, felt that the plight of African Americans was close to his heart, and this became the focus of most of his philanthropy for the rest of his life. Rosenwald read Booker T. Washington's 1901 autobiography, Up From Slavery (1901), reached out to the author, and began a very close friendship with Washington. Ultimately, Rosenwald endowed the Tuskegee Institute and also match-funded a number of projects devoted to the education of African Americans. Rosenwald was asked to serve on the Board of Directors of the Tuskegee Institute in 1912, and remained a board member until he died. One biographer notes (Zinsmeister 2018):

> In 1912, Rosenwald…announced he would be celebrating his 50th birthday by giving away close to $700,000 (about $16 million in current dollars)…"Give While You Live," was his slogan. One of Rosenwald's birthday gifts was $25,000 to Washington's Tuskegee Institute. Soon Rosenwald and Washington were ramping up their program, [leading to] schools all across the South over more than 20 years.

In 1917, after Booker Washington's death, Rosenwald started the Rosenwald Fund, for "the well-being of mankind", which became the chief sponsor of several southern African American school projects (called Rosenwald Schools) prior to the *Brown* decision (see earlier and Chap. 9). The fund also offered $1000 "Fellowship Grants" from 1928–48 to various African American artists and intellectuals for their personal growth; recipients ranged from Marion Anderson to social scientists Kenneth and Mamie Clark, of the "Clark Doll Study" fame, which was ultimately used to argue the *Brown* decision, and was a "Who's Who of black America in the 1930s and 1940s" (Schulman 2009).

Finally, the Rosenwald Fund expanded into "medical economics" (Jones 1993)—later termed community medicine, and named Michael Davis as its medical director. The Fund was eager to sponsor public health initiatives benefiting the African American community and was reviewing proposals. In 1929, the same year as the stock market crash, the Rosenwald Fund partnered with the PHS to help sponsor the first "major serological survey and treatment program for syphilis" (Reverby 2009). The Rosenwald Fund provided matching funds to the PHS to help find and treat syphilis in the African American population of six rural states/counties whose populations typically could not be served even by free clinics (Jones 1993), which were already overextended. The study "for demonstrations of the control of venereal disease in the rural South, in cooperation with the United States Public Health Service and with the state and local authorities" (Jones 1993) was approved in November 1929 by the Rosenwald Fund Board. Drs. Thomas Parran, director of the Division of VD and Taliaferro Clark, Advisor to the PHS (appointed by Hugh S. Cumming, the Surgeon General of the USPHS) selected the treatment sites; one of them was Macon County, Alabama, where there were eight African Americans for every white American. The PHS would use staff from the Tuskegee Institute, as the Rosenwald Fund required them to use African American healthcare professionals. Nurse Rivers was one of the nurses that had assisted in this project as well (Jones 1993). The goal of the treatment study was to find out the extent of syphilis incidence and to prove that containment with treatment was possible. The treatment study began in Macon County around February 1930, and continued until around September 1931. Oliver C. Wenger, who handled a similar project in Mississippi, and a venereal disease expert, was put in charge of the study.

The syphilis case-finding and treatment project was a high-quality research initiative even by today's standards, with the exception of problems with informed consent, which was not yet a standard. The Rosenwald Fund sent outside observers and evaluators (African American physicians, Drs. Frost and H.L. Harris Jr.) to review onsite conditions, provide oversight and report back various concerns and problems; it also commissioned a follow-up qualitative study of the patients treated, which was the Johnson study discussed earlier. As for an outside syphilologist, Dr. Thomas Parran, and E.L. Keys, former president of the American Social Hygiene Association, acted as reviewers and had only praise for the study.

There were lots of technical and methodological problems with sampling and enrollment (going house to house vs. blood sample drives in one setting); adequate amount of therapies; melting mercury; as well as medical and clinical ethics issues: drug side-effects; poor conditions in overcrowded shack-schoolhouses, and very poor consent protocols with the poor and illiterate population, including coercion through power relations, as noted by Dr. Frost (Jones 1993), as well as participants "who were entirely ignorant of the character of the disease for which they were being treated" because they were confused by the term "bad blood" (Jones 1993). Later, Johnson would note that not one of the 612 interviews he conducted understood the connection between sex and syphilis, and also stated that "bad blood" was a non-specific term that typically referred to a whole host of ailments (Jones 1993).

In the final analysis, the Rosenwald Fund treatment study could best be described as a "successful failure" in that it indeed found a very high incidence of syphilis using a "Wasserman test dragnet" approach (Reverby 2009) where they essentially tested all at-risk African Americans in the county; they also treated them. To the PHS, the findings seemed to assure continuation of the treatment project.

The Rosenwald Fund saw things differently. The issue according to their outside reviewers was that they found so many problems with barriers to consent and compliance due to poverty, illiteracy, co-morbidities, treatment contraindications—including starvation from malnutrition—that resource allocation became confounding and morally distressing. What they found, and could realistically address with their available Fund was essentially "too big to fund" and not feasible, regardless of the Wall Street crash. After a careful review and analysis, the Rosenwald Fund suggested to the PHS that *primary care*—and not solely syphilis—was a much bigger problem, and the best way to deal with the syphilis problem, given the extraordinary amount of chronic disease and malnutrition they found, was to consider funding overall healthcare and nutrition, not just case-finding and treating syphilis. The PHS was open to transitioning to a more comprehensive healthcare program. But they also felt the study's goals were achieved: they were able to effectively demonstrate that they *could* establish the prevalence of syphilis and treat contagion.

Pulling Funding

The year the Rosenwald Fund treatment project began was the same year as the Wall Street crash in October 1929. The Crash was not just a one-time event; between

October 1929 and July 1932, there were small fits and starts of the stock market rebounding. By 1931, roughly 2300 banks had failed, $1.7 billion in deposits were lost, over 100 businesses were closing per day. Julius Rosenwald died January 6, 1932 before things got even worse: on July 8 1932, the stock market closed at $41.22—its lowest level that century, and stocks lost 89 percent of their value across the board (Kenton 2018). That period marked the beginning of the worst financial crisis in U.S. history and the official start of the Great Depression. As today, study funding was volatile, but it is inaccurate to say that the Rosenwald Fund pulled funding from the syphilis treatment study entirely due to the crash, but it was certainly a factor. The Rosenwald Fund mainly pulled funding because of their analysis of the data and needs assessment of the region: continuing to focus solely on treating syphilis wasn't a good use of their resources, given the extent of the health inequities they found; addressing primary care and basic nutrition was what their data told them was the best use of resources. The other factor in pulling their funding was that the Rosenwald Fund required matching state funds from Alabama, which was indeed reeling from the Depression and was not prepared to match funding at that time for a major free healthcare initiative. What the treatment study also revealed would become the foundation for the observational protocol. Johnson noted in his qualitative interviews after the treatment study: "The tradition of dependence and obedience to the orders of authority, whether these were mandatory or not, helps to explain the questionless response to the invitation to examination and treatment" (Jones 1993).

Thus, the now infamous observational protocol designed to follow African American men with latent and/or late stage syphilis *was an afterthought* that arose from the data analysis phase of the well-intentioned Rosenwald Fund *treatment* study.

The Second Tuskegee Observational Study Protocol (1932–1972): What We Mean When We Say "Tuskegee Study"

After funding for the Rosenwald treatment study ended, the PHS investigators found the data too rich to simply "drop" and so they began to think about what they *could* study given their limited funding. Talliaferro Clark felt that the incidence of syphilis uncovered in Macon County was an ideal setting in which to study the "natural history" of late stage syphilis in African Americans. Since there was a dearth of published studies about syphilis (with the exception of the Oslo study), Clark wanted to build on the treatment study data and design an observational protocol to track morbidity and mortality in late stage syphilis "to assess the extent of medical deterioration" (Gray 1998). Since there was no funding to continue the treatment study anyway, the PHS could still follow the infected men, and examine them clinically to see what really happened to them when the disease was left untreated. When Clark reviewed the data from 1930–31, and considered the extent of syphilis prevalence in the Macon County region, he felt the region was an excellent site for an African American version of the Oslo study since leading syphilologists believed that "Syphilis in

the negro is in many respects almost a different disease from syphilis in the white."
But instead of a retrospective study, it would be an improved *prospective study*. To
make it easier to stage, he would select only men because their genital sores in early
syphilis would be far easier to see and stage than women. The study Clark had in
mind would be for about a year. At that point, Clark named 35 year-old Raymond
Vondehler, a PHS Officer, to direct the new prospective observational study of late
stage syphilis, and asked Wenger to continue to be involved.

Clark next met with all the stakeholders: the Alabama Board of Health, the private
practitioners in the region, and the Tuskegee Institute. The Alabama Board of Health's
director, J.N. Baker, made it clear that Clark's study design would be approved by
the State on condition that every person diagnosed with syphilis would be treated;
as it would only be a short trial of 6–8 months, no one could get the current recommended standard of care, but they agreed to minimal treatment of eight doses of
neoarsphenamine and some additional mercury pills unless it was contraindicated.
As these doses were regarded by syphilologists to be wholly insufficient, or even
useless to treat late stage syphilis, the men they followed were considered, for their
purposes, to have *"untreated syphilis"* (Jones 1993; Gray 1998; Reverby 2009).

Clark approached Dibble with his plan, and Dibble briefed Moton by letter about
the "worldwide significance" of the Clark PHS study and how the opportunity could
benefit Tuskegee Institute nurses and interns (Reverby 2009), while the Tuskegee
Institute would "get credit for this piece of research work …the results of this study
will be sought after the world over. Personally, I think we ought to do it" (Jones
1993). Both Moton and Dibble agreed it was definitely worthwhile, and offered the
Tuskegee Institute and John A. Andrew Memorial Hospital resources for this "new
deal"—the observational study. The plan was for Dibble to have some of his interns
in the field help deliver the minimal treatment protocol agreed upon with the State
Health Board. Dibble and Moton fully understood the practical purpose of Clark's
PHS study—that it could identify how much treatment was really necessary, since
most of their community had no access to it at all. (This same design would later be
invoked in the 1990s in African AIDS protocols, discussed in Chap. 2.)

The initial Clark PHS study was just one version away from what would become
the *notorious* "Tuskegee Study of Untreated Syphilis in the Male Negro" (or "Negro
Male" in some documents) that lasted 40 years.

Dibble and Moton undoubtedly viewed Clark's study design within a "global
health ethics" framework—justifying a project that could potentially benefit the most
people in the long-term where there are no resources at all. Even Wenger had noted
that there were greater health resources in China and the Phillipines "where better
medical services are rendered to the heathens than we find right here in [Macon
County]" (Jones 1993:76). There were no defined ethics guidelines for human subject
research in 1932 (see under Healthcare Ethics Issues), so informed consent was not on
anyone's mind. The country was still plunged into economic chaos, as "Hoovervilles"
(Shanty towns for millions of white Americans who were homeless) were becoming
the norm as the unemployment rate rose to over 25% for white Americans. The
thinking went: Without the Clark study, the men would have died with late stage
syphilis anyway with no treatment at all; what harm was there in assessing them,

with the benefit of giving them some minimal treatment, a hot meal on exam days and medical attention to other smaller ailments?

In 1933, just as President Franklin Roosevelt had begun his first term, and more socialist "New Deal" programs would begin, Clark wrote in a PHS annual report that: "the ideal method of therapy is seldom possible of attainment, and the vast majority of infected people receive treatment that is generally regarded as inadequate or no treatment at all. It is highly desirable, therefore to ascertain, if possible, the relative benefits accrued from adequate and inadequate treatment" (Reverby 2009).

But with the goal of observation and *some undertreatment*, Jones (1993) noted that: "The Tuskegee Study had nothing to do with treatment. No new drugs were tested; neither was any effort made to establish the efficacy of old forms of treatment. It was a nontherapeutic experiment, aimed at compiling data on the effects of the spontaneous evolution of syphilis on black males."

Methods: 1932–1933

The plan was to do another Wassermann testing sweep to enroll men similarly as they did in the Rosenwald study by inviting them for testing and examination. This time, they would enroll only males in the latent or tertiary phase, and follow them by doing clinical exams to take a good look at what was going on. The minimal heavy metal treatment they agreed to provide would be furnished to anyone found with early stage syphilis to prevent contagion. Again, the men with latent or tertiary syphilis also received the minimal metals treatment, and was less than half the dosage recommended by the PHS to cure syphilis; it was thus considered so inadequate and ineffective, the study investigators still defined it as "*untreated syphilis.*" The exams would comprise continued Wassermann blood tests, urine tests, cardiac exams (fluoroscopy and x-rays), lumbar punctures to test spinal fluid for neurosyphilis, and if they died, an autopsy.

The goal was to study the "natural history" of late stage syphilis in African Americans to (1) help settle debates over racial differences in syphilis pathogenesis; (2) determine whether treatment in later stages was really necessary, given the side-effects of the heavy metal treatments in general. The social significance to the Tuskegee Institute was that if there were significant morbidity, the data could convince the government that funding public health and syphilis disparities in the African American population was an important health priority.

Regarding the consent procedures, as in the first study, the men would be told that they were being checked by "government doctors" for "bad blood," and would receive "treatment" for their ailments (in addition to the minimal heavy metal doses, they would receive aspirin, tonics and vitamin pills). The men would also be served a hot meal on exam days. But there would be no "informed" consent as it was. not yet a standard: there was no disclosure they were being followed for syphilis specifically, and no disclosure for the enrolled men that their syphilis was not being adequately treated. There was ongoing deception as to the purpose of the study, deceptive practices surrounding lumbar puncture described as "spinal shots" and "special free treatment" and its potential risks. (See Reverby 2009, pg. 45 for the actual study letter.)

As one of the men later said when questioned in the 1970s about his 40 years of participation, "I ain't never understood what this study was about" (Jones 1993).

In terms of study personnel roles: Vondelehr, Wenger, and some of Dibble's interns at the Tuskegee Institute would work in the field; Dibble would be in charge of the team at the Tuskegee Institute, and John A. Andrew Memorial Hospital would serve as the local teaching hospital site for the cardiology tests, lumbar punctures, acute medical needs, and any autopsies to be performed, which meant that when the men were close to death, they were brought to Dibble's hospital so they could die there to improve the autopsy results (Jones 1993). Nurse Rivers, suggested by Dibble, would be hired at this point as the study nurse ("Scientific Assistant") at $1600 per annum ($1000 per annum, and $600 for transport expenses) to serve as a liaison who would interact with the men, ensure they made it to their exams, and offer any primary nursing consultation or care.

By the end of the recruitment phase, they enrolled 399 men in the disease group, did their exams, and Vonderlehr performed the spinal taps on 296 of them—often incompetently (Jones 1993; Gray 1998; Reverby 2009), which caused many side-effects; the men would stay in the hospital overnight after the spinal taps. As the team started to write everything up in 1933, Vonderlehr suggested to Clark that they should consider *extending* the study between 5 and 10 years as the data was so rich. Before Clark could decide, he retired and Vonderlehr was named as his replacement as director of the Division of Venereal Diseases. And so, the Clark design now turned into the *Vonderlehr design*—which ultimately lasted for 40 years. In the 1933 Vonderlehr design of the PHS study, they added 201 men without syphilis as controls, who were not informed that they were actually a "control" in a study (Jones 1993), for a total of 600 men. They stopped doing the spinal taps to improve retention (the men were actively refusing them), and so simply stopped diagnosing neurosyphilis. They also decided that all the men in the syphilis group should have autopsies at death, as that yielded the best data. Wenger noted in a letter to Vonderlehr: "As I see it, we have no further interest in these patients *until they die*" (Reverby 2009). They would still provide small amounts of "treatment" in the form of aspirins, tonics, and oral mercury compounds "if asked" (Reverby 2009). Vonderlehr reached out again to the Tuskegee Institute—Moton and Dibble. The Institute would continue as the teaching hospital site for the exams, and Dr. Jesse Jerome Peters, an African American pathologist, would handle the autopsies at the VA hospital onsite at the Tuskegee Institute (Jones 1993). Dibble would handle treatment of any study subjects (using the minimal treatment), while local African American doctors were to refer any study subject who wanted syphilis treatment back to Dibble should they encounter one of the enrolled men. At this point, Dibble was named as one of the Tuskegee study's official government consulting doctors ($1/annum); and they would rehire Rivers as the study nurse for $1200.00 per annum ($1000 per year, and $200 for expenses). The PHS would next appoint John R. Heller to the study site to direct the field work. By 1934, the PHS secured $50 per family for burial costs as an incentive to agree to autopsy; the funder was the Milbank Memorial Fund (the Rosenwald Fund was approached first).

Vonderlehr had now shaped the study design into what is meant by the "cultural shorthand of Tuskegee Study or Tuskegee Experiment…it was never really an experiment in the sense of a drug, biologic or device being tested. It was supposed to be prospective study of what the doctors called the 'natural history' of late latent syphilis…Its name varies in published articles" (Reverby 2009). As for the issue of withholding adequate treatment from the enrolled men, Jones notes (1993):

> Apologists for the Tuskegee study contended that it was at best problematic whether the syphilitic subjects could have been helped by the treatment that was available when the study began…The [heavy metal treatments] were highly toxic and often produced serious and occasionally fatal reactions in patients… was painful and usually required more than a year to complete. As one CDC officer put it, the drugs offered 'more potential harm for the patient than potential benefit'…PHS officials argued that these facts suggested that the experiment had been conceived in a moral vacuum. For if the state of the medical art in the early 1930s had nothing better than dangerous and less than totally effective treatment to offer, then it followed that, in the balance, little harm was done by leaving the men untreated…

Yet undertreatment totally corrupted the data from the very beginning. Gray (1998) notes:

> Nearly all these patients were given some treatment with arsenicals or mercury, and often with both in the course of the initial 1932–33 [Clark design]. The amount was believed too small to have effect on the disease, but it ruined the study as one of "untreated syphilis". The doctors botched the sample doing the short-term study, secured it by going through the motions of treatment, before a long-term study was even contemplated. Rather than call it quits, the doctors falsified the sample selection procedure in their initial papers by arbitrarily defining the little treatment as no treatment at all. Although it baffled later doctors how so many patients had gotten some, albeit obviously inadequate, treatment, no one had read the files.

The Shift to Lifetime Follow-Up

The standard one-slide summary of the observational study on almost every "Introduction to Research Ethics" course typically reads: "From 1932–1972, the U.S. PHS followed ~400 black males with syphilis but didn't treat them and didn't tell them they were not being treated." But the devil is in the details: They enrolled 600 men, when you count the controls—several of whom wound up with syphilis, too, and were then moved into the syphilitic arm; most men in the syphilitic arm received some heavy metals treatment or protoiodide pills (dispensed by Heller), but the medication dispensed highly varied among the men, and was often left up to practitioner discretion (Jones 1993). This created a quagmire of confusing data for anyone trying to analyze a study of "untreated" syphilis.

As the study ticked on into subsequent calendar years, it began to dawn on everyone that the "pot of gold" was really the data obtained from the autopsies so they could best validate their clinical findings, and ascertain *to what extent syphilis was really a factor in the death.* Since many African American men died in Macon County from a range of untreated chronic diseases, it was difficult to figure out whether they were actually dying from syphilis, or to what extent it was a contributing factor. The team thus shifted its goals by 1936 to *lifetime follow up*. Jones (1993) uncovered that:

> The PHS was not able to locate a formal protocol for the experiment. Later it was learned that one never existed; procedures, it seemed, had simply evolved…but the basic procedures called for periodic blood testing and routine autopsies to supplement the information that was obtained through clinical examinations.

In addition to the burial stipend, another incentive for the men to stay enrolled was lifetime *access to Nurse Rivers*, who formed deep relationships with the men followed, drove them into town for their periodic exams, responded to their questions, and also provided them and their families with basic public health hygiene and nursing regarding primary care, reminiscent of her Movable School days. Nurse Rivers had other incentives to getting to know the men's families: she was the one who consented the family for autopsy after each man died. A year before the shift to lifetime follow-up was made, Moton retired.

In 1938, Austin V. Diebert, a Johns-Hopkins trained PHS syphilologist, joined the study to work in the field. Diebert reviewed the data so far, and had major problems with its corruption due to "some treatment" in a supposed study of *no treatment*. He re-tweaked the protocol again to ensure that fresh men were recruited into the syphilitic arm who were to receive no treatment for syphilis at all; he also noted that undertreatment of several of the men may explain why they weren't seeing as severe complications in the men as they would predict (Jones 1993). This finding, alone, should have been cause to end the study. No new men were added to the study after 1939 (Reverby 2009).

Moton died in 1940, and was honored when an air field nearby was named after him: Moton Field. Things pretty much ticked along through the rest of the Depression era until the bombing of Pearl Harbor in December 1941, when the United States entered World War II. At this point, needing as many troops as possible, Moton Field became the site for training the first African American fighter pilots to be sent overseas. They were called the "Tuskegee Airmen," and the initiative was known as the "Tuskegee Experiment" because the prevalence of sickle cell disease in African Americans had been associated with barring them from flying or becoming pilots. Since the Tuskegee Airmen completed many successful bombing missions in Europe, the so-called "Tuskegee Experiment" was associated with *this* event 30 years before Moton's legacy would be linked to the notorious "Tuskegee Syphilis Study". More germane to this time frame, however, was another "experimental" miracle drug introduced into the military to combat syphilis: penicillin. *The Henderson Act* was passed in 1943, "requiring tests and treatments for venereal diseases to be publicly funded" which eventually included penicillin (McVean 2019).

The Penicillin Era and Beyond

Penicillin did not become a standard of care for syphilis treatment until 1946. By 1947, the *Henderson Act* led the USPHS to open Rapid Treatment Centers in which syphilis was being treated with penicillin (McVean 2019). When penicillin was first dispensed to U.S. soldiers during the war, *it was still experimental*, and bioethicists have raised many questions about that study design, too. Around that time, a completely secret and egregious penicillin experiment was also going on in Guatemala in which the USPHS was *actually infecting men with syphilis using prostitutes they enlisted* and then treating them with penicillin to test doses and efficacy (Reverby 2013). This study's fallout is discussed more under Healthcare Ethics Issues (further).

Providing inadequate heavy metal treatment was not considered medically or ethically problematic in the pre-penicillin era considering its risks and questionable efficacy beyond latency. But the withholding of penicillin started to become an ethical issue, despite clinical equipoise. To deal with the risk of "curing" their enrolled men, a "do not treat list" of study subjects was circulated to the local physicians as well as a "do not enlist/do not treat" list to the army recruiters in the area (Jones 1993; Reverby 2009). A local physician in Macon County recalled that at that point in the study's history, the men "were being advised they shouldn't take treatment [with penicillin] or they would be dropped from the study" which meant that they would lose the benefits being promised (Jones 1993).

When interviewed by historian James H. Jones in 1977, Rivers relayed: "I never told somebody *not* to take any medication." When asked how she was able to keep the men from getting penicillin, she said "I don't know that we did…I was never really told not to let them get penicillin, and we just had to trust that to those private physicians" (Reverby 2009). In another interview in the 1970s, Rivers stated: "I never told anybody that you couldn't get treatment. I told them "So who's your doctor? If you want to go to the doctor go and get your treatment…that [the white doctors] had to…have an excuse so they put it on me that I wouldn't let the patients get treatment." (Reverby 2009:179, note 85)

Jones (1993) surmised:

> Discrediting the efficacy of mercury and slavarsan helped blunt the issue of withholding treatment during the early years, but public health officials had a great deal more difficulty explaining why penicillin was denied in the 1940s. PHS spokespersons noted later that withholding penicillin was "the most critical moral issue…one cannot see any reason that they could not have been treated at that time…" [and] "I don't know why the decision was made in 1946 not to stop the program…" (Jones 1993).

When penicillin became widely available in the later 1940s, it was known to cure early stage syphilis, but no one knew if it was effective on latent or tertiary syphilis; there were theories that it would do more harm than good because of reactions to the drug. It took about a decade to understand that penicillin could cure syphilis at the latency stage if organ damage had not yet occurred (Reverby 2009), but since the men in the study were beyond that, the investigators were convinced "the men could not be helped at this point" (Reverby 2009). Jones (1993) refers to "therapeutic nihilism" in that there was a belief that "penicillin was a new and largely untested drug in the 1940s [and that] the denial of penicillin was a defensible medical decision."

It is historically inaccurate to state that penicillin was a "standard of care" in latent or tertiary syphilis until at least the mid-1950s. On the contrary, there was *clinical equipoise* over that question (see under Healthcare Ethics Issues). Penicillin only became a standard of care for late stage syphilis by the later 1950s, and there was still debate as late as 1964 (Reverby 2009). It took years of its use for its benefits in latent or late stage syphilis to be clear; part of the clarity was due to incidental use of penicillin. Many men in the Tuskegee study had indeed been given penicillin incidentally for other problems if they saw local physicians, and so the data again was becoming polluted. Vonderlehr remarked in 1952 to one of the field doctors at the time: "Hope that the availability of the antibiotics has not interfered too much with this project" (Reverby 2009).

During the penicillin era, investigator bias surrounding the Tuskegee study caused the team to "double down" on the original aims of the study. Thus, rationalizing the ongoing study dominated the paper trail from that point on.

Sidney Olansky, John C. Cutler and Stanley Schuman, became the observational study's new leaders in the 1950s (Reverby 2009). In a memo by Olansky to Cutler in 1951, he wrote: "We have an investment of almost 20 years of Division interest, funds and personnel; a responsibility to the survivors both for their care and really to prove that their willingness to serve, even at risk of shortening of life, as experimental subjects. And finally, a responsibility to add what further we can to the natural history of syphilis" (Reverby 2009:69).

Hiding in Plain Sight: Data Sharing and the Road to Moral Outrage

As Miss Evers states, the study "wasn't no secret; everybody knew what was going on". Indeed, the study investigators were pretty good about data sharing over the years. The first paper on the second study was published in 1936 by Vonderlehr, Clark, Wenger and Heller to report on their findings, followed by 11 more papers over 40 years (See Jones 1993). Notes Reverby (2009):

> The PHS was not hiding the Study, and medical journals published the articles and reports. Unless a physician picking up the articles thought deeply, it would have been easy to see the men as 'volunteers', and indeed in one report they were referred to as 'people [who] responded willingly.' Reading between the lines would have been necessary since the reports did not make clear that the aspirins, tonics, and spinal punctures were being described to the men as 'treatment'. The language and titles of the articles—'untreated syphilis in the Male Negro'—easily distanced the medical reader.

From a data integrity standpoint, the Tuskegee study was almost as egregious as the ethical abuses; none of the published papers really reflected what the investigators were doing, and the papers began to raise questions from peers. Any rigorous reading of the articles revealed badly compromised data, particularly when articles referred to "some treatment". It was very sketchy to decipher who got treated and who did not, how much treatment each man in the study received, and how it affected clinical findings. With such murky data, the ethical questions were more pronounced because it made it harder to justify why the men had been enrolled in this research to begin with, and why the study was continuing. As to the clinical findings, it was considered "messy science" (Reverby 2009) because it was not clear what the team was ever really observing when documenting complications of syphilis; every problem in someone with syphilis seemed to be attributed to syphilis when this was not the case (Reverby 2009).

In 1948, the new director of the Division of VD, Theodore J. Bauer, was alerted to problems with the study by Albert P. Iskrant, chief of the Office of Statistics in his division. Iskrant questioned the methods, data contamination, and whether the investigators had followed Alabama law for testing and treatment, Iskrant concluded that the data could be salvaged if it were analyzed as a study of "inadequately treated" syphilis (Jones 1993).

In 1951, Olansky and Shuman undertook a major internal review of the data to determine how to improve the data analysis, and even if the study should continue. They arranged for an outside reviewer to take a look at the scientific merits. Norwegian syphilologist, Trygve Gjestland, who had also re-examined the Oslo data, came to Atlanta in 1952 to meet Olansky and Cutler. Gjestland thought the entire observational study was a "scientific mess because of unclear categories and uncertain criterion for much of the diagnoses" (Reverby 2009). However, Gjestland suggested how to improve the data instead of suggesting the study end. Additionally, by 1952, the investigators thought the study data could also be used to study aging in their aging men; Peters was also furnished with better equipment to improve autopsies, which he continued until the early 1960s (Jones 1993).

The first documented ethical concern by peers surfaced in 1955. Count D. Gibson, an Associate Professor at the Medical College of Virginia heard Olansky give a lecture about the observation study, presumably reviewed the literature, and wrote to Olansky:

> I am gravely concerned about the ethics of the entire program…The ethical problem in this study was never mentioned…There was no implication that the syphilitic subjects of this study were aware that treatment was being deliberately withheld…It seems to me that the continued observation of an ignorant individual suffering with a chronic disease for which therapeutic measures are available cannot be justified on the basis of any accepted moral standard: pagan (Hippocratic Oath), religious (Mainmonides, Golden Rule), or professional (AMA Code of Ethics)…Please accept this letter in a spirit of friendliness and an intense desire to see medical research progress in its present remarkable rate without any blemish on its record of service to humanity. (Reverby 2009: 71).

Reverby documents Olansky's response, as follows:

> He acknowledged Gibson's worries that the Study was 'callous and unmindful of the welfare of the individual.' But he explained: 'I'm sure it is because all of the details of the study are not available to you.' He confessed that when he had started with the study in 1950 'all of the things that bothered you bothered me at the time…Yet after seeing these people, knowing them and studying them and the record I honestly feel that we have done them no real harm and probably have helped them in many ways… No females were selected so that the question of congenital [syphilis] could be eliminated.' Olansky repeated what would

become the mantra: that the men had no chance to get treatment elsewhere when the Study started, that they 'knew that they had syphilis and what the study was about, [and that] only those with latent syphilis were chosen. They got far better medical care as a result of their being in the study than any of their neighbors' and a nurse 'takes care of all their needs.' " (Reverby 2009: 71)

Olansky also inadvertently revealed to Gibson that the data was truly compromised. He wrote that "some of the patients did receive penicillin, but we are continuing to follow them…" Gibson was "deeply disturbed" by Olansky's response (Reverby 2009).

On the 30th anniversary of the study in 1964, when the *Archives of Internal Medicine* published the paper on untreated syphilis by Rockwell et al., it prompted a letter to the editor by a young Canadian physician, Irwin Schatz. According to the *New York Times* (Roberts and Irwin 2015):

> Dr. Schatz sent his letter, comprising three sentences, to the study's senior author, Dr. Donald H. Rockwell. He wrote: "I am utterly astounded by the fact that physicians allow patients with potentially fatal disease to remain untreated when effective therapy is available. I assume you feel that the information which is extracted from observation of this untreated group is worth their sacrifice. If this is the case, then I suggest the United States Public Health Service and those physicians associated with it in this study need to re-evaluate their moral judgments in this regard."

> The letter was passed to a co-author, Dr. Anne R. Yobs of the Centers for Disease Control, who wrote in a memo to her bosses: "This is the first letter of this type we have received. I do not plan to answer this letter."

Years later, Schatz learned that William B. Beam, editor of the *Archives of Internal Medicine* in 1972, noted then that he regretted that the journal had published the 1964 report, and said the journal had "obligations to apply moral and ethical standards" (Reverby 2009).

In 1965, Rivers and Dibble retired, but Rivers continued to see the men until her replacement could be found (see further). No medical historian has identified any replacement role for Dibble per se, but autopsies continued to be performed by Dr. Peters until the study was closed. A prominent polio expert at the Tuskegee Institute, Dr. John Hume, appears to have served as "Acting Medical Director" sometime after Dibble had left until 1971 (Jet 1971), when Dr. Cornelius L. Hopper became the new Medical Director of the John A. Andrew Memorial Hospital and Tuskegee Institute's Vice President of Health Affairs.. Hume was then named chief of staff and director of the orthopedic services at John A. Andrew Memorial Hospital (AAOS 2018). There are no specific documents that describe what Drs. Hume or Hopper knew about the study.

The Whistleblowing Phase

In 1966, a year after the *Civil Rights Act* was passed, and the same year that a major whistleblowing paper in the *New England Journal of Medicine* had been published by Henry Beecher (see further below), Peter Buxton, a 27 year-old Jewish PHS social worker (hired in 1965) and based in San Francisco, started to look at internal documents and records about the ongoing Tuskegee study. Buxton was born in Prague, and brought to the United States as a baby when his parents fled the Holocaust. Buxton heard "watercooler" chatter about the study while lunching with some of his co-workers and found it hard to believe it was going on. As part of his job, he was expected to do short papers on venereal disease, and so he decided to focus on the Tuskegee study, and review what was available on it. He was so concerned about the moral issues with the trial, he sent a registered letter to the director of the Division of Venereal Disease, William J. Brown, with his own ethical analysis of the study in the context of the Nuremberg Trials. Brown did not respond immediately to the letter but asked a CDC colleague based there to meet with Buxton and explain the study to him. Buxton was not placated. At a conference a few months later, Buxton was summoned to a PHS meeting with Brown and Cutler to review the goals of the study in light of his concerns. The meeting did not go well; Cutler "dug in" about the value of the study and dismissed Buxton's concerns. Buxton resigned in 1967 and went to law school, as the turbulent late 1960s raged on. In 1968, after the assassination of Martin Luther King, when racial tension exploded in riots across the country, Buxton wrote a second letter to Brown. In June 1968, Dibble died. During late 1968, Brown showed the letter to David Sencer, the director of the CDC, and it prompted the PHS to have a major interdisciplinary medical team meeting, inviting Olansky who had left. The meeting took place on February 6, 1969 (no African Americans were present); its purpose was to review the study to date, and decide whether to end or continue the study. One attendee agreed that the study should end and there was a moral obligation to treat the men, but he was outnumbered, and the rest decided to continue the study but re-seek stakeholder approval. They discussed hiring Rivers replacement, and they also discussed the potential that the study could be viewed as "racist" by some, but dismissed it as an impediment.

Stakeholder approval meant they needed to seek support for continuing the study from local African American physicians who now made up the majority membership of the Macon County Medical Society, as well as the medical professionals and new leadership at the John A. Andrew Hospital. John Caldwell, now in charge of the Tuskegee study briefed all the stakeholders on the study, and the local African American physicians agreed to continue to support it. In 1970, the PHS renewed its partnership with the John A. Andrew Memorial Hospital at the Tuskegee Institute for handling the x-rays, and hired Elizabeth Kennebrew to replace Rivers. During this period, Brown wrote back to Buxton to let him know there was overwhelming consensus to continue the study and that its design was sound.

Meanwhile, Buxton, continued to have considerable moral distress over the knowledge that the study was ongoing. In July 1972, one month after the Watergate news was breaking, he was introduced at a social event to an *Associated Press* reporter (Edith Lederer) and told her all about the study, and provided documentation. Lederer took the story to her editors, who assigned it to Jean Heller. The story broke July 25, 1972. A new era had begun in the Tuskegee Syphilis Study (See Heller 1972, for the link to the original *Associated Press* article).

Fred Gray, an accomplished civil rights attorney who lived not far from Macon County, Alabama, first read about the Tuskegee study in the newspaper like everyone else (Gray 1998). Gray had represented Rosa Parks in 1955, and was involved in several major civil rights cases. He was approached a few days later by Charles Pollard, who also read Heller's story, didn't realize he was part of an experiment until then, and told Gray he was one of the men in the study and wanted to sue. Thus *Pollard et al. v. United States et al.* was filed[3] which led in 1975 to a multi-million-dollar settlement for the surviving men and their families; the estates and heirs of the deceased men. That settlement still provides reimbursement to heirs.

Fallout from the Heller article led the Assistant Secretary for Health and Scientific Affairs to appoint a panel to review the study; The Tuskegee Syphilis Study Ad Hoc Advisory Panel was chartered August 28, 1972, and was charged to investigate the study, and make recommendations on a defined set of questions[4], including whether the study should end. The multidisciplinary subcommittee was chaired by Brodus Butler, and was a diverse committee. Astonishingly, the Ad Hoc Panel indeed interviewed Rivers, but the Chair decided to destroy her interview tapes (Washington 2007), clearly concerned that they may be problematic for Rivers. Ultimately, the Ad Hoc Panel wound up becoming bitterly divided, but did make the recommendation for the study to end immediately when they submitted their first report on October 27, 1972[5], The final report was dated April 28, 1973.[6] But this protocol thereafter would become forever "studied" by medical historians and ethics scholars as one of the most notorious chapters in medical research history and health disparities.

In February and March 1973, the Senate Subcommittee on Health of the Committee of Labor and Public Welfare, led by Senator Ted Kennedy, held hearings on medical research, and discovered the Tuskegee study was not yet closed. Rivers never testified, but Buxton was among several witnesses who testified. The study finally closed after the Kennedy hearings insisted the men get treated. Hopper was involved in negotiating with the federal government over how best to treat the patients, but there is no evidence he participated in the study.

Coming full circle to the film, as many began to read about the study in subsequent decades, the decision to silence Rivers in 1972 motivated Feldshuh to allow her to speak her truth:

Evers: We proved there was no difference between how whites and blacks respond to syphilis.

Senator: If they were white would they have been treated as anyone?

---

[3]Pollard v. United States, 384 F. Supp. 304 (M.D. Ala. 1974) may be viewed here: https://law.justia.com/cases/federal/district-courts/FSupp/384/304/1370708/.

[4]See Footnote 1.

[5]See Footnote 1.

[6]The committee report can be viewed as this link: http://www.research.usf.edu/dric/hrpp/foundations-course/docs/finalreport-tuskegeestudyadvisorypanel.pdf.

Evers: You wouldn't have dared. You wouldn't have voted for it for forty years, if they were white. Somebody would have said something about this before now. Everybody knew what was going on. It wasn't no secret. But because they were black nobody cared.

Consequently, a new phase of the Tuskegee study would begin: the postmortem *Bioethics Phase* (see under Healthcare Ethics Issues further), which blossomed in the 1990s as a direct result of Jones' and Feldshuh's works. The Bioethics Phase comprised the commissioning of a new committee to recommend "moral reparations" that culminated into a Presidential Apology, a national endowment and revitalization of the Tuskegee Institute to become a University and major Center for Bioethics, as well as more revelations about unethical syphilis studies conducted by the U.S. government on vulnerable populations.

Other Observation and Non-treatment Studies

When teaching from the history of medicine context of the Tuskegee study, it's important to note that troubling non-therapeutic, or "under-therapeutic" observational protocols abound with problematic consent in both the United States. and other countries. In the AIDS era, "protocol 076" was essentially a Tuskegee study, prompting the Editor of the *New England Journal of Medicine* to resign (see Chap. 2). In India, cervical cancer screenings were denied to 138,624 women to see if the unscreened women developed cervical cancer more frequently; 254 died (Suba 2015). In 1994, observing prostate cancer and withholding treatment had become a standard in men over 65 (Wilt et al. 2017), but the data shows that significant numbers of men died prematurely, consequent to no treatment (Frandsen et al. 2017). Debates are currently raging over observation-only of early breast cancers—ductal carninoma insitu (O'Connor 2015; Lagnado 2015; Drossman et al. 2015; Bath 2016; Rosenthal 2015); and observation-only of biopsied confirmed thyroid cancer without informed consent, echoing the Tuskegee study (Rosenthal et al. 2017). In essence, once race is removed from the story, the Tuskegee study protocol is ongoing in many different disease contexts and remains a problematic theme in medicine, where observation is motivated by the desire to save money with less treatment. As for current treatment protocols for syphilis, we have come full circle, too. Although antibiotics have long been the treatment since the penicillin era began, due to some emerging antibiotic-resistant strains in underserved populations, some practitioners suggest re-exploring the metals treatments again (Ambrose 2016).

Healthcare Ethics Issues

When examining the plethora of healthcare ethics issues entangled with the Tuskegee study, and the film *Miss Evers' Boys*, there are four areas to consider for discussion.

First, there are research ethics considerations, which surround the study design within the prevailing research ethics standards or guidelines, as well as the "reparations" phase of the Tuskegee study, which involved compensation to the harmed subjects and their families. Second, there are clinical ethics considerations: did the healthcare providers involved deliver the standard of care, and did they knowingly violate the prevailing clinical and professional ethical standards of care? Third, there are distributive justice considerations that surround health disparities in this specific population vs. health disparities overall; public health considerations surrounding the study's impact on syphilis. Finally, there is the "Bioethics Phase" of this study surrounding "preventative ethics" guidelines; moral corrections, and accountability. The staging of the play, "Miss Evers' Boys" as well as the making of the film, is considered by some to be part of this final phase—letting the public see what occurred as part of an accounting of the harms.

Research Ethics Considerations

In the field of bioethics, invoking the noun "Tuskegee" is code for "egregious study design". As discussed in the History of Medicine section above, the initial Rosenwald treatment study is not what we mean when we say "Tuskegee." What we really mean is the observational study that resulted from the "Vonderlehr design".

Although even the Rosenwald treatment study had problems with informed consent, it was debatable at that stage whether consent standards were any better in other populations at the time. What made the Rosenwald treatment study ethical were its specific aims, which indeed met the beneficence standard for both the individuals treated and the population at large. When the Rosenwald study ended, the "Clark design" of the study fails the informed consent test, but had some ethically defensible aims in that it indeed provided the standard of care immediately to men in the earlier phases of syphilis, with the goal of treating all study subjects within a six-month period. It's clear that if the Clark design ended as planned, nobody would have ever heard of the Tuskegee study at all.

What makes the Tuskegee study infamous was its morphing into the "Vondelehr design": deliberate *misinformed* consent and deceit; withholding the pre-penicillin standard of care; and finally, withholding penicillin in the penicillin era, which could have cured the subjects.

Research Ethics Guidelines 1932–1972

To ethically locate the research ethics issues initially, it's important to first consider the research ethics guidelines in 1932. When the Tuskegee study began in 1932, the obvious problems with informed consent were not unique, and public health studies were judged by the quality of the generalizable data that led to improved public health. Between 1932 and 1947, there were no codified "research ethics" guidelines at all, and "research ethics" was not a consideration. However, as discussed earlier, the Tuskegee study's data by even these standards was a bad study because its sloppy design and methods polluted the data.

After the Nazi medical experiments were revealed in 1945, and the subsequent Nuremberg Trials took place, The Nuremberg Code was created by the American military. It was the first codified Research Ethics guideline and formally published by the U.S. government in 1949 (NIH 2020). It is clear to medical historians that by 1947 the Tuskegee study was in violation of The Nuremberg Code because there was no informed consent. But so were dozens of other American studies going on at that time, many of which involved vulnerable populations such as children. In the 1950s, for example, the polio vaccine was tested on children without consent (Welner 2005). American practitioners did not seem to recognize that The Nuremberg Code was applicable in the United States, and many assumed it was intended to define Nazi medical experiments specifically; undoubtedly, the word "Nuremberg" confused U.S. practitioners, and they did not seem to understand its purpose.

In 1964, the World Medical Association announced The Declaration of Helsinki in the July 18 issue of the *British Medical Journal*. This set of guidelines largely echoed the Nuremberg Code, but helped to define distinctions between therapeutic and nontherapeutic clinical research. By the new 1964 standards, the Tuskegee study fell into the category of nontherapeutic clinical research without informed consent, now in violation of the Declaration of Helsinki. It's possible that the Declaration of Helsinki may have been influenced by an infamous 1962 study truly resembling Nazi medical experiments. In this case, Chester Southam from Memorial Sloan Kettering, partnered with the Jewish Chronic Disease Hospital to inject live HeLa cells into some of its elderly patients without informed consent. Three Jewish doctors refused to participate in the study, but many took part. This experiment was exposed in 1964 when the front-page headline of the New York *World-Telegram* on Jan. 20, 1964 read: "Charge Hospital Shot Live Cancer Cells Into Patients." (Hornblum 2013).

There were no other codified U.S. guidelines at this time, but by 1966, a whistle-blowing article in the *New England Journal of Medicine* by Henry Beecher was published that "outed" 22 clinical trials in violation of the Nuremberg Code, which included the Brooklyn Jewish Chronic Disease Hospital.

Informed Consent Versus Beneficence

In the post-war years, once the study continued past the Nuremberg Code (1947), lack of informed consent started to become viewed as a serious ethical issue as awareness of Nazi medical experiments became more widely known. Even by 1932 standards, informed consent problems were noted. According to Jones (1993):

> Dr. J.W. Williams, who was serving his internship at Andrews Hospital at the Tuskegee Institute in 1932 and assisted in the experiment's clinical work, stated that neither the interns nor the subjects knew what the study involved. 'The people who came in were not told what was being done...we told them we wanted to test them. They were not told, so far as I know, what they were being treated for, or what they were not bring treated for...and the subjects'

'thought they were being treated for rheumatism or bad stomachs...We didn't tell them we were looking for syphilis. I don't think they would have known what that was'.

Charles Pollard, one of the subjects, clearly recalled the day in 1932:

when some men came by and told him that he would receive a free physical examination if he appeared the next day at a nearby one room school." 'So I went on over and they told me I had bad blood...and that [is] what they've been telling me ever since. They come around from time to time and check me over and they say Charlie, you've got bad blood...All I knew was that they just kept saying I had the bad blood—they never mentioned syphilis to me, not even once. (Jones 1993)

But in 1932, informed consent was not the standard, especially when it came to patients with literacy issues. Even in more educated patient populations, few doctors provided "informed consent"; in fact, "therapeutic misconception" was much more the standard. Clearly, the Nuremberg Code's first requirement was violated in that the voluntary consent of the subject was not obtained, and coercive incentives to participate were used, which included deception. However, pre-Nuremberg, by 1930s standards, what did "coercion" really mean?

Jones observes (1993):

The press quickly established that the subjects were mostly poor and illiterate, and that the PHS had offered them incentives to participate. The men received free physical examinations, free rides to and from the clinics, hot meals on examination days, free treatment for minor ailments, and a guarantee that [fifty dollars for] burial stipends would be paid to their survivors.

Post-Nuremberg, the incentives continued: in 1957, in honor of the 25th year of the Study, each man also received a special certificate singed by Surgeon General Leroy E Burney for "completing 25 years of active participation in the Tuskegee medical research study" and a dollar for every year of 'service'....Rivers stated the men were "thrilled with their cash awards" (Reverby 2009). Rivers acknowledged the consent issues in interviews, but she also considered the Tuskegee study as "beneficent" because the men got more attention than they would have otherwise. She was fully aware of the deception, but given the research standards of the time, did not think anything of it. She states (Smith 1996):

I got with this syphilitic program that was sort of a hoodwink thing, I suppose....Honestly, those people got all kinds of examinations and medical care that they never would have gotten. I've taken them over to the hospital and they'd have a GI series on them, the heart, the lung, just everything. It was just impossible for just an ordinary [black] person to get that kind of examination...they'd get all kinds of extra things, cardiograms and...some of the things that I had never heard of. This is the thing that really hurt me about the unfair publicity. Those people had been given better care than some of us who could afford it....They didn't get treatment for syphilis, but they got so much else".

In terms of meeting the standard of "beneficence", it's a stickier question in the pre-penicillin era because of clinical equipoise over the metals therapy, but the concept of "clinical equipoise" would not be articulated until 1987, which refers to clinical disagreement among the community of experts over which therapy is better or more effective (Freedman 1987). Freedman argued in 1987 that when there is clinical equipoise, a randomized controlled trial can ultimately disturb equipoise so long as there is enough statistical power to settle the question. It can be argued that the observation trial was the presumed rationale to attempt to disturb equipoise, but the fact that the men got variable amounts of (under)treatment ruled out any potential for disturbing equipoise. What about the argument that there was "some benefit" to the subjects that transpired in the pre-penicillin years given the health disparities in the region, as most would have had "no treatment" otherwise. Under-treatment where "usual care" is nothing has been used as an ethical justification to enroll subjects in many clinical studies. But once penicillin became a standard of care, there is no "beneficence" argument that could justify continuing the study.

It's possible that some of the men may still have consented to the Tuskegee study had they been told the true nature of the study, and its potential benefits and risks. For example, had they been told that they will receive less treatment than the standard of care, but more treatment than not seeing a doctor at all given the regional health disparities, informed consent would have become the "honest broker" as to whether the study was advantageous to them in some way.

Rivers stated: "Now a lot of those patients that were in the Study did get some treatment. There were very few who did not get any treatment" (Reverby 2009). Historians have tried to analyze Rivers' views on beneficence-based grounds:

> [Rivers] knew that the 'iron tonics, aspirin tablets and vitamin pills' that she gave out were not treatments for syphilis. But she described these drugs, as well as the physical exams, as being part of the treatment. She knew the aspirins helped with the pains of arthritis and the iron tonics gave the men 'pep' in the spring. She said: 'This was part of our medication that they got and sometimes they really took it and enjoyed it very much. And these vitamins did them a lot of good. They just loved those and enjoyed that very much very much.' ... Nurse Rivers seemed more troubled when she talked in her interviews about what penicillin had meant for the treatment of syphilis after the 1940s…She communicated in a 'just following orders' nursing voice…But, as with any of the doctors, she also emphasized the dangers of the Herxheimer reactions and her memory of someone dying from anaphylactic shock from a penicillin allergy (Reverby 2009).

By 1950, the PHS investigators clearly started to question whether the study was, in fact, "beneficent". They mused that maybe the medical activities performed to date on this population weren't so benign after all. The inadequate heavy metal treatments provided through the years may have caused morbidity and sped up some of the men's deaths (had treatment been adequate with a cure intended, the risks could be justified); many invasive diagnostic tests, including lumbar punctures, had significant risks (had the tests' purpose been to determine treatment, those risks could have been justified), and finally, the issue of medical neglect overall through the years took a toll on the population. Thus, O.C. Wenger concluded in a 1950 paper he delivered at a venereal disease conference: "[T]hese patients received no treatment on our recommendation. We know now, where we could only surmise before that we have contributed to their ailments and shortened their lives. I think the least we can say is that we have a high moral obligation to those that have died to make this the best possible study" (Gray 1998; Reverby 2009). In essence, stopping the study seemed "unethical" at a certain point to the study investigators.

The wording of The Nuremberg Code, however, specifically references that the code applied to an "experiment" (NIH 2020). It is not clear whether the PHS investigators viewed the Tuskegee study as an "experiment" per se, and the term "study" had different connotations. Since they were "following" a disease, but not testing any new drug or therapy, and even providing some nominal amounts of treatment, the Tuskegee study did not fit neatly into traditional categories of "experiment".

By 1964, when the first version of the Declaration of Helsinki was unveiled, the Tuskegee study could more clearly be defined as "non-therapeutic research" and it is clear that it violated most of the guidelines set forth in that document (WMA 2020), including all of Sections II and III. But by this period, the Tuskegee study published its findings on its 30th anniversary, and it is not clear—even now—how any investigator in this time frame would have applied "new" guidelines to old studies, in a pre-institutional/ethics review board (IRB) era. Were these new guidelines to be retroactive to 30 year-old studies? This question, even now, remains unclear.

John R. Heller, who was the decision-maker regarding penicillin from 1943–48 said in 1972 there was "nothing in the experiment that was unethical or unscientific" (Jones 1993). A CDC spokesperson defended the Tuskegee study that same year: "We are trying to apply 1972 medical treatment standards to those of 1932…At this point in time, with our current knowledge of treatment and the disease and the revolutionary change in approach to human experimentation, I don't believe the program would be undertaken" (Jones 1993). It's important to note that Heller called it an "experiment" while the CDC spokesman referred to the Tuskegee study as a "program". There is complete inconsistency in how the investigators viewed the Tuskegee study overall, and thus, it is difficult to assess what they understood to be applicable to either The Nuremberg Code or The Declaration of Helsinki.

Where there *is* consistency is in the contemporaneous peer assessment of when the Tuskegee study became definitively "unethical". Because there was clinical equipoise over the use of penicillin in latent or tertiary syphilis until about the mid-1950s (see earlier), the withholding of penicillin beyond that time frame is when the PHS was seen as truly crossing a bright moral line for medical practitioners reviewing the study, which is what led to someone finally blowing the whistle.

Clinical and Professional Ethics Issues

One of the core themes in *Miss Evers' Boys* surrounds the moral actions of the African American healthcare providers in enabling the Tuskegee study and collaborating with the PHS investigators. In evaluating the clinical and professional ethics issues surrounding the Tuskegee study, it's important to try to uncover what the healthcare providers involved believed was the prevailing clinical and professional ethical standards of care at the time, and whether they were in violation with those standards. The central problem in evaluating this question is germane to clinical trials in general: when patients are research subjects, what standard of patient care should apply? (For a prolonged discussion, see Clinical Ethics on Film, Chap. 1.) Even today, these questions are difficult to nail down, but in the years the Tuskegee study ran, particularly in the Plessy era, the complexity of holding African American health-care providers to the same patient care standards applicable today are not realistic. At the same time, dismissing the African American healthcare providers involved as having no moral agency is not responsible, either. In 1974, Fred Gray had taken a deposition from Nurse Rivers. He decided that she, along with all other Tuskegee Institute personnel, were merely "victims" of a racist protocol, with absolutely no power or control to stop the study, and had no choice but to go along (Gray 1998). Gray's perspective reduced the Tuskegee Institute professionals to non-autonomous, childlike beings, who had no moral agency whatsoever; it is a position that can be understood in the context of the 1970s, but became an increasingly problematic lens from which to view Dibble, Rivers and Moton in future years, given their training and positions (see earlier sections). The real answer is that these healthcare professionals agreed with the study aims at the time from a *global health ethics perspective*, in much the same way that we look upon studies in developing countries today where the standard of care is nothing. To these providers, the Tuskegee study was a "racial uplift" project (see earlier) they thought would address an important question for science regarding how much treatment was necessary in late stage syphilis in a population that had no access to medical care. They also thought the Tuskegee study was providing some benefit to the men, who were otherwise a completely medically neglected population. What they could not control were the wider health disparities issues that existed, which is a wider macro-ethics problem for which we cannot hold them responsible. Because most important of all, they saw the study as a vehicle for drawing more resources to the cause of African American healthcare—to poten-tially reduce health disparities. As for their professional ethics duties and the issue of deception—they were modeling the paternalistic attitudes that prevailed throughout American medicine in that time frame. Medical paternalism was equally applied to white patients, too.

In 1957, Rivers received the Oveta Culp Hobby Award for her "selfless devo-tion and skillful human relations…had sustained the interest and cooperation of the subjects of a venereal disease control program in Macon County Alabma" (Reverby 2009). She continued to take pride in her role right up until the Associated Press article outed the trial in 1972. Unfortunately, we don't know what Dibble would have said about this study if he were still living in the 1970s, but I have provided reasonable speculative analysis in earlier parts of this chapter.

Rivers was blindsided by the criticism over her participation in the Tuskegee study. According to Jones (1993): "Once her identity became known, Nurse Rivers excited considerable interest, but she steadfastly refused to talk with reporters. Details of her role in the experiment came to light [through] an article about the Tuskegee Study that appeared in *Public Health Reports* in 1953." In that article, she was noted to be the liaison between the researchers and subjects and to live in Tuskegee. She apparently stated in 1973 to a *Jet* magazine reporter (Smith 1996): "I don't have any regrets. You can't regret doing what you did when you knew you were doing right. I know from my personal feelings how I felt. I feel I did good in working with the people. I know I didn't mislead anyone".

It's possible that Rivers, over time, reconsidered her role. Reverby (2009) notes that James Jones was "sure [Rivers] had no moral uncertainty during the Study years, [but] he recalled that she told him to shut off the audiotape he was using [when interviewing her]. She then turned to him and said: 'We should have told those men they had syphilis. And God knows we should have treated them.' "

Smith (1996) observes:

> Black professionals faced a dilemma imposed by American racism in how to provide adequate health services to the poor within a segregated system. Furthermore, the gendered nature of public health work meant that the nurse, invariably a woman, was at the center of public provisions, both good and bad. As her actions show most starkly, black professionals demonstrated both resistance to and complicity with the government and the white medical establishment as they attempted to advance black rights and improve black health. Rivers and other black professionals counted on the benefits of public health work to outweigh the costs to the poor…but there were dire consequences when they were wrong.

In my view, the "moral complicity" argument is not applicable when given the lived context of these African American healthcare providers, living in the Plessy era in the Jim Crow South. Remember: these African Americans did not see any substantive changes to segregation and their civil liberties for most of their natural lives. The following scene from *Miss Evers' Boys* perhaps best captures the issue they faced with respect to withholding treatment.

> Evers: I'm a nurse. I'm not a scientist.

> Brodus: There is no difference. Not here. Not now. Not for *us*…You serve the race your way. I serve it in mine…I can't rock the boat while I'm trying to keep people from drowning. There are trade-offs you can't even imagine. Don't you see that?

Distributive Justice and Health Disparities

The original sin in the Tuskegee study are the conditions that created a vulnerable population and health disparities to begin with: mainly racism and apartheid as federal and social policy. The concept of social and racial justice in the Plessy era was not on speaking terms with American law at that time, while the *Brown* decision (see earlier) was specific to education. Healthcare access would remain largely inaccessible for impoverished Americans until the passage of Medicaid. The concept of "vulnerable population" was not properly articulated until 1978, when it first appeared in The Belmont Report (HHS 2020) which was largely informed by the Tuskegee study, but was also responding to the studies revealed in the Beecher paper in 1966. The 1973 Kennedy hearings surrounding protections for medical research subjects were indeed prompted by the Tuskegee study as well as other egregious studies such as Willowbrook and the Jewish Chronic Disease Hospital. These hearings led to the *National Research Act* (1974) and the formation in 1975 of the National Commission for the Protection of Human Subjects; its first report, Research With Children (National Commission 1977), addressed the Willowbrook experiments, followed by The Belmont Report in 1978 (HHS 2020), which addressed the Tuskegee study, and which specifically articulates the meaning of "vulnerable population" and distributive justice considerations as a requirement in research, codifying that such populations cannot be exploited in research, and that the burdens and benefits of research must be evenly distributed across populations.

The Bioethics Phase

The bioethics legacy of the Tuskegee study was to craft future preventative ethics guidelines that more clearly defined universal ethical principles in clinical research, codified in The Belmont Report: Respect for Persons, Beneficence, and Justice. All three were reverse-engineered principles designed to prevent another Tuskegee study from recurring. The Principle of Respect for Persons articulated that even those who cannot consent must be protected in research; the Principle of Beneficence makes clear that research must ensure that there is a greater balance of benefits than harms and that researchers must maximize benefits and minimize harms; and finally, the Principle of Justice dealt with preventing exploitation of vulnerable populations.

When Jones published the first definitive medical history of the Tuskegee study, it was just after The Belmont Report was published, and also informed by the first definitive scholarly text on medical ethics: Principles of Biomedical Ethics (Beauchamp and Childress 1979). But there were few bioethics courses available, and the field, per se, did not yet exist. As one bioethicist would later note, however, "Tuskegee gave birth to modern bioethics and James H. Jones was the midwife" (Reverby 2009). Coming full circle to the beginning of this chapter, one could say, too, that James H. Jones was the midwife to the film, *Miss Evers' Boys*.

As bioethicists began to populate academic medical centers, and the teaching of the Tuskegee study led to further analysis, there was a renewed sense of morally unfinished business. One piece of unfinished business, partly informed by this notorious study, was the articulation of a fourth bioethics principle.

The Principle of Non-Maleficence and the Duty to Warn

The Principle of Non-Maleficence spelled out the specific duty not to "intentionally" harm patients or their beneficiaries by performing known harmful therapies with no benefit, as well as knowingly withholding beneficial therapies.

In 1976, an obscure health law case would become relevant to public health ethics in particular. *Tarasoff v. Regents of the University of California* established the "duty to warn" third parties in imminent danger. (See Chap. 2.) By the 1980s, the "duty to warn" began to be a standard applied to the public health context, which meant that the duty to warn identifiable third parties that they may have contracted a serious sexually transmitted disease, including syphilis, was another factor to consider in the laundry list of harms done to the Tuskegee study enrollees. Family members who were known to be at-risk during the study years, and contracted syphilis, or who gave birth to children with congenital syphilis, were also considered harmed by the Tuskegee study.

The 1973 class action lawsuit filed by Fred Gray[7] was essentially based on the violation of the Principle of Non-Maleficence. Compensation packages were not just to the men, but to the families connected to this study who were not informed or warned they were at risk for syphilis, as well as those born with congenital syphilis because their fathers were never treated.

The Tuskegee Committee

A major outgrowth of the bioethics phase was to reconvene a bioethics committee to re-examine the Tuskegee study with more historical distance. Thus, the Tuskegee Syphilis Study Committee was formed in 1996, and issued its report, which included a recommendation for a formal Presidential Apology.[8] The report stated (Tuskegee Syphilis Study Committee 1996):

> Because the Tuskegee study is a starting point for all modern moral reflection on research ethics, a meeting of the NBEAC at Tuskegee in conjunction with a Presidential apology would be an ideal new beginning.

The apology finally came to fruition May 16, 1997,[9] also following Fred Gray's reaction to the HBO airing of *Miss Evers' Boys* in February of that year.

President Clinton's formal apology was an important enhancement to the film's release—still the only feature film made about the Tuskegee study as of this writing. With the film's release, and the apology, a new generation of scholarship and journalism surrounding the Tuskegee study began to flourish, which informed deeper insights into health disparities in the African American population, as well as trust issues that scholars now coin the "legacy of Tuskegee" (Reverby 2009). Harriet Washington in her book, Medical Apartheid (2007), pointed out that the Tuskegee study was fully representative of the treatment of African American patients received, but was not at all unique. Then an unexpected event occurred. Historian Susan Reverby, while doing research for her book, Examining Tuskegee (2009), discovered a *second* unethical syphilis study conducted by the U.S. government that *had* been kept secret, known as The Guatemalan Syphilis Study (1946–8). Here, the U.S. government enrolled human subjects in Guatemala without informed consent, deliberately infected them with syphilis, and then treated them with penicillin (Reverby 2012). The purpose of the Guatemalan study was to optimize penicillin dosages for the treatment of syphilis—the drug treatment withheld from the Tuskegee study subjects at the time. This time, a second "Clinton Apology" was needed. Hillary

---

[7]See Footnote 3.

[8]The entire report can be viewed here: http://exhibits.hsl.virginia.edu/badblood/report/.

[9]The Apology transcripts can be reviewed at this link: https://www.cdc.gov/tuskegee/clintonp.htm.

Clinton in the role of Secretary of State, serving the first African American U.S. President, Barack Obama, issued the formal apology about the second egregious syphilis study conducted by the U.S. government that harmed Guatemalans and their families and American medical research integrity in general (Hensley 2010).

Conclusions

Ultimately, the infamous Tuskegee Syphilis Study dramatized in *Miss Evers' Boys* spans multiple generations, lasted more than the average life span of any of its enrolled human subjects, was rooted in twentieth century systemic racism, but continues to harm patients and their families well into this century because of lasting disparities and mistrust of the medical community, and the American government institutions that are supposed to provide protections. Echoes of this study can be seen in the twenty-first century from the wake of Hurricane Katrina (2005), to the poisoning of African American citizens in Flint Michigan (2013), to the burden of the COVID-19 deaths in 2020 (see Afterword). When teaching *Miss Evers' Boys* today—as an enhancement to any number of courses—the first question we must address is whether we have made any progress at all in racism and health disparities. We need to be transparent about the fact that the United States is still a racist country—just "less racist" than it was when the Tuskegee study was running. Within this context, no one should be shocked that this study not only ticked along for 40 years, but that Congress continued to vote for it year after year—as Miss Evers reminds us in the film. The Tuskegee study's protocol, as Nurse Evers correctly states, "wasn't no secret; everybody knew what was going on…we gave the best care that was in our power to give to those men."

Theatrical Poster

Miss Evers' Boys (1997)

> Director: Joseph Sargent
> Producer: Robert Benedetti, Laurence Fishburne, Derek Kavanagh, Kip Konwiser, Kern Konwiser, Peter Stelzer
> Screenplay: Walter Bernstein
> Based on: "Miss Evers' Boys" (1992 stage play) by David Feldshuh
> Starring: Alfre Woodard, Laurence Fishburne, Craig Sheffer, Joe Morton, Obba Babatunde, and Ossie Davis
> Music: Charles Bernstein
> Cinematography: Donald M. Morgan
> Editor: Michael Brown
> Production Company: HBO NYC Productions and Anasazi Productions
> Distributor: HBO
> Release Date: February 22, 1997
> Run time: 118 min

References

Academy Awards. (2019). *Speech for 1939 actress in a supporting role.* http://aaspeechesdb.oscars.org/link/012-2/.

Alliance for Human Research Protection. (2014). *1962: Dr. Chester Southam injected live cancer cells into 22 elderly patients.* http://ahrp.org/1962-dr-chester-southam-injected-live-cancer-cells-into-22-elderly-patients-at-jewish-chronic-disease-hospital-in-brooklyn/.

Ambrose, C. T. (2016). Pre-antibiotic therapy of syphilis. *NESSA Journal of Infectious Diseases and Immunology, 1*(1), 1–20. https://uknowledge.uky.edu/microbio_facpub/83.

American Academy of Orthopedic Surgeons (AAOS). (2018). John F. Hume. http://legacyofheroes.aaos.org/About/Heroes/stories/hume.cfm.

Anne Frank.org. (2018). www.annefrank.org.

Bath, C. (2016). Is observation without surgery a viable strategy? *The ASCO Post,* November 10, 2016. http://www.ascopost.com/issues/november-10-2016/is-observation-without-surgery-a-viable-strategy-for-managing-ductal-carcinoma-in-situ/.

Beauchamp, T. L., & Childress, J. F. (1979). *Principles of biomedical ethics.* New York: Oxford University Press.

Beecher, H. K. (1966). Ethics and clinical research. *New England Journal of Medicine, 274*(24), 1354–1360.

Brandt, A. M. (1978). Racism and research: The case of the Tuskegee Syphilis study. *The Hastings Center Report, 8,* 21–29.

Brownstein, R. (2014). How Brown v. Board of education changed—and didn't change American education. *The Atlantic,* April, 25, 2014. https://www.theatlantic.com/education/archive/2014/04/two-milestones-in-education/361222/.

Bruusgaard, E. (1929). The fate of syphilitics who are not given specific treatment. *Archiv tur Dermatologie and Syphilis, 157,* 309.

Canby, V. (1972, July 20). Irving wallace's 'the man': Political movie stars James Earl Jones. *New York Times.*

Centers for Disease Control (CDC). (1981). *Pneumocystis pneumonia, 30*(21), 1–3. https://www.cdc.gov/mmwr/preview/mmwrhtml/june_5.htm.

Coaxum, J. (2018). Historically black colleges and universities. *State University Webpage.* http://education.stateuniversity.com/pages/2046/Historically-Black-Colleges-Universities.html.

Conner, E. (2019). https://www.ranker.com/list/whoopi-goldberg-and-ted-danson-dated/eric-conner.

Cornell University. (1993). Documentary. *Susceptible to Kindness.*

Cornell University. (2018). David Feldshuh faculty page: http://pma.cornell.edu/david-m-feldshuh.

Dibble, E., Rabb, L., & Ballard, R. (1961). John A. Andrew Memorial Hospital. *Journal of the National Medical Association, 53,* 104–118. https://www.ncbi.nlm.nih.gov/pmc/articles/PMC2641895/pdf/jnma00690-0004.pdf.

Dougherty, M. (1987). Playing South African Activist Winnie Mandela, Alfre Woodard Captures the soul of a nation. *People,* September 28, 1987. https://people.com/archive/playing-south-african-activist-winnie-mandela-alfre-woodard-captures-the-soul-of-a-nation-vol-28-no-13/.

Downs, M. L. (2015). *Great depression in Alabama.* April 21, 2015. http://www.encyclopediaofalabama.org/article/h-3608.

Drake, S. (1990). Stage review: Subject propels Miss Evers' Boys. *Los Angeles Times.* July 20, 1990. http://articles.latimes.com/1990-07-20/entertainment/ca-18_1_miss-evers-boys.

Drossman, S. R., Port, E. R., & Sonnenblick, E. (2015). Why the annual mammogram matters. *New York Times,* October 29, 2015. http://www.nytimes.com/2015/10/29/opinion/why-the-annual-mammogram-matters.html?_r=1.

Encyclopedia.com. (2018). Laurence Fishburne, 1961. Contemporary black biography. *Encyclopedia.com.* July 17, 2018. http://www.encyclopedia.com/education/news-wires-white-papers-and-books/fishburne-laurence-1961.

Encylopedia.com (2019). Charles Spurgeon Johnson (1893–1956). Encyclopedia of world biography. *Encyclopedia.com.* July 12, 2019. https://www.encyclopedia.com/people/history/historians-miscellaneous-biographies/charles-s-johnson.

Fairclough, A. (1997). Civil rights and the Lincoln Memorial: The censored speeches of Robert R. Moton (1922) and John Lewis (1963). *The Journal of Negro History, 82*(4), 408–416. https://www.jstor.org/stable/2717435?seq=1#page_scan_tab_contents.

Ferris State University. (2018). *Jim Crow Museum or Racist Memorabilia.* https://www.ferris.edu/HTMLS/news/jimcrow/who/index.htm.

Frandsen, J., Orton, A., Shreive, D., & Tward, J. (2017). Risk of death from prostate cancer with and without definitive local therapy when gleason pattern 5 is present: A surveillance, epidemiology, and end results analysis. *Cureus, 9*(7), e1453. https://www.ncbi.nlm.nih.gov/pmc/articles/PMC5590810/.

Freedman, B. (1987). Equipoise and the ethics of clinical research. *New England Journal of Medicine, 1987*(317), 141–145.

Frith, J. (2012). Syphilis—Its early history and treatment until penicillin and the debate on its origins. *Journal of Military and Veterans' Health*, 20, 4. https://jmvh.org/article/syphilis-its-early-history-and-treatment-until-penicillin-and-the-debate-on-its-origins/.

Furman, M. (2012). 90th Anniversary of the dedication of the Lincoln memorial. *Ranger Journal*, 30 May, 2012. https://www.nps.gov/nama/blogs/90th-anniversary-of-the-dedication-of-the-lincoln-memorial.htm.

Gebreyes, R. (2016). Actress Mary Steenburgen's friendship with the Clintons goes way back. *Huffington Post*, Oct. 27, 2016. https://www.huffingtonpost.com/entry/mary-steenburgen-hillary-clinton_us_581110a8e4b064e1b4b04957.

Geiger, J. H. (1981). An experiment with lives. *Book Review. New York Times*, June 21, 1981. https://archive.nytimes.com/www.nytimes.com/books/98/12/06/specials/jones-blood.html?mcubz=1.

Gjestland, T. (1955). The Oslo study of untreated syphilis; an epidemiologic investigation of the natural course of the syphilitic infection based upon a re-study of the Boeck-Bruusgaard material. *Acta Derm Venereol Suppl (Stockh), 35*(Suppl 34), 3–368; Annex I-LVI. https://www.ncbi.nlm.nih.gov/pubmed/13301322.

Gray, F. D. (1998). *The Tuskegee Syphilis study.* Montgomery, AL: New South Books.

Harvard Library (2018). *Rivers interview by BWOHP.* https://sds.lib.harvard.edu/sds/audio/443302359.

Harvey, W. B., Harvey, A. M., & King, M. (2004). *The impact of Brown v. Board of Eduation on post-secondary participation of African Americans.* https://www.jstor.org/stable/4129615?seq=1#page_scan_tab_contents.

Health and Human Services (HHS). (2020). *The Belmont Report.* https://www.hhs.gov/ohrp/regulations-and-policy/belmont-report/index.html.

Heller, J. (1972). Syphilis victims in U.S. study went untreated for 40 years. *New York Times*, July 26, 1972. https://www.nytimes.com/1972/07/26/archives/syphilis-victims-in-us-study-went-untreated-for-40-years-syphilis.html.

Hensley, S. (2010). U.S. apologizes for syphilis studies in Guatemala. *NPR*, October 1, 2010. https://www.npr.org/sections/health-shots/2010/10/01/130266301/u-s-apologizes-for-medical-research-that-infected-guatemalans-with-syphilis.

Hill, R. E. (Ed.). (1991). The black women oral history project: From the Arthur and Elizabeth Schlesinger library on the history of women in America, Radcliffe College. In R. E. Hill (Ed.). Westport, CT: Meckler.

Holt, P. (1996). Alice Walker on the making of the film 'The Color Purple'. *San Francisco Chronoicle*, January 7, 1996. https://www.sfgate.com/books/article/Alice-Walker-on-the-Making-of-the-Film-The-Color-3000001.php.

Hornblum, A. M. (2013). NYC's forgotten cancer scandal. *New York Post*, December 28, 2013. https://nypost.com/2013/12/28/nycs-forgotten-cancer-scandal/.

Jet (1971). People: Dr. John Hume. *Jet*, Mar 25, 1971.

Jet (1997). Laurence Fishburne and Alfre Woodard star in HBO movie about Tuskegee experiment on syphilis. *Jet*, February 24, 1997.

Journal of the National Medical Association (JNMA). (1962). Dr. Eugene Heriot Dibble, Jr., Distinguished Service Medalist for 1962. *Journal of the National Medical Association, 54*, 711–712.

Isenberg, B. (1990). Recreating a night of good intentions: Dr. David Feldshuh's 'Miss Evers' Boys' examines a dark hour in medicine: the Tuskegee syphilis study. *Los Angeles Times*, July 15, 1990. http://articles.latimes.com/1990-07-15/entertainment/ca-372_1_miss-evers-boys.

Johnson, C. (1931). *The Negro in American civilization*. London: Constable and Company Ltd.

Johnson, C. (1934). *In the shadow of the plantation*. Chicago: University of Chicago Press.

Jones, J. (1981, 1993). (First edition, 1981; Revised edition, 1993). *Bad blood*. New York: Free Press.

Kagan, J. (2004). *Visual History with Joseph Sargent. Interviewed by Jeremy Kagan, Directors Guild of America*. Retrieved from https://www.dga.org/Craft/VisualHistory/Interviews/Joseph-Sargent.aspx.

Katz, R. V. (2008). The legacy of Tuskegee. *Journal Healthcare Poor Underserved, 19*(4), 1168–1180. (https://www.ncbi.nlm.nih.gov/pmc/articles/PMC2702151/.

Kenton, W. (1929). Stock market crash of 1929. *Investopedia*, Apr 17, 2018. https://www.investopedia.com/terms/s/stock-market-crash-1929.asp.

King, S. (1997). A government study gone bad. *Los Angeles Times*. Retrieved from http://articles.latimes.com/1997-02-16/news/tv-29113_1_miss-evers.

Lagnado, L. (2015). Debate over early stage cancer. *Wall Street Journal*, October 19, 2015. http://www.wsj.com/articles/debate-over-early-stage-cancer-to-treat-or-not-to-treat-1445276596.

Legal Defense Fund. (2019). Landmark: Brown v. Board of Education. http://www.naacpldf.org/case/brown-v-board-education.

Levitt, S. (1993). Changing partners. *People,* November 22, 1993.

Lewis, N. A. (1995). Ex-colleague says Clinton nominee knew of Tuskegee study in 1969. *New York Times*, February 28, 1995. https://www.nytimes.com/1995/02/28/us/ex-colleague-says-clinton-nominee-knew-of-syphilis-study-in-1969.html.

Marriott, M. (1997). First, do no harm: A nurse and the deceived subjects of the Tuskegee Study. *New York Times*, February 16, 1997. https://www.nytimes.com/1997/02/16/tv/first-do-no-harm-a-nurse-and-the-deceived-subjects-of-the-tuskegee-study.html.

McVean, A. (2019). 40 years of human experimentation in America: The Tuskegee Syphilis Study. *McGill Office for Science and Society*, January 25, 2019. https://www.mcgill.ca/oss/article/history/40-years-human-experimentation-america-tuskegee-study.

Meyer, H. S. (1981). Bad blood: The Tuskegee syphilis experiment. *JAMA, 246*(22), 2633–2634. https://jamanetwork.com/journals/jama/article-abstract/365193.

Mills, B. (1997). It takes a special kind of villain to sacrifice herself for … *Chicago Tribune*, February 16, 1997. http://articles.chicagotribune.com/1997-02-16/entertainment/9702160182_1_tuskegee-syphilis-study-miss-evers-boys-untreated-syphilis.

Moton, R. R. (1913). *Some elements necessary to race development*. Press of the Hampton Normal and Agricultural Institute. https://www.amazon.com/Some-elements-necessary-race-development/dp/B00086ZGJG/ref=sr_1_13?s=books&ie=UTF8&qid=1533951749&sr=1-13&refinements=p_27%3ARobert+Russa+Moton.

Moton, R. R. (1916). *Racial goodwill*. An address reprinted by Ulan Press, 2012. https://www.amazon.com/Racial-Robert-Russa-1867-catalog/dp/B00ABUU1MQ/ref=sr_1_6?s=books&ie=UTF8&qid=1533950894&sr=1-6&refinements=p_27%3ARobert+Russa+Moton.

Moton, R. R. (1920). *Finding a way out: An autobiograpy*. Double Day Books and Co. Scanned edition: https://docsouth.unc.edu/fpn/moton/moton.html.

Moton, R. R. (1922). *The Negro's debt to Lincoln*. Unknown binding. https://www.amazon.com/Negros-Lincoln-Robert-Russa-Moton/dp/B000882EZS/ref=sr_1_12?s=books&ie=UTF8&qid=1533950894&sr=1-12&refinements=p_27%3ARobert+Russa+Moton.

Moton, R. R. (1929). *What the Negro thinks*. Doubleday, Doran and Company.

National Commission for the Protection of Human Subjects of Biomedical and Behavioral 1569 Research. (1977). *Report and recommendations: Research involving children. (O77-0004).* Federal Register: U.S. Government.

National Institutes of Health. (2020). *The Nuremberg Code.* https://history.nih.gov/research/dow nloads/nuremberg.pdf.

National Medical Association (NMA). (2018). https://www.nmanet.org/page/History.

National Park Service. (2019). *Lincoln memorial webpage.* https://www.nps.gov/linc/learn/histor yculture/lincoln-memorial-important-individuals.htm.

O'Connor, S. (2015). Why doctors are rethinking breast cancer treatment. *Time*, October 1, 2015. http://time.com/4057310/breast-cancer-overtreatment/.

Olenin, A. & Corcoran, T. F. (1942). *Hours and earnings in the United States 1932–40.* U. S. Department of Labor, Bulletin No 697. https://fraser.stlouisfed.org/files/docs/publications.bls/ he_bls_1942.pdf.

Padgett, M. (2018). *Black face minstrel shows.* http://black-face.com/minstrel-shows.htm.

Reverby, S. M. (2000). Dibble interview: [Helen Dibble, Daniel Williams, "An Interview with Nurse Rivers," *Tuskegee's Truths*, ed. Reverby, p. 327.].

Reverby, S. M. (2009). *Examining Tuskegee.* University of North Carolina Press.

Reverby, S. M. (2012). *Ethical failures and history lessons: the U.S. Public Health Service Research Studies in Tuskegee and Guatemala.* https://publichealthreviews.biomedcentral.com/track/pdf/ 10.1007/BF03391665.

Roberts, S., & Irwin, S. (2015). 83, Rare critic of Tuskegee study is dead. *New York Times*, April 18, 2015. https://www.nytimes.com/2015/04/19/health/irwin-schatz-83-rare-critic-of-tuskegee-study-is-dead.html?_r=1.

Robertson, C. (2018). A lynching memorial is opening. *New York Times*, April 25, 2018. https:// www.nytimes.com/2018/04/25/us/lynching-memorial-alabama.html.

Rockwell, D., et al. (1964). The Tuskegee study of untreated syphilis: The 30th year of observation. *Archives of Internal Medicine, 114*, 792–798.

Rosenthal, M. S. (2015). Boo: Scary new guidelines for breast cancer. *Endocrine Ethics Blog*, October 30, 2015. http://endocrineethicsblog.org/2015/10/.

Rosenthal, M. S., Ain, K. B., Angelos, P. A. et al. (2017). *Problematic clinical trials in thyroid cancer.* Published online at: https://doi.org/10.2217/ije-2017-0008.

Schulman, D. (2009). *A force for change: African American art and the Julius Rosenwald Fund.* Evanston: Northwestern University Press.

Skloot, R. (2010). *The immortal life of Henrietta Lacks.* New York: Crown Publishing.

Smith, S. L. (1996). Neither victim nor villain: Nurse Eunice Rivers, the Tuskegee syphilis experiment, and public health work. *Journal of Women's History, 8*, 95–113. https://muse.jhu.edu/art icle/363745/pdf.

Smithonian National Museum of History. (2019). *Lincoln Memorial webpage.* https://americanh istory.si.edu/changing-america-emancipation-proclamation-1863-and-march-washington-1963/ 1963/lincoln-memorial.

Staples, B. (2018). How the Suffrage movement betrayed black women. *New York Times*, June 28, 2018. https://www.nytimes.com/2018/07/28/opinion/sunday/suffrage-movement-racism-black-women.html?action=click&module=RelatedLinks&pgtype=Article.

Suba, E. J. (2015). India cervical cancer testing is Tuskegee 2.0. *Alabama.com*, June 17, 2015. https://www.al.com/opinion/2015/06/india_cervical_cancer_testing.html.

Tuskegee Syphilis Study Committee Report. (1996). http://exhibits.hsl.virginia.edu/badblood/rep ort/.

Washington, H. (2007). *Medical apartheid.* New York: Doubleday.

Wellner K. L. (2005). Polio and historical inquiry. *OAH Magazine of History, 19* (5), Medicine and History (Sep., 2005), 54–58.

Wilt, T. J. et al. (2017). Follow up of prostatectomy versus observation for early prostate cancer. *New England Journal of Medicine 377*, 132–142. https://www.nejm.org/doi/full/10.1056/NEJ Moa1615869.

Winer, L. (1989). Patients sacrificed in the name of research. *New York Times*, December 21, 1989. https://www.nytimes.com/1989/12/21/theater/review-theater-patients-sacrificed-in-the-name-of-research.html.

World Medical Association (WMA). (2020). *Declaration of Helsinki*. https://www.wma.net/policies-post/wma-declaration-of-helsinki-ethical-principles-for-medical-research-involving-human-subjects/.

Zinsmeister, K. (2018). Philanthropy round table website, 2018. https://www.philanthropyroundtable.org/almanac/people/hall-of-fame/detail/julius-rosenwald.

Zorthian, J. (2016). This is how February became Black History Month. *Time*, January 29, 2016. http://time.com/4197928/history-black-history-month/.

Chapter 2
"How Many Dead Hemophiliacs Do You Need?" *And the Band Played on* (1993)

The early and shameful history of HIV/AIDS[1] in the United States, and the decimation of the U.S. gay population throughout the 1980s and early 1990s, is not well known to the majority of healthcare practitioners who were licensed in the twenty-first century. But it must be, and this film is the conduit. *And the Band Played On* is a faithful representation of the 1987 630-page book of the same name, written by journalist Randy Shilts. As an openly gay reporter for the *San Francisco Chronicle*, in a city that had the largest population of gay males at that time—roughly 17% of the city's population was gay (KQED 2009), Shilts tracked the deadly pandemic from its first appearance around 1979 through the disease's infancy and early childhood, signing off in 1987, when he completed the book. In "AIDS Years" 1987 was a timeframe that was considered a turning point in HIV/AIDS, when it finally began to be treated like a twentieth century infectious disease instead of a seventeenth century plague, and the year that azidothymidine (AZT), the first anti-retroviral medication for AIDS, was introduced. Shilts refrained from taking the HIV antibody test until he completed the book (CBS 1994); he tested positive and died of AIDS in 1994 at age 42, two years before protease inhibitors were discovered. His greatest contribution to society and medicine was writing down what occurred so that the tragic ethical violations and medical harms that characterized this time period can be taught to future generations of practitioners who now regard AIDS as casually as type 2 diabetes, and who may even see a cure for AIDS in their professional lifetime. While some may never see a patient die from full-blown AIDS or an opportunistic infection, they may be on the front lines of the next pandemic, such as Ebola or COVID-19 (see Chap. 6 and Afterword). Shilts' work on this critical early period of AIDS is comparable to Daniel Dafoe's A Journal of the Plague Year (1772), which was about the account of a 1665 outbreak of the bubonic plague in London, as witnessed by one man. Like Dafoe, Shilts tracks AIDS chronologically, across multiple "ground zero" settings, and bears witness to a time and place that is unique in the social and medical history

---

[1] See under the History of Medicine section for the history of nomenclature for both the Human Immunodeficiency Virus (HIV) and Acquired Immune Deficiency Syndrome (AIDS).

© Springer Nature Switzerland AG 2020
M. S. Rosenthal, *Healthcare Ethics on Film*,
https://doi.org/10.1007/978-3-030-48818-5_2

of the twentieth century. This is a story about intentional medical harms; denial; greed; and the political climate of misusing AIDS to weaponize homophobia and health disparities. For many living Americans who were homophobic in the 1980s, gay males were considered as dispensable a patient population as African Americans in Macon County, Alabama in the 1930s, when the Tuskegee study was underway (see Chap. 1). When the first AIDS cases appeared to be isolated to homosexuals, the prevailing attitude was that it was not a public health problem, but endemic in the gay population due to high-risk sexual behaviors or lifestyles, which mirrored similar misconceptions about syphilis in the African American population in the 1930s. In the early 1980s, most heterosexuals were not interested in the plight of gay males unless they were personally connected to one.

When the first cases of AIDS were reported among hemophiliacs in 1982 it served as evidence that the public blood supply was contaminated by the AIDS virus through blood donors who were infected. Yet the blood industry refused to implement any screening of the public blood supply prior to 1985, when it was much too late. "How many dead hemophiliacs do you need?" is the iconic question that characterizes how AIDS became a public health catastrophe. The question is shouted in the film by Don Francis, a Centers for Disease Control (CDC) physician (portrayed in the film by Matthew Modine), out of frustration and recognition of ethical violations by the blood industry's failure to act on clear evidence that Factor VIII (clotting factor made from donor blood) was contaminated with the AIDS virus. But because the AIDS virus at this early point was mainly killing gay males and drug addicts—patients few actually cared about at the time—the "handful of hemophiliacs" who contracted AIDS through the public blood supply were considered inconsequential compared to the high cost and inconvenience of screening the blood supply. (This is particularly ironic, considering that Ronald Reagan's own campaign manager for his 1984 re-election—Roger Ailes—was a hemophiliac.) Moreover, despite growing numbers of transfusion-AIDS cases—particularly in the neonatal, pediatric and post-partum populations—donor screening did not occur until 1985.

Ultimately, *And the Band Played On* is a "maleficence blockbuster" that harshly, but accurately, judges competent and well-trained health policymakers and some researchers who had the tools and knowledge to contain the AIDS virus, but didn't because it was too personally and professionally inconvenient.

This chapter discusses the origins of the book and film's production within the sociological context of the gay liberation movement of the late 1970s in the United States, which abruptly ended in 1981; social misconceptions about AIDS; early AIDS activism and activists such as Larry Kramer; and key historical figures in the social and political history of AIDS, ranging from Rock Hudson to Ryan White. Next, this chapter discusses this film within a History of Medicine context from 1981 to 1993, highlighting early AIDS epidemiology and the facts about Gaetan Dugas (controversially known as "patient zero" in an early cluster study, discussed further); pioneering AIDS researchers, clinician-activists such as Mathilde Krim (co-founder of AMFAR) and Linda Laubenstein (the female doctor in the play "The Normal Heart"); funding issues and failure to screen the blood supply; the discovery of HIV and AZT, as well as how to frame the film in a post-"AIDS cocktail" period (1996 and beyond). Finally, this chapter will discuss the healthcare ethics issues raised in this film: clear violations of the Principle of Non-Maleficence; health disparities; discrimination and fear, which led to the abandonment of patients; and conflicts of interest that led to delays in prevention and treatment.

The Life and Times of Randy Shilts: From the Book to the Film

Randy Shilts, author of And the Band Played On, was born in 1951; he was a classic baby boomer who came of age at the height of the Civil Rights movement. Three years before he was born, Alfred Kinsey had published his groundbreaking book, Sexual Behavior in the Human Male (1948), which first introduced the Kinsey Scale, and identified homosexuality as part of the "normal" spectrum of male sexual behavior. Kinsey's work, though controversial at the time, planted the seeds for the next generation of gay males to live openly and begin to "come out". Shilts was just 18 years old when the Stonewall Riots took place on June 28, 1969—a few weeks shy of Apollo 11. The Stonewall Riots marked the first major "gay civil rights" protest. When the police raided The Stonewall Inn, a gay club in Greenwich Village, its patrons fought back and rioted against the police. The incident marked the beginning of the Gay Liberation Movement, which flourished and peaked in the 1970s, and which also established large gay communities in particular cities, such as San Francisco and New York. In fact, the first HIV infection identified in American gay males was traced to a sexual transmission during a bi-centennial celebration in New York City harbor in 1976 (Shilts 1987).

The philosophy of the Gay Liberation Movement was to celebrate sexual freedom by making sexual activity a focal point in the community, and encouraging gay males (open or closeted) to embrace sexual promiscuity as a banner. One historian recalls (Larsen 2015):

> In 1981, the New York gay community was still living in the heady days of post-Stonewall gay liberation. Gays and lesbians were still fighting for basic legal rights, such as freedom from police harassment, and the current successes of the gay marriage lawsuits were barely a fantasy for most of them. The Gay Pride movement offered gays an ideology of basic self-worth as an antidote to a society that largely ostracized them and viewed them as either mentally ill or morally degenerate and therefore a threat to society. Many in the gay community embraced a hyper-sexualized culture based on free love, partying, and drug use, thereby unwittingly creating ideal conditions for a disease that had been lurking unnoticed in the American population since at least the 1960s because it lacked the opportunity to spread easily.

By the 1990s, the sexualized lifestyle would be labeled "high risk behaviors". Shilts notes, for example (Shilts 1987: 19):

> This commercialization of gay sex was all part of the scene, an aspect of the homosexual lifestyle in which the epidemics of venereal disease, hepatitis, and entric disorders thrived. The gay liberation movement of the 1970s had spawned a business of bathhouses and sex clubs. The hundreds of such institutions were a $100 million industry across American and Canada, and bathhouse owners were frequently gay political leaders as well, helping support the usually financially starved gay groups.

Dr. Lawrence D. Mass, a New York City physician stated this in 1982 about the spread of AIDS: "gay people whose life style consists of anonymous sexual encounters are going to have to do some serious rethinking."(Altman 1982)

Upon the book's release in 1987, the *New York Times* noted this about the gay community's promiscuity (Geiger 1987):

> With few exceptions, they denied that the epidemic existed except as a homophobic fantasy, fiercely labeled attempts to modify behavior as "sexual facism" and an infringement on civil liberties and failed to mobilize effectively for more funding for research and treatment.

Shilts lived as an openly gay male by age 20 (Weiss 2004). Born in Iowa, and raised in Illinois, he came out while still in college, while he was earning his journalism degree. He worked from 1975 to 1977 for the gay press. Although he struggled to find work as a mainstream journalist due to his lifestyle, he eventually began to cover gay issues for the mainstream press in cities that had large gay communities, including common health problems such as Hepatitis B, which was sexually transmitted.

The title of the book is a nod to the 1912 Titanic disaster, in which the "band played on" as the ship was sinking the night of April 14 and early morning of April 15 until about 1:55 AM, when the last lifeboat was lowered. Until Shilts, the phrase was always connected to the Titanic disaster, based on eyewitness accounts. Musician Wallace Hartley led a string quartet that played on the upper deck all throughout the sinking to calm passengers. The last songs heard were "Nearer My God to Thee" and "Autumn." Hartley's last words were apparently: "Gentlemen, I bid you farewell" (Kopstein 2013).

From 1977 to 1980, Shilts worked for an Oakland television news station, covering gay issues, which led to his chronicling of the 1978 assassination of the first openly gay politician in San Francisco, Harvey Milk, who championed the first piece of U.S. legislation to ban gay discrimination. Shilts' reporting led to his first book, a biography of Harvey Milk, entitled The Mayor of Castro Street: The Life and Times of Harvey Milk (Shilts 1982). In 1984, a documentary entitled *The Times of Harvey Milk* aired, and in 2008, the film *Milk*, based on Shilts' book, was released with Sean Penn (see Chap. 7) in the title role. Shilts began to gain recognition as a superb journalist and next landed a plum job at the *San Francisco Chronicle* with the assignment of covering the "gay beat"; by 1982, the main story in the gay community was a new "plague" that seemed to be targeting gay males, which was initially labelled GRID—Gay-Related Immune Deficiency—in a May 11, 1982 *New York Times* article (Altman 1982). Although the article also noted that some were calling the disease "A.I.D." for Acquired Immune Deficiency, the disease would not be known as AIDS until September 1982 (see under History of Medicine). Thus, Shilts officially became the San Francisco "chronicler" of the early days of AIDS, which turned into a 630-page book, And the Band Played On: Politics, People, and the AIDS Epidemic (1987). Shilts stated: "Any good reporter could have done this story, but I think the reason I did it, and no one else did, is because I am gay. It was happening to people I cared about and loved" (Grimes 1994). The book was explosive in its condemnation of policy makers and decision-makers. Shilts states in his prologue (Shilts 1987):

The bitter truth was that AIDS did not just happen to America – it was allowed to happen. ...
From 1980, when the first isolated gay men began falling ill from strange and exotic ailments,
nearly five years passed before all these institutions - medicine, public health, the federal
and private scientific research establishments, the mass media, and the gay community's
leadership - mobilized the way they should in a time of threat. The story of these first five
years of AIDS in America is a drama of national failure, played out against a backdrop of
needless death.

The book (and later, film) tracks five main political theaters of early AIDS history: (1)
the gay community: patients, partners, caregivers, the gay "libertarians" and sexual
freedom fighters, as well as their opponents—gay advocates for risk reduction and
safe sex; (2) the clinicians on the ground dealing with lack of resources for AIDS
patients; (3) the public health professionals [from local to the CDC] embroiled in risk
reduction strategies and "screening battles" to protect the public blood supply; (4) the
Reagan Administration response; and (5) the battle over credit for the discovery of
the AIDS virus, which has become known as one of the most infamous research ethics
stories. The book tells these stories simultaneously in chronological, not thematic
order. At the time, the book created a fast-paced, dizzying adventure story with
multiple subplots and characters. The film version is faithful to this style of narrative,
and mirrors the same pace with the same effects.

In a November 1987 *New York Times* review (Geiger 1987), it was clear the book
was being published in the center of an uncontrolled epidemic:

We are now in the seventh year of the AIDS pandemic, the worldwide epidemic nightmarishly
linking sex and death and drugs and blood. There is, I believe, much more and much worse
to come...And so acquired immune deficiency syndrome is not only an epidemic; it is a
mirror, revealing us to ourselves. How did we respond? [The book] is at once a history and
a passionate indictment that is the book's central and often repeated thesis... A majority of
the anticipated tens of thousands of 1991 New York City AIDS patients will be black and
Hispanic intravenous drug users, their sexual partners and their babies.

However, to a 1987 reader, the multiple political theaters unfolding in the book were
difficult to grasp; the book review also noted the following (Geiger 1987):

There is also the clinical story of physicians struggling both to treat and care for AIDS
patients - desperately comparing notes, searching the medical journals, fighting for hospital
beds and resources. There is the story of the scientific research that led at last to a basic
understanding of the disease, the identification of the virus, the test for antibodies. And,
finally, there is the larger political and cultural story, the response of the society, and its
profound impact on all the other aspects of the AIDS epidemic.

Mr. Shilts tells them all - but he tells them all at once, in five simultaneous but disjointed
chronologies, making them all less coherent. In the account of a given month or year, we
may just be grasping the nature of the research problem - and then be forced to pause to read
of the clinical deterioration of a patient met 20 or 40 or 60 pages earlier, and then digress to a
Congressional hearing, and then listen to the anxious speculations of a public health official
and finally review the headlines of that month. The threads are impossible to follow.

The reader drowns in detail. The book jacket says that Mr. Shilts - in addition to his years
of daily coverage of the epidemic - conducted more than 900 interviews in 12 nations and

dug out thousands of pages of Government documents. He seems to have used every one of them. Reading <u>And the Band Played On</u> sometimes feels like studying a gigantic mosaic, one square at a time.

Indeed, it does. That is why the 1993 film version, faithful to the book's pace and narrative, suffices for new generations of learners who may not have the three months that it takes to read the entire book. However, Shilts' 630 pages age very well, and have become the "go to" for any medical historian wishing to document this period of history. With only a few exceptions noted further on, Shilts' historical account holds, and also led to a scathing analysis of organizational decision-making by the Institute of Medicine in 1995 (see under History of Medicine).

An AIDS Book in the Reagan Era

When Shilts' book is published in 1987, Ronald Reagan, one of the most popular Republican Presidents to date, was finishing his second term, had survived an assassination attempt in his first term (March 30, 1981 by John Hinckley Jr., who was found not guilty by reasons of insanity), and would be successful in being elected for a "third term" of sorts, when voters chose his Vice President, George H. W. Bush (aka "Bush 41") in 1988 to continue his agenda, including the "slow walk" on AIDS policy. As mentioned earlier, considering that both Reagan and Bush 41 were very dependent on infamous Republican strategist Roger Ailes—a hemophiliac—for getting elected, the disconnect of ignoring AIDS is notable, since Ailes no doubt was probably taking clotting factor VIII before the blood supply was screened. Reagan was not a supporter of gay rights, and made that clear in his 1980 bid for the presidency: "My criticism is that [the gay movement] isn't just asking for civil rights; it's asking for recognition and acceptance of an alternative lifestyle which I do not believe society can condone, nor can I" (Scheer 2006).

Prior to 1985, AIDS was not a major topic of concern for most Americans. As hundreds of gay males in San Francisco began to drop dead between 1981 and 1984, another part of the Bay Area was flourishing as the PC Revolution took shape; Microsoft and Apple started, and the famous "Macintosh 1984" commercial aired December 31, 1983. Most Americans have memories during this period not of AIDS, but of Madonna, Michael Jackson's "Thriller" album, and Bruce Springsteen's "Born in the USA" album as the soundtrack to a time frame of shameless promotion of materialism, opulence and greed, demonstrated in the film, *Wall Street* (1986). The biggest fears amongst most heterosexual American liberals between 1981 and 1984 centered on nuclear holocaust as Reagan poured billions into increased defense, and ramped up tensions with the Soviet Union. An entire genre of 1980s nuclear war films abounded around Reagan's first term: *The Day After* (1983); *Testament* (1983); *War Games* (1983); the horrific British film, *Threads* (1984); and the animated film, *When the Wind Blows* (1986). Reagan actually changed his nuclear policies as a result of watching *The Day After*.

In October 1985, when actor Rock Hudson died of AIDS (see further), a flurry of heightened mainstream media coverage of the AIDS epidemic followed. Shilts, in fact, divides early AIDS history into "before Rock" and "after Rock" (see further). Reagan famously avoided all discussion of AIDS during the entire first term of his presidency; he mentioned it for the first time in 1985 in the aftermath of Hudson's death, after approximately 4000 Americans had died from AIDS (a greater number than those killed on 9/11 and roughly the same number of American troops killed in the Iraq war), and roughly 8000 had been infected (Avert 2018) by then. Reagan began to mention AIDS in several speeches in 1987, which was likely due to the popularity of Shilts' book.

By 1987, AIDS was beginning to penetrate into heterosexual lives more significantly. The third top-grossing film of that year was *Fatal Attraction*, which was not about AIDS per se, but about how sex with the wrong person can literally kill your whole family. *Fatal Attraction* resonated with heterosexuals on a range of levels, but ultimately was a film that demonstrated the depths to which having unprotected sex can harm you. At the same time, another high-grossing film, *Three Men and a Baby*, which featured three heterosexual men living together and raising a baby as a result of a one-night stand of one of them—also resonated culturally, as "macho" heterosexual men were portrayed here as maternal, capable of caregiving, and taking ownership of consequences of their sexual partners. Young adults coming of age in this period—demonstrated in sitcoms such as *Family Ties*, for example—had the opposite experiences of their parents, who grew up during the Sexual Revolution. Yet another 1987 film, inspired by misplaced "nostalgia" for easier sexual times, was *Dirty Dancing* (see Chap. 5). This film was yet another cautionary tale about unprotected sex, in which the entire plot revolves around an unwanted pregnancy and unsafe abortion. Viewers of *Dirty Dancing* are also aware of a growing epidemic of maternal transmission of AIDS, and transfusion AIDS in the neonatal setting. In essence, three major films in 1987 that are not *specifically* about AIDS, are still about the consequences of unprotected sex, reflecting the cultural anxieties over AIDS. If you were in college at this time, HIV testing and condoms became part of the college scene. By December 1, 1988, the first Annual AIDS Day was launched, marking the end of the Reagan era, and the start of the Bush 41 era, which was also noted for an "appalling" record on AIDS, discussed further (Thrasher 2018).

Ultimately, Reagan's first term coincided with the "Moral Majority" movement, in which conservative Christians dominated the political agenda, openly voicing fear and loathing of homosexuals and anyone else with AIDS. However, by the end of Reagan's second term, the Moral Majority had begun to dissolve as a movement (later re-emerging as the religious right base of the Republican Party), while AIDS was no longer considered a "gay disease" but a political one, largely due to Shilts' book. Shilts recalled in an interview in 1987 (Geiger 1987):

If I were going to write a news story about my experiences covering AIDS for the past five years, the lead would be: In November of 1983, when I was at the San Francisco Press Club getting my first award for AIDS coverage, Bill Kurtis, who was then an anchor for the 'CBS Morning News,' delivered the keynote speech…He started with a little joke…In Nebraska the day before, he said he was going to San Francisco. Everybody started making AIDS jokes and he said, 'Well, what's the hardest part about having AIDS?' The punch line was, 'Trying to convince your wife that you're Haitian.' …[THIS] says everything about how the media had dealt with AIDS. Bill Kurtis felt that he could go in front of a journalists' group in San Francisco and make AIDS jokes. First of all, he could assume that nobody there would be gay and, if they were gay, they wouldn't talk about it and that nobody would take offense at that. To me, that summed up the whole problem of dealing with AIDS in the media. Obviously, the reason I covered AIDS from the start was that, to me, it was never something that happened to those other people.

Shilts' book remained on the *New York Times* best-seller list for five weeks, and was nominated for a National Book Award. Unbeknownst to his readers, Shilts got an HIV test the day he turned in his manuscript to his publisher, and discovered he was HIV-positive in March 1987 (Michaelson 1993). He chose not to disclose he was HIV positive because "Every gay writer who tests positive ends up being an AIDS activist. I wanted to keep on being a reporter" (Grimes 1994). He progressed to AIDS in 1992, and finished his next book on gays in the military, Conduct Unbecoming (1993), from his hospital bed (Schmalz 1993). That book inspired the opening act of the Clinton Administration—the "don't ask, don't tell" policy (see further). In February 1993, when Shilts was 41, he publicly disclosed he had AIDS, stating: "I want to talk about it myself rather than have somebody else talk" (Michaelson 1993). Shilts lived to see *And the Band Played On* made into a film.

The Life and Times of Larry Kramer

Shilts' book also covered extensively the "life and times" of writer, playwright and activist, Larry Kramer (1935–2020), which the film, does not document, due to too much complexity. For these reasons, I would suggest that when screening *And the Band Played On* as part of a comprehensive course on the history of AIDS, one should also assign as companion works, the HBO feature film, *The Normal Heart* (2014) and/or the HBO documentary, *Larry Kramer: In Love and Anger* (2015), which thoroughly covers Kramer's rightful place in early AIDS history. I provide a more concise recap here.

Kramer's early activism career started with a controversial novel, Faggots (1978), which was autobiographical in nature, and critically questioned the Gay Liberation movement, and the unhealthy lifestyle sexual promiscuity encouraged, including the activities that went on in gay bathhouses (see earlier). The book warns that such a lifestyle may be dangerous, and suggests that monogamy, and not promiscuity, should be what gay males strive for. The book was attacked by the gay press for daring to "preach" about sexual liberation. When the first cases of AIDS began to proliferate in New York City, Kramer was an early believer that it was likely sexually transmitted, and advocated for risk reduction through reducing sexual activity. In 1981, eighty men gathered in Kramer's apartment to discuss raising awareness and money for "gay cancer"—before the term GRID was used; that meeting led to the formation of the Gay Men's Health Crisis (GMHC) in 1982, with founding co-members, Nathan

Fain, Larry Mass, Paul Popham, Paul Rapoport, and Edmund White. The organization functioned as a social and healthcare services organization: "An answering machine in the home of GMHC volunteer Rodger McFarlane (who will become GMHC's first paid director) acts as the world's first AIDS hotline—it receives over 100 calls the first night" (GMHC.org). That same year, GMHC produced the first informative newsletter about the AIDS crisis that went to roughly 50,000 readers comprising doctors, hospitals, clinics and the Library of Congress. GMHC also introduced the "Buddy program" which is modeled in countless organizations today; volunteers would visit and care for the sick in their community, and often would go to their hospital rooms to care for them when nurses refused to touch them.

In March 1983, Kramer authored an essay in a gay magazine titled: "1,112 and Counting" (Kramer 1983), which I discuss further on (see under Healthcare Ethics issues). The essay remains a masterpiece of advocacy journalism but can be categorized as one of the first clinical ethics articles documenting discrimination and moral distress in the AIDS crisis. Shilts notes this about "1112 and Counting" (Shilts 1987: 244):

> With those words, Larry Kramer threw a hand grenade into the foxhole of denial whre most gay men in the United States had been sitting out the epidemic. The cover story of the New York Native, headlined "1,112 and Counting" was Kramer's end run around all the gay leaders and GMHC organizaers worried about not panicking the homosexuals and not inciting homophobia. As far as Kramer was concerned, gay men needed a little panic and a lot of anger.

AIDS was handled differently in New York City, where Ed Koch is mayor, than in San Francisco, where progressive Dianne Feinstein is mayor at the time. Koch was a closeted gay male (Shilts 1987; Kramer 1983), who refused to address AIDS in his community. Koch represented a genre of closeted American gay politicians—still seen today—who display a public "anti-gay" ultra-Conservative face, in contrast to their private lives. Ultimately, Kramer's aggressive style of activism was so politically incorrect and unpopular, it led to his being ousted by his own GMHC Board of Directors. Kramer next wrote the powerful play, "The Normal Heart", which debuted off Broadway in 1985—while Shilts is still chronicling the early years of the epidemic. "The Normal Heart" told the incredible story—in real time—of men dying of AIDS in New York City completely alone and abandoned, and the frustrations of the healthcare workers and eyewitnesses at "ground zero", as well as Kramer's experiences with GMHC. Shilts' recounts the opening night of the play like this (Shilts 1987: 556):

> A thunderous ovation echoed through the theater…True, The Normal Heart was not your respectable Neil Simon fare, but a virtually unananimous chorus of reviewers had already proclaimed the play to be a masterpiece of political drama…One critic said Heart was to the AIDS epidemic what Arthur Miller's The Crucible had been to the McMcarthy era. New York Magazine critic John Simon, who had recently been overheard saying that he looked forward to when AIDS had killed all the homosexuals in New York theater, conceded in an interview that he left the play weeping.

The reviews were critical. From the *New York Times*, with Bruce Davison cast as the "Kramer" character, Ned Weeks (Rich 1985):

The blood that's coursing through "The Normal Heart," the new play by Larry Kramer at the Public Theater, is boiling hot. In this fiercely polemical drama about the private and public fallout of the AIDS epidemic, the playwright starts off angry, soon gets furious and then skyrockets into sheer rage. Although Mr. Kramer's theatrical talents are not always as highly developed as his conscience, there can be little doubt that "The Normal Heart" is the most outspoken play around - or that it speaks up about a subject that justifies its author's unflagging, at times even hysterical, sense of urgency.... The trouble is not that the arguments are uninteresting, but that Mr. Kramer is not always diligent about portraying Ned's opponents...The more the author delves into the minutiae of the organization's internecine politics, the more "The Normal Heart" moves away from the larger imperatives of the AIDS crisis and becomes a parochial legal brief designed to defend its protagonist against his political critics.

From the *Los Angeles Times*, with Richard Dreyfus in the role: (Sullivan 1985).

Beneath the social concerns of "The Normal Heart" is the story of a man who must learn to adjust his expectations of other men downward if he hopes to do any good among them. This is interesting. So is the play's message that having sex with as many partners as desired is actually a kind of addiction. What play in the liberated '70s and '80s has dared to say that?" As an AIDS documentary, it is also already something of a period piece, thank God: The causes of the disease have been more clearly pinpointed now.

The play's revival on Broadway in 2011, featured an open letter in the playbill to the audience that read, in part (Gans 2011; Nuwer 2011):

Thank you for coming to see our play.

Please know that everything in The Normal Heart happened. These were and are real people who lived and spoke and died, and are presented here as best I could. Several more have died since, including Bruce, whose name was Paul Popham, and Tommy, whose name was Rodger McFarlane and who became my best friend, and Emma, whose name was Dr. Linda Laubenstein. She died after a return bout of polio and another trip to an iron lung. Rodger, after building three gay/AIDS agencies from the ground up, committed suicide in despair. On his deathbed at Memorial, Paul called me (we'd not spoken since our last fight in this play) and told me to never stop fighting.

Four members of the original cast died as well, including my dear sweet friend Brad Davis, the original Ned....

Please know that AIDS is a worldwide plague...Please know that no country in the world, including this one, especially this one, has ever called it a plague, or acknowledged it as a plague, or dealt with it as a plague.

Despite the reviews, which seem to prefer another AIDS play—"As Is", which opened a month before Kramer's—"The Normal Heart" play was a big hit, and Kramer next founded AIDS Coaltion to Unleash Power (ACT UP), which staged a number of groundbreaking activist-led "performance art" that literally sped up AIDS drug development, treatment, and AIDS patients' access to experimental protocols and clinical trials (see under History of Medicine). ACT UP formed in March 1987, the same year *And the Band Played On* was published. Both Shilts and Kramer played integral, if not, symbiotic roles in helping to make AIDS a political priority. ACT UP was particularly active during the Bush 41 era, staging all kinds of creative protests in front of the White House and around D.C. Kramer was one of the few AIDS survivors of his generation, who lived long enough to take advantage of AZT and then protease inhibitors (see under History of Medicine).

Paper Covers Rock: From Rock Hudson to amfAR

The story of the early days of AIDS is incomplete without discussing actor Rock Hudson (whose real name was Roy Harold Scherer, Jr. until it was changed by his agent), the first celebrity to admit that he was dying from AIDS, blowing his carefully concealed cover that he was gay. Shilts writes about the "before Rock" and "after Rock" timeframes, as after Hudson came forward, media coverage about AIDS abounded, and the American Foundation for AIDS Research (amFAR) was founded.

Rock Hudson (1925–85) was a closeted 6-foot 4-inch gay actor who was the picture of heterosexual machismo, and "tall, dark and handsome" leading man. He made a number of romantic comedies with actress Doris Day; they were reunited on July 16, 1985 to promote a cable show Day was doing for the Christian Broadcasting Network on pets. Hudson's appearance and demeanor were so alarming, it led to rumors about his health; many in Hollywood who knew he was gay actually suspected cancer (Collins 1992). The footage of Hudson and Day was repetitively aired, and is one of the few presentations of Hudson with late stage AIDS. Hudson was a patient of Dr. Michael Gottlieb at UCLA, who authored the first paper on AIDS in 1981 (see under History of Medicine). Hudson was one of the first AIDS patients to be treated with experimental HPA-23 antiretroviral drug therapy at the Pasteur Institute in Paris (see under History of Medicine); he was diagnosed in 1984 while he was starring in the television series, *Dynasty*. When Hudson collapsed in his Paris hotel lobby on July 25, 1985 and was flown back to UCLA, his publicist admitted he was in Paris receiving treatment for AIDS. This was groundbreaking, and big news. When AIDS began to kill off many celebrities in the early 1980s, most of the AIDS-related Hollywood obituaries would not openly state the cause of death other than "long illness" and sometimes "pneumonia" (Shilts 1987). Hudson was the first to make the morally courageous and consequential decision to announce he was undergoing treatment for AIDS. Hudson's close friend, Elizabeth Taylor, came to visit him when he was in the hospital that August, and met his doctor, Gottlieb. She approached Gottlieb after her first visit to help get her to the hospital in a less publicized fashion, and he picked her up and took her to the hospital, using a side-entrance typically reserved for faculty and staff. Taylor and Gottlieb formed a friendship.

AIDS Project L.A. and the Commitment to Life Dinner

While Hudson was ill, a former gay Xerox executive, Bill Misenhimer, had formed an AIDS patient services organization called AIDS Project Los Angeles (APLA), which was planning a big fundraising dinner for September 19, 1985. Misenhimer contacted Taylor to chair and host the gala dinner which they were calling the "Commitment to Life" dinner. Taylor's interest in helping was also motivated by the fact that her own daughter-in-law (Aileen Getty) had been diagnosed with AIDS that same year (1985), and was presumed to be infected through IV drug use (Dulin 2015). As Gottlieb and Taylor got to know one another, they had an impromptu dinner with Misenhimer and Taylor's publicist (Chen Sam) to discuss the gala event, and then transitioned to discussion of Taylor and Gottlieb forming a national AIDS foundation they would call National AIDS Research Foundation (NARF). Next, Hudson told Gottlieb and Taylor that he would donate $250,000.00 to help NARF get off the ground. On September 19, 1985, the Commitment to Life charity dinner was a huge success, netting $1 million for APLA. Hudson prepared a statement to be read by Burt Lancaster to the 2500 persons in attendance at the dinner (Higgins 2015):

> I am not happy that I am sick; I am not happy that I have AIDS. But if that is helping others, I can, at least, know that my own misfortune has had some positive worth.

At the same dinner, Burt Reynolds read a telegram from Ronald Reagan, which the *Los Angeles Times* reported the next morning as follows (Oates 1985):

> A page-long statement from Reagan, read by Reynolds, said "remarkable progress" had been made in efforts to conquer the disease, but "there is still much to be done."

> Scattered hissing broke out in the audience when Reynolds read a line that began: "The U.S. Public Health Service has made remarkable progress …"

> Reynolds stopped reading and told the audience that "I don't care what your political persuasion is, if you don't want the telegram read, then go outside."

> There was applause and Reynolds continued to read the statement.

> Reagan, who spoke out publicly on acquired immune deficiency syndrome for the first time at a news conference earlier this week, said the fight against the disease is a "top priority" of his Administration but told supporters of the AIDS Project that "we recognize the need for concerted action by organizations like yours, devoted to education, support services and research."

Birth of amfAR

When AIDS researcher, Mathilde Krim (see amfar.org) heard about Hudson's donation, she contacted Gottlieb to discuss joining forces (Collins 1992). Krim was a basic AIDS researcher at Memorial Sloan Kettering in New York City. In 1983, Krim and several of her colleagues had formed the AIDS Medical Foundation. Gottlieb got Krim and Taylor together, and they decided to pool their efforts and resources and merge into one organization they renamed the American Foundation for AIDS Research (amfAR), with Taylor as the famous face of the organization as its National Chair and spokesperson. Ultimately, Hudson's initial donation, combined with individual donations after his death for AIDS research, is what led to *independent funding* of AIDS research beyond the limitations and constraints of federal agencies. On September 26, 1985, Taylor, Gottlieb and Krim announced the formation of amfAR at a press conference.

Hudson's Death

Rock Hudson died on October 2, 1985, and AIDS was announced as the cause of death. Hudson's death helped destigmatize AIDS into a disease worthy of charitable donations—something that was not possible before. Through amfAR, Taylor used her fame to secure millions of dollars in fundraising for AIDS research; she also coached Ronald Reagan to speak at one of amfAR's first major fundraisers in 1987, in which Reagan made controversial comments about the need for "mandatory testing", which was contrary to expert recommendations (see under History of Medicine). By 1992, Taylor's efforts led to amfAR raising $20.6 million in funding (Collins 1992), which was a staggering amount for one organization; she was featured on the cover of *Vanity Fair* with the title "Liz Aid" which stated:

> No celebrity of Taylor's stature up to that point had had the courage to put his or her weight behind a disease that was then thought to be the province of gay men. Elizabeth Taylor brought AIDS out of the closet and into the ballroom, where there was money—and consciousness—to be raised.

In teaching about the early history of AIDS, it's important to discuss Hudson's and Taylor's roles as countless extramural grants were funded because of them. Without Hudson and Taylor, AIDS funding at the levels needed at that time would not have occurred, nor would thousands of basic AIDS research careers.

An easy way to showcase Hudson and Taylor together is by presenting some clips from the film *Giant* (1956).

From Band to Bush: AIDS in the Bush 41 Era

In 1988, Reagan's Vice President, George H. W. Bush, was the Republican nominee for President, facing off against Massachusetts Governor Michael Dukakis, who was leading by 15 points. Bush approached one of the most heralded Republican strategists, Roger Ailes (later CEO of Fox News), who had also worked on Reagan's 1984 campaign, and Nixon's 1968 campaign. Ailes was also in a high-risk group for being infected with HIV because he had hemophilia. Ailes worked with Bush's campaign manager, Lee Atwater, playing the "race card" with the infamous Willy

Horton ad, which was a negative ad about Governor Dukakis' prison reform policies, featuring an African American inmate, Willy Horton, who committed crimes
while on a weekend furlough (Bloom 2018). The ad played to white voters' fears
of African Americans raping them. It was the ad that led to Bush being elected, and
a continuation of tepid AIDS policies, coinciding in 1990 with the death of Ryan
White (see further). In 1992, Elizabeth Taylor was asked about Bush's AIDS policies
at the Eighth International Conference on AIDS in Amsterdam. She stated: "I don't
think President Bush is doing anything at all about AIDS. In fact, I'm not even sure
if he knows how to spell 'AIDS.' "

On September 18, 1989, the National Commission on AIDS (created November
4, 1988—a week before Vice President George H. W. Bush would be elected as the
41st President) met for the first time.

The Bush 41 era was also the timeframe in which Shilts' book was being optioned
and considered for a film. The 10th year of AIDS was marked during the Bush 41 era,
in June 1991, a decade after the first article appeared on Kaposi's sarcoma (see under
History of Medicine). But a decade into the deadly epidemic with no cure began to
take its toll on the American psyche and healthcare infrastructure, and was no longer
a "gay disease". During the Bush 41 era, three prominent heterosexual cases of HIV
infection would begin to shift public opinion about AIDS, and public opinion about
Bush, who seemed stuck in the early 1980s.

The Ryan White Story

Ryan White (1971–1990) was a child in Indiana with hemophilia who was diagnosed
with AIDS in 1984; he had been infected through a blood supply product, Factor
VIII. His story was widely publicized during Reagan's second term, which revolved
around his battle with the school boards to be allowed to attend his public school.
During the Reagan era, he became the face of the social story of pediatric AIDS, as he
endured intense discrimination. But Ryan's "superstardom" as an AIDS activist hit
its stride during the Bush 41 years due to the ABC television film that aired in 1989:
The Ryan White Story, which made him into a moral hero for pediatric AIDS patients,
as he used his cause celeb to champion AIDS public education in countless public
forums. White attracted a lot of famous friends, including Elton John and Michael
Jackson, who gave him a car in 1988. Just prior to Bush's inauguration, a *New York
Times* review on the White film was published, which stated the following (O'Connor
1989), hinting at barriers faced by producers working with Shilts' material for film:

> The vast majority of AIDS patients are homosexuals and drug addicts, but television appar-
> ently is not ready to explore these groups with any degree of compassion. Innocent youngsters
> trapped by circumstances beyond their control provide far easier dramatic hooks for uplift
> exercises.

> Still, these stories are indeed heartbreaking and do serve as vehicles for exposing public
> ignorance and prejudice about AIDS. "The Ryan White Story" at 9 o'clock tonight on ABC
> is a good case in point. It is the true story of a remarkably gutsy 13-year-old whose battle to
> continue going to school made national headlines. A hemophiliac, Ryan discovered he had

AIDS a little over four years ago. At the time, he was given three to six months to live. He is still active today, occasionally visiting schools around the nation to tell his story

It is a story not only about ignorance but also about an almost total lack of enlightened community leadership in the city of Kokomo. Residents are understandably concerned and frightened, but panic is allowed to take over. Petitions to keep Ryan out of school are signed not only by neighbors but also by the local police. A radio station broadcasts vicious bilge The print and electronic press push for sensationalism. The Whites find a bullet hole in their living-room window. Ryan is systematically isolated and ostracized…. It is not a pretty story Worse, it is a story that didn't have to happen.

Bush 41 was inaugurated just days after ABC's White film begins to change hearts and minds, but Bush is still stuck in the Reagan era on the policy front. He is blindsided by a shift in public sentiment calling for greater action on AIDS. His administration was faulted for underfunding the growing epidemic as the magnitude of transfusion AIDS cases was unfolding, as well as maternal transmission cases. Transfusion AIDS cases were not just due to a failure to screen the blood supply prior to 1985, but a failure to warn patients needing transfusions of these risks when the blood supply was unsafe (see under Healthcare Ethics Issues), a fact that the film, *And the Band Played On*, highlights for its 1993 audience. Additionally, White's story of discrimination resonates within the African American community, which by now is experiencing an explosion of AIDS cases (see further). By this time frame, homophobia and cruel discrimination against AIDS patients persisted; health disparities regarding screening and testing persisted; and the costs of treatment mounted. Bush was unable to contain anger and frustration over a lack of basic infrastructure for AIDS patients, who were often unable to get, or maintain: health insurance, adequate housing, education or employment, or even adequate end of life care or hospice. By the time Ryan White died in 1990, Bush had inherited national rage regarding general underfunding of AIDS healthcare, while there were real limits to the efficacy of AZT. The *Ryan White Comprehensive AIDS Resources Emergency (CARE) Act* passed in 1990 (see under History of Medicine), which was a healthcare "payor of last resort" for HIV/AIDS patients; in retrospect, it was one of the most significant pieces of AIDS legislation passed by Congress, but its benefits did not kick in until 1991, and its reach would not be completely realized until Bush was out of office, as resource allocation issues abounded after passage (see further). The legislation was seen as too little too late: by 1990 over 100,000 Americans had died of AIDS—nearly twice the number of Americans who died in the Vietnam War, while AIDS had become the number one cause of death for U.S. men ages 25–44 (GMHC 2018).

If White was one transfusion AIDS case bookend to the start of Bush's presidency, Arthur Ashe (see further) was the transfusion AIDS case bookend at the end of his term. But Magic Johnson would bridge the two, and he ultimately humiliated Bush by calling him out on his AIDS policy failures.

Earvin "Magic" Johnson: From NBA to the National Commission on AIDS

On November 7, 1991, the famous basketball player, Magic Johnson, publicly announced that he had tested positive for HIV and was retiring from the L.A. Lakers team and the NBA. Johnson traced his infection to unprotected heterosexual activities when he had been single and promiscuous, and made clear that his wife and daughter were not infected. Johnson was one of the first famous figures infected through heterosexual contact, and always denied being bisexual or gay (Friend 2001).

In response, Bush appointed Johnson to the National AIDS Commission his administration initiated by federal statute, which consisted of 15 members appointed by Congress and the White House. The AIDS Commission had recommended expansion of Medicaid to cover low income people with AIDS and universal healthcare to deal with healthcare access, disparities and the growing number of those infected with HIV as well as AIDS patients. Johnson was acutely aware of disparities in the African American community, echoing the same problems found with the Tuskegee issues (see further).

But none of these recommendations were taken up by the administration. Eight months later, Johnson quit, stating that the Bush Administration had "dropped the ball" on AIDS (Hilts 1992a, b):

> As I think you know, along with my fellow commission members I have been increasingly frustrated by the lack of support, and even opposition, of your Administration to our recommendations – recommendations which have an urgent priority and for which there is a broad consensus in the medical and AIDS communities...I cannot in good conscience continue to serve on a commission whose important work is so utterly ignored by your Administration.

Johnson went on to be an HIV/AIDS "super-educator", which continues today, and also played professionally again by 1996, when the drug cocktail became available (see further). Ultimately Johnson's role on the Bush AIDS Commission was a microcosm of AIDS policy in the Bush 41 era: all talk—an improvement over Reagan's refusal to even talk about it—but alas, no action that translated into meaningful healthcare delivery for AIDS patients. If the Reagan years were about committing to funding for AIDS research, the Bush years were about getting a commitment to fundamental AIDS healthcare delivery, which did not occur; "last resort" healthcare was not the same thing.

The Arthur Ashe Case

Arthur Ashe (1943–93) was an African American tennis player and intellectual; he was the first African American to win Wimbledon. He had been active in a variety of human rights causes, including protesting apartheid in South Africa (Barker 1985). In the scholarly world, he was known for a three-volume history of black athletes in the United States (Ashe 1993).

In 1979, Ashe suffered a heart attack, even though he was fit and in his 30s. This forced him to retire in 1980. He wrote the following about his heart disease in the *Washington Post* (Ashe 1979):

Why me?

The why-me question was more than just a complaint from someone who felt cheated by fate. It was a serious question.

I'm not your typical heart attack victim. My blood pressure is below normal. I don't smoke or take drugs. I'm thin. My serum cholesterol count is low. I'm not hypertense. I have no trouble absorbing sugar. And with all the tennis I play, I'm about as physically fit as a 36-year-old man can be.

Maybe all of this explains why I didn't pay much attention at first to the chest pains. The first four incidences of pain lasted two minutes apiece and disappeared – I thought for good.

Ashe had quadruple bypass surgery in 1979. He then needed a double bypass surgery in 1983, which is when he became infected with HIV from a blood transfusion during that surgery. He didn't realize he was infected until he was hospitalized for bizarre symptoms in 1988, and he was diagnosed with AIDS when he was found to have an opportunistic disease. He wanted to keep his illness private, but when a *USA Today* reporter learned of it from an unnamed hospital source—making the eventual case for the privacy rule of the *Health Insurance Portability and Accountability Act* (HIPAA)—Ashe was given the opportunity to disclose it himself before they published the story. Ashe contacted the *USA Today*'s sports editor, Gene Polincinski, to request 36 hours for him to pull together a press conference and announce it himself. In other words, Ashe was forced to disclose he had AIDS due to confidentiality violations. Ashe's case helped to pave the way to the HIPAA privacy rules that were passed in 2003. Ashe disclosed he had AIDS April 8, 1992 (Walton 2017):

Beginning with my admittance to New York Hospital for brain surgery in September 1988, some of you heard that I had tested positive for HIV, the virus that causes AIDS. That is indeed the case…I am angry that I was put in a position of having to lie if I wanted to protect my privacy…I did not want to have to go public now because I'm not sick. I can function very well and I plan to continue being active in those things I've been doing all along—if the public will let me.

He also stated (Reimer 1992; Parsons 1992):

There is no good reason for this to have happened now. I didn't commit any crime. I'm not running for public office. I should have been able to keep this private… Still, I didn't commit any crimes and I'm not running for public office. I should be able to reserve the right to keep things like that private. After all, the doctor-patient relationships are private…There was certainly no compelling medical or physical necessity to go public with my medical condition.

On April 12, 1992, Ashe penned a scathing editorial in the *Washington Post* (Ashe 1992) entitled "Secondary assault of AIDS spells the public end to a private agenda", in which he wrote the following:

I'm pissed off that someone would rat on me, but I'm even more angered that I would have had to lie to protect my privacy... I wasn't then, and am not now, comfortable with being sacrificed for the sake of the "public's right to know." ...

"Going public" with a disease such as AIDS is akin to telling the world in 1900 that you have leprosy. Fortunately, this general reaction is abating somewhat as people such as Magic Johnson show that one need not shun an AIDS victim...Since I was not sick, I felt no compelling need to tell anyone who didn't need to know. ... Keeping my AIDS status private enabled me to choose my life's schedule. That freedom now has been significantly eroded. Will I be able, for instance, to fulfill my duties as an HBO broadcaster at Wimbledon? I assume so, but I'm not sure. When I give a tennis clinic for kids somewhere, will they shy away for no good reason except parental caution?

After going public (Freeman 1992), he formed the Arthur Ashe Foundation for the Defeat of AIDS. As an African American, Ashe was continuously misrepresented as having contracted AIDS through sex (like Johnson), but like Johnson, he was also an unintentional spokesperson for health disparities in the African American communities who were becoming the fastest growing population of AIDS patients due to the "legacy of Tuskegee" (see further). Ashe was also arrested while protesting the Bush Administration's treatment of Haitian refugees fleeing a military coup at the time (Finn 1993). Haitians who were HIV-positive were either refused entry into the United States, or held in quarantine if they were at high risk of being infected with HIV.

The Arthur Ashe story and saga followed Bush through an election year. Although Bush was praised for leading the end of the Cold War, as well as his handling of the Gulf War, he was unable to win approval for his handling of AIDS, or his domestic policy achievements. He seemed unable to demonstrate empathy (by 1992 standards), which did not look good next to Bill Clinton, who oozed empathy and was the Democratic nominee for President in the 1992 election. At a debate with Clinton and Ross Perot, Bush actually said this about AIDS when defending his Administration's policies:

It's one of the few diseases where behavior matters. And I once called on somebody, "Well, change your behavior! If the behavior you're using is prone to cause AIDs, change the behavior!" Next thing I know, one of these ACT UP groups is saying, "Bush ought to change *his* behavior!" You can't talk about it rationally!

Clearly, this response indicated that Bush had no real understanding or appreciation of transfusion AIDS cases or pediatric AIDS, which had nothing to do with anyone's behavior but the blood industry's decision-makers. Ashe died February 6, 1993, just a few weeks after Bill Clinton was inaugurated.

The Legacy of Tuskegee: AIDS and African Americans

The story of the Tuskegee study, exhaustively discussed in Chap. 1, was only comprehensively brought to light in the time of AIDS through James Jones' groundbreaking book, Bad Blood: The Tuskegee Syphilis Experiment, which was released in June 1981, coinciding with the June 5, 1981 article in *Morbidity and Mortality Weekly Report (MMWR)*, entitled "Pneumocystis pneumonia" (CDC 1981a), the very first published medical article about HIV/AIDS. The CDC followed up with a second article a month later in *MMWR* on the outbreak of Kaposi's sarcoma in a similar population (CDC 1981b). As Jones' book is being read, the AIDS crisis is unfolding, and begins to make its way into the African American communities—particularly through intravenous (IV) drug use and needle sharing—by the mid-1980s. By 1988: "For the first time, more new AIDS cases in NYC are attributed to needle sharing than to sexual contact. The majority of new AIDS cases are among African Americans; people of color account for more than two thirds of all new cases" (GMHC 2018). Although there was a catastrophic rise in HIV/AIDS cases, because of what would be called the "legacy of Tuskegee," distrust of "government doctors" from the CDC, NIH, and other public health agencies, led to resistance by the African American communities to get tested, screened, or practice preventative measures. Fool me once, but not twice, was the thinking.

As detailed in Chap. 1, the Tuskegee study involved deceiving poor African American males by telling them that if they agreed to be screened for syphilis, they would receive treatment. Instead, they received no treatment or inadequate treatment, and were even harmed, with African American healthcare providers lying to them, too. This time, aware of the Tuskegee history, African American males were not interested in hearing about any "voluntary" testing or screening programs conducted by mainstream medicine. Additionally, due to misinformation about the Tuskegee study that abounded throughout the 1970s (Reverby 2009), a large percentage of African Americans actually believed the U.S. government deliberately infected them with syphilis. Similarly, when HIV/AIDS began to proliferate through African American communities, there was suspicion that it was a "secret" genocidal virus the government cooked up to kill them, along with gay males. Misinformation about HIV/AIDS, complicated by the legacy of Tuskegee, became a "wicked problem". As I discussed in Chap. 1, the situation became so bad, that Jones felt compelled by 1993 to release a second edition of Bad Blood with a new chapter entirely devoted to AIDS, in which he states (Jones 1993: ix):

> In this expanded edition of Bad Blood, I have traced the Tuskegee's Study's legacy in the age of AIDS. Briefly, I have shown why many blacks believe that AIDS is a form of racial genocide employed by whites to exterminate blacks, and I have examined how the Tuskegee Study has been used by blacks to support their conspiracy theory.

Even with funding for HIV/AIDS research in the African American community, research became more challenging because of new ethical guidelines for research, published by the National Commission for Protection of Human Subjects, which formed because of the Tuskegee study (see Chap. 1). These guidelines mandated protections for vulnerable populations in medical research, but major trust issues between African American patients, medical research, and medical institutions became barriers. For example, "[a]t the first CDC conference on AIDS and minorities in 1987, members of a black coalition caucus argued that the 'black clergy and church' needed to assure African Americans that 'AIDS testing and counseling initiatives are not just another Tuskegee tragedy being perpetrated on the black race'" (Reverby 2009). In fact, there was a *valid* belief in this community that the government was "giving them AIDS" because of the U.S. government's failure to protect the public blood supply, which Shilts' reveals in his book. Nonetheless, by 2010, the consequences were stunning; although African Americans represented approximately 12% of the U.S. population, they accounted for almost half (46%) of people living with HIV in the United States., as well as nearly half (45%) of new infections each year (CDC 2013). Among female adults and adolescents diagnosed with HIV infection in 2008, 65% were African American compared to 17% who were white and 16% who were Hispanic/Latino. Thus, the HIV incidence rate for African American women is nearly 15 times as high as that of white women.

Band Width: Getting Shilts' Story on Film

Almost as soon as Shilts' book was published, interest in purchasing the rights for a film began. Aaron Spelling, who was a long-time television producer, met with Shilts to pitch the film version to ABC television, and was turned down because of the content. Next, a producer with NBC, Edgar Scherick, purchased the rights to the book, but the project was put on hold for two years and then it was dropped because of the "gay content" (Michaelson 1993). Network advertisers did not want to support any program that featured homosexuals (Michaelson 1993). Spelling then purchased the rights again, but they were bought by HBO, which was not under the same pressure from advertisers. Bob Cooper, an executive at HBO recalled (Michaelson 1993):

> The day the project was dropped is the day we picked it up—immediately...in that it has a ticking clock...How does society cope with a huge crisis? Or not cope?...[It was]a great David and Goliath drama: little Don Francis and little Bill Kraus fighting every group, constituency, agency and being rejected.

To tackle the 630-page book and make it into a workable script, screenwriter Arnold Schulman delivered 17 drafts as there were continuous changes with directors, legal reviews, and Shilts' reviews of the script. The first director was hired in April 1991—Joel Schumacher—who wanted to change the project into a documentary that was similar to PBS' "The Civil War.", which had just aired the previous year (Michaelson 1993). Nobody liked that idea. Next, director, Richard Pearce, who had worked on sensitive Civil Rights material with *The Long Walk Home* (1990) was promising, but left the project to work on the film *Leap of Faith* (1992) for Paramount (Michaelson 1993). The final director, who completed the project was Roger Spottiswoode, who made critical decisions about the content that made the film work. He recounts (Michaelson 1993):

> When Dick Pearce left, I moved in quickly...I knew it was out there; I knew the book. It's an absolutely fascinating subject. A dramatic detective story, a story about politics, a morality

tale about how we all behaved and whether we behaved properly, which I don't think we did. … We all have to accept some responsibility for the fact that for the first five years people in this country believed that for some reason this was a gay plague, that it was brought about by gay people, that it was their problem and that somehow it would go away.

HBO film projects, which I discuss more in Chap. 1, were never intended to be "made for TV" productions. Instead, the network's independent films were produced using feature film quality and production standards. From the start of the project, the producers wanted an ensemble cast of well-known actors to "offset any reluctance that some viewers might have about AIDS and the movie's predominantly compassionate portrayal of the plight of the gay community" (Michaelson 1993). Several feature film directors and A-list actors were contemplated, including Penny Marshall, (Michaelson 1993) who had directed *Awakenings* (1990), which I discuss in Clinical Ethics on Film (Rosenthal 2018). Actors discussed included Richard Gere (who plays choreographer Michael Bennet of "A Chorus Line", who died of AIDS). Gere is heterosexual although there was tabloid speculation about his orientation. He was eager to join the cast, stating (Michaelson 1993):

As you know and I know, there is nothing wrong with being gay. Now if you start with that premise, the rest is all kid stuff and silliness, isn't it?…And they beseeched me…There wasn't really a part in there for me.

Although Gere was already booked up for months with other projects, he accepted the role as Bennet in a cameo, who in the film, gives money to the CDC to help fund research—potentially a nod to Rock Hudson (see earlier).

Shilts' book does not revolve around any single figure as the main character, but screenwriter Schulman chose to focus the script around the CDC's Don Francis, who shouts "How many dead hemophiliacs do you need?" at a contentious meeting with blood industry officials. Recalled Schulman (Michaelson 1993):

I made [Francis] the hero because here's a man who had no other agenda. He worked in the field. If you want to stop an epidemic, you do this, this and this. And people's toes get stepped on–'Too bad; we're here to save lives.' And suddenly he gets into this situation.

Matthew Modine was cast as Francis, who was praised for his work in *Memphis Belle* (1990). Modine had very clear memories of AIDS, stating (Michaelson 1993):

When I moved to New York, I was a chef in a restaurant, and to the best of my knowledge all the waiters are dead. They were some of the first to go before they called it gay-related immune deficiency. They died of colds? That was in '79-'80-'81. If we think of ourselves as tribes, there's a tribe of media people, a tribe of hospital people, a tribe of businessmen. Well, my tribe, the artistic tribe–actors, dancers and jugglers and storytellers–they've been decimated.

One character that was banned as a "focus" was so-called "Patient Zero" from the first epidemiological cluster study (see under History of Medicine), who many screenwriters may have wanted to emphasize. However (Michaelson 1993):

Shilts had already stipulated in his "deal memo" with HBO that the handsome young French Canadian airline steward Gaetan Dugas–better known as "Patient Zero" because [a large

number of the first infected men] found to have AIDS in the United States either had had
sex with Dugas or with someone else who had had sex with him–not be "a major plot line."

The first drafts of the script had been approved to focus, instead, on a number
of public health characters, while later drafts of the script omitted unnecessarily
"added" characters, such as any of the spouses of the public health officials to avoid
the "Sissy Spacek problem" we see in films such as *JFK* (1991), where she plays
the neglected wife of the prosecutor and merely detracts from the plot (Michaelson
1993). Thus, decisions were made to focus on the science and healthcare protagonists
in the story—which is one reason this film can be used to teach healthcare trainees and
students today. The film's vast, all-star cast delivers more of a Robert Altman-style of
direction—quick-paced, high quality (the stuff of 1976's *Nashville* fame)—instead
of a sappy 1970s disaster feel of an Irwin Allen's style of direction seen in *Earthquake*
(1974) or *Airport 1975* (1975). The film version of Shilt's book also mirrors the book's
chronological pace, featuring the work—not the personal lives—of: Don Francis,
Mary Guinan (played by Glenne Headly), Selma Dritz (played by Lily Tomlin),
San Francisco Director of Public Health, Merv Silverman (played by David Dukes),
pediatrician, Art Ammann (Anjelica Huston, playing a female version), Marc Conant
(played by Richard Jenkins) the key Pasteur Institute researchers, and Robert Gallo
(played by Alan Alda), who was accused of research misconduct (see under History
of Medicine). It also covers Bill Kraus (played by Ian McKellen), the San Francisco
politician and activist, with only small glimpses of his personal life for 1993 viewers
to understand the "family systems" aspects of gay male partnerships. Patients who
were either instrumental in helping to fill in knowledge gaps in the early days of
the epidemic, such as Dugas, transfusion-AIDS patient, Mary Richards Johnstone
(played by Swoosie Kurtz), as well as a range of actors playing real characters from
Shilts' panorama of afflicted gay males who have now died. The film also discusses
the tense battle over closure of the San Francisco bath houses, with Phil Collins
appearing as one of the owners of these establishments. Ian McKellen, who plays
Bill Kraus, recalled when his friend, Scottish actor, Ian Charleson (who starred in
1984's *Chariots of Fire*), had died from AIDS in 1990 (Michaelson 1993), and was
who he thought of when doing Kraus' deathbed scenes in the same hospital as actual
dying AIDS patients (Michaelson 1993):

> I saw (him) in his last performance as Hamlet at the National Theatre, about two or three
> months before he died... I remember Ian when he was very weak at the end, putting all his
> energies into his body, sometimes very bright, and (other) times going away from you. ...
> You know, you (wear the makeup) of the disease Kaposi's sarcoma, and it's just a bit of
> plastic. And just across the ward there are people who are marked for life, and now facing
> the possibility of dying. That's when acting and life come right up close against each other.

Documentary Versus Docudrama

Shilts' book spent three years in development as a film; from 1989 until 1991, the
film went through multiple changes in direction. Schumacher recalled: "In the well-
intentioned efforts to dramatize this (story), we have veered from journalism toward

fiction, and I think that this would be an immoral, irresponsible way to present the material" (Michaelson 1991). But Shilts didn't want a documentary because he felt it would alienate much of the audience (Michaelson 1991). Schumacher was also contemplating having eye witness talking heads interspersed, similar to Warren Beatty's masterpiece, *Reds* (1981).

There was also discussion as to whether the film should cover any part of Ryan White's story, who had died in 1990—after the book was published. Shilts did not really cover White in the book, and instead, focused on Hudson's death as a turning point. But White appears in the montage of AIDS victims after the film. Schumacher said in 1991: "When this innocent child died, I think everybody realized [AIDS] was something that could come into your own home" (Michaelson 1991).

By August 1991, creative differences in the direction of the film led to Schumacher leaving the project, and he signed on, instead to a "toxic L. A." film called *Falling Down* (1993–I discuss the "toxic L. A" genre of films in Clinical Ethics on Film). Schumacher had told HBO that he didn't want to make a drama but a documentary, and so they parted ways. At that point, Spottiswoode was brought in, who had no problem with the docudrama form. He said (Michaelson 1993):

> I don't think you should lie. I don't think you should attack somebody who shouldn't be attacked. And I don't think you should intentionally portray a falsehood. However, to portray what you hope is the truth, you sometimes have to (alter) for clarity's sake, for dramatic purposes. There are times you have to take certain liberties with details.

In the final script, key lines in Shilts' book deliver the truth about early AIDS, such as the line delivered by Dr. Marc Conant: "You know damn well if this epidemic were killing grandmothers, virgins and four-star generals instead of gay men, you'd have an army of investigators out there." Spottiswoode also used actual AIDS patients as extras in critical hospital scenes, and had Bill Mannion, chairman of the Shanti Foundation, visit the set to educate the crew about risks of working with extras who were people with AIDS (Michaelson 1993).

Script Approval and Fact Checking

Considering that Shilts was HIV-positive during this timeframe (prior to the development of protease inhibitors), getting his book translated onto the screen was a race against time, as he began to get sick around 1992. The film's accuracy and quality were dependent upon Shilts' availability to review and approve the final script. Screenwriter, Arnold Schulman, also did his own research and fact-checking when translating Shilts' book to the screen. He met repeatedly with Shilts to discuss several facts and aspects of the script. He met several of the key figures he was writing about, including Francis, CDC doctors and researchers. Before writing about Gallo, Schulman also read up on reports of his scientific misconduct, which I discuss further below (see under History of Medicine). Schulman also reviewed other works about AIDS, which were published after Shilts, including Dominique Lapierre's 1991 book, Beyond Love (Michaelson 1993). Ultimately, Shilts reviewed all of the drafts of the script, and also insisted on scenes that accurately depicted the gay community's buddy program, started by the GMHC, for example (Michaelson 1993).

Synopsis

And the Band Played on is a fairly literal screen adaptation of the 1987 book, which tracks a cross-section of the early AIDS epidemic theaters in chronological order: healthcare providers, researchers, patients, and public health officials on the front lines. It captures the most essential details of the 630-page book without falling victim to maudlin and unnecessary subplots. The film focuses on specific characters at the CDC, such as Don Francis; the work of Mary Guinan; as well as local San Francisco figures such as Selma Dritz and Marc Conant. It presents critical meetings that took place accurately, particularly the meeting in which the virus is finally labeled AIDS. The film does not cover the New York City theater of AIDS, or any of the content surrounding Larry Kramer and the GMHC. This film ages well, and can be used as a History of Medicine film about the early history of HIV/AIDS, which medical students frequently are not taught. It can also be taught as a research ethics film with respect to the discovery of HIV, and spends considerable time on NIH scientist Robert Gallo, played by Alan Alda. This film is also a public health ethics film, and illustrates one of the most egregious examples of ethical violations by healthcare providers who failed to warn patients of the risks of transfusion AIDS and who engaged in discriminatory practices by abandoning patients; research ethics misconduct with respect to translational research; and public health ethics violations with respect to protecting the public blood supply. With a tremendous cast of actors in both major and small parts, the film is gripping and fast-paced and makes a tremendous contribution to AIDS education. Finally, airing in 1993, it ends with scenes of one of the first AIDS quilt demonstrations and a moving montage of notable people and celebrities who had died from AIDS as of 1993, including Rock Hudson, Ryan White and Arthur Ashe.

The Picture of AIDS in the Clinton Era: 1993

In 1993, *And the Band Played On* premiered on HBO on September 11—a strangely ominous date that in eight years would completely redefine the United States, and become synonymous with catastrophic loss of lives. It's important to note, however, that on September 11, *1993* the number of American civilians who died from AIDS was *57 times* the number of civilian casualties on 9/11, not including the deaths from dust exposure (see Chap. 4). Yet the government response was different. As the *New York Times* noted [emphasis mine] when the film premiered (Michaelson 1993):

At a glance, one might surmise that the long journey from book to movie might have drained some controversial punch. After all, the Reagan Administration is long gone; the blood banks have been cleaned up…Meanwhile, the numbers keep piling up. The AIDS body count, or "butcher's bill," is the grim thread running through Shilts' book. In July, 1985, the month the world learned that Rock Hudson had AIDS, the number of Americans with AIDS had surpassed 12,000 and the toll of dead was 6,079. **Through the end of 1992, the Centers for Disease Control reported that 244,939 people in the United States had contracted AIDS, of whom 171,890 had died**.

Let that sink in. But unlike 9/11 (conspiracy theories aside), those who contracted AIDS through the U.S. blood supply could literally say it was an "inside job" and the U.S. government actually had blood on its hands. By 1993, the politics in the United States had shifted and the Clinton era had begun, which promised progressive policies such as universal healthcare, with a proposed "Healthcare Security Act", (see Chap. 4).

By this point, Randy Shilts was dying of AIDS, and completed his third book from his hospital bed (Schmalz 1993), entitled, Conduct Unbecoming: Lesbians and Gays in the U.S. Military, Vietnam to the Persian Gulf (Shilts 1993). Shilts referred to it as the "definitive book on homophobia." By exposing the U.S. military's treatment of homosexual service members, the film, *And the Band Played On* resonated. Clearly, thousands of heterosexual and pediatric deaths could now be traced to deep homophobia within the U.S. government, based on its initial response to "GRID". Shilts' book covered the history of persecution of homosexuals in the military, and became a bestseller at 700 pages. In 2014, Shilts' book resonated with the film, *The Imitation Game*, which was about the treatment of Alan Turing, who is not only responsible for turning World War II around due to his code cracking, but who was shamefully discharged and treated by the British military with harmful hormone therapies for his homosexuality (Casey 1993).

This time, the President of the United States responded. Indeed, the opening act of the Clinton administration was to address "gays in the military". Clinton had promised during the 1992 campaign to reverse military policy banning homosexuals from service. He kept his promise, and pushed for reform. Just 10 days after his inauguration, President Clinton announced his plan to change the policy, which was reported by the *New York Times* [emphasis mine] like this (Ifil 1993):

After days of negotiation and rancor in the military and Congress, Mr. Clinton temporarily suspended the formal discharge of homosexuals from the military and announced that new recruits would no longer be asked if they are homosexuals. He ordered the Pentagon to produce an order for him to sign by July 15 [1993]. …In the meantime, the military will continue discharge proceedings against avowed homosexuals or those it believes to be gay. While acknowledged homosexuals will not actually be ousted from the service, Mr. Clinton was forced to agree to a plan that will place them in the unpaid standby reserves – in effect, putting their military careers in limbo – and require that they petition for reinstatement if the ban is permanently lifted.

Mr. Clinton called his announcement today a "dramatic step forward"; it represented the first time a President had taken steps to aid gay men and lesbians in the armed services. But he acknowledged that he had yielded on important issues. "This compromise is not everything I would have hoped for, or everything that I have stood for, but it is plainly a substantial step in the right direction," **Mr. Clinton said during the first news conference of his Presidency,**

which was entirely devoted to the question of homosexuals in the military and not the economic issues that Mr. Clinton has sought to emphasize.

Although this was not a complete lifting of the ban, the new policy would make a distinction between banning gay service members for "conduct" such as wanted, or unwanted homosexual acts, versus their sexual status or orientation: being a homosexual.

The *New York Times* explained the changes this way (Ifil 1993):

> Questions about sexual orientation will be removed from the induction form; Homosexuality will no longer be grounds for outright discharge from the military; Court proceedings against those discharged for being homosexual will be delayed; Rules on sexual conduct will be explained to recruits; The Pentagon will draft an executive order banning discrimination against homosexuals.

On July 19, 1993—just 6 months after he was sworn in—President Clinton announced the official new policy on the treatment of homosexuals in the military, allowing them to serve so long as they kept their sexual orientation and preferences to themselves. This became known as the "Don't Ask, Don't Tell" policy (see further) that would continue until the Obama era in 2011. The new policy was formally signed on December 21, 1993. The next day, the second major Hollywood film about AIDS (following *And the Band Played On*) was released to a limited number of theaters: *Philadelphia* (1993), starring Tom Hanks and Denzel Washington. *Philadelphia* was essentially about "Don't Ask, Don't Tell" in the workplace, and was based on the 1987 true story of attorney, Geoffrey Bowers, who was fired because he was discovered to have AIDS due to visible Kaposi's sarcoma. The attorney filed a wrongful dismissal case and sued his law firm, which became the first known lawsuit over HIV/AIDS discrimination (Navarro 1994). In the film, the plaintiff's name is Andrew Beckett, played by Tom Hanks, who deliberately concealed he was gay and that he had AIDS. His lawyer, played by Denzel Washington, is a homophobic African American attorney who reluctantly takes on the case, rightfully locating it as a civil rights case. Ironically, Bowers' case only settled in December 1993—just as *Philadelphia* was released (Navarro 1994). Bowers had died in September, 1987—while Reagan was still President, coinciding with the release of Shilts' book. Bowers' companion also died of AIDS in 1988, and court hearings on the case occurred between July 1987 and June 1989.

It is not known if Shilts' ever saw *Philadelphia*, as he died February 17, 1994—but did live to see the policy "Don't Ask, Don't Tell" which his work helped to inspire. Bowers said in court: "In light of the fact that I was dealing with my AIDS and my Kaposi's sarcoma, I merely felt as though they had taken the last thing in the world that meant anything to me" (Navarro 1994). When *Philadelphia* was released, before the Bowers case was resolved, his surviving family members sued Scott Rudin, the producer of *Philadelphia* as the film duplicated so much of the Bowers' case and family experience without their consent to use their case for a film. They settled in 1996 as Rudin conceded that much of the script was based on public records of the case, although he had also interviewed the family and lawyers. Ultimately, Bowers case, *Philadelphia*, and Shilts' third masterpiece on gay issues,

Conduct Unbecoming, highlighted the intersection between gay civil rights and AIDS discrimination. There were a host of HIV/AIDS films that began to burst forth around this time frame, including the cinema-verite, "reality" documentary, *Silverlake Life* (1993), filmed by the partner of a dying man with AIDS, about their everyday lives in Los Angeles (Ehrenstein 1993).

Don't Ask, Don't Tell

The phrase "Don't Ask, Don't Tell" was coined by a military sociologist, Charles Moskos in 1993. No one could foresee on September 11, 1993—the day HBO aired *And the Band Played On*—how important "Don't Ask, Don't Tell" would become to national security eight years later, in which the American military would become embroiled in two wars as a result of 9/11. In the post-9/11 years, the U.S. military would be stretched beyond the capacity of its all-volunteer military. It would be highly dependent on its gay service members, who would go on to fight in Afghanistan and Iraq.

Reversing the ban on homosexuals in the military occurred during one of the most peaceful times in the twentieth century, as well as the most hopeful. Initially, President Clinton wanted to completely lift the ban, but once he was inaugurated, there was considerable resistance to that plan by his Joint Chiefs and his Democratic-led Congress; Barney Frank, who would later come out as the first openly gay congressman, actually voted to keep the ban. The argument then was that openly gay service members would "undermine 'unit cohesion' and threaten combat effectiveness" (De La Garza 2018). After a six-month review, Clinton would go for a compromise, which meant that service members could no longer be asked about their sexual orientation, *and they were not required to disclose it.* However, they could still be dismissed if they chose to disclose it, which "prompted outrage from many gay rights advocates who argued the new policy was simply a repacked version of the old ban that was put into place" (La Garza 2018). In fact, the debate in 1993 resembles many of the same issues surrounding the controversial ban on transgender troops in the Trump Administration. However, given the intense homophobia that still prevailed in 1993, on the heels of *And the Band Played On* and *Philadelphia*, "Don't Ask Don't Tell" was an important bridge to the future, particularly in a timeframe when there was still no sign of a cure for AIDS.

For example, in August 1992, a Marine Corps chaplain wrote a position paper that stated: "In the unique, intensely close environment of the military, homosexual conduct can threaten the lives, including the physical (e.g. AIDS) and psychological well-being of others"(Schmitt 1992). In 1993, Senator Strom Thurmond stated to a gay service member when he toured a military base—to applause no less (De La Garza 2018): "Your lifestyle is not normal. It's not normal for a man to want to be with a man or a woman with a woman." He then asked the service member if he ever sought out medical or psychiatric aid (De La Garza 2018). Around the same time, a Navy Commander stated: "Homosexuals are notoriously promiscuous" (a fact made clear in Shilts' book) declaring that heterosexuals would feel uncomfortable in shared quarters or showers (Schmitt 1993). There were noted Republicans who favored lifting the ban, including Dick Cheney (Secretary of Defense in the Bush 41 Administration), and Barry Goldwater, who wrote an opinion piece in 1993 where he stated that "You don't have to be straight to shoot straight" (Bull 1998).

Despite President Clinton's promise to completely lift the ban, in 1993 it was indeed a bridge too far, but the policy opened the door to at least open dialogue surrounding gay Americans. "There was a vast sea change of opinion about this over a decade and a half, and a big part of that was keeping the dialogue going and presenting incredibly patriotic individuals who were serving openly" (De La Garza 2018). In 1994, 45% of Americans wanted to keep the ban, but by 2010, only 27% wanted the ban. On December 21, 1993, Military Directive 1304.26 was issued with the "Don't Ask, Don't Tell" policy codified:

> E1.2.8.1. A person's sexual orientation is considered a personal and private matter, and is not a bar to service entry or continued service unless manifested by homosexual conduct in the manner described in subparagraph E1.2.8.2., below. Applicants for enlistment, appointment, or induction shall not be asked or required to reveal whether they are heterosexual, homosexual or bisexual. Applicants also will not be asked or required to reveal whether they have engaged in homosexual conduct, unless independent evidence is received indicating that an applicant engaged in such conduct or unless the applicant volunteers a statement that he or she is a homosexual or bisexual, or words to that effect.

Gay advocates continue to argue that "Don't Ask, Don't Tell" was only a marginal step forward (De La Garza 2018). A 1993 RAND report (Rostker et al. 1993) actually supported lifting the ban, but the report was apparently shelved. The argument was that privacy about being gay was not considered the same as serving openly gay, as gay service members could still be discharged for demonstrating homosexual behaviors, while the same culture of harassment of gay service members persisted. For example, if a service member didn't tell, he was still vulnerable for being "outed" if his license plates were spotted at a gay bar, for example (De La Garza 2018).

Military Directive 1304.26 went into effect February 28, 1994—11 days after Shilts had died of AIDS, and two days after the first anniversary of the 1993 World Trade Center Bombing of February 26, 1993, an event that foretold September 11, 2001 (Wright 2006). It remained the official policy of the U.S. Department of Defense until it was finally repealed in 2011 by the Obama Administration, which completely lifted the ban once and for all—in the 30th year of AIDS, when there were close to 70,000 gay service members. President Obama said on the day the ban was lifted: "As of today, patriotic Americans in uniform will no longer have to lie about who they are in order to serve the country they love" (De La Garza 2018). By 2019, Mayor Pete Buttigeig, an openly gay veteran, who honorably served in the post-9/11 wars, announced himself as one of the Democratic Presidential candidates for the 2020 Presidential election, and was open about being married to a male, in light of the Supreme Court ruling that made same sex marriage the law of the land in 2015. He went on to win the 2020 Iowa Caucus, Shilts' home state.

Although "Don't Ask Don't Tell" seemed to many at the time to be an odd "first battle" of President Clinton's Administration, in the context of delayed action on

AIDS in the 1980s due to homophobia, his initial gesture to begin to integrate gay Americans into the military was symbolic of gay integration into American life overall, which was necessary in the next phases of battling AIDS. Even President Clinton did not foresee that his bridge would lead 26 years later to an openly gay veteran as candidate for President in 2020.

Ultimately, the premise of "Don't Ask, Don't Tell" and continuing HIV/AIDS discrimination in housing, the workplace, and elsewhere, would form the basis for the 1996 *Health Insurance Portability and Accountability Act* (HIPAA), which President Clinton would sign the same year protease inhibitors would be made available (see under History of Medicine). When it comes to AIDS, the most important aspect of HIPAA would be its Privacy Rules, which would take effect April 14, 2003 (see under Healthcare Ethics Issues)—only three weeks after thousands of troops (many gay) would head to Iraq. Being fired for concealing AIDS in the workplace—as told in the *Philadelphia* story—would no longer be possible, as Protected Health Information (PHI) could no longer be shared with employers or anybody else without explicit permission of the patient.

History of Medicine Context: From "GRID"-Lock to Cocktails

The purpose of teaching *And the Band Played On* in the healthcare ethics context is to use the early history of AIDS as a case study to primarily demonstrate violations of the Principle of Non-Maleficence. However, many educators may be using the film to simply teach history of medicine, or as part of the social history in LBGT studies. This section will focus on the early medical history, research and treatment of AIDS from 1981 until 1996, when protease inhibitors were approved. This section will review key HIV/AIDS practitioners and researchers during this timeline; the battle to screen the blood supply; the first era of HIV testing and treatment; and finally, the cocktail era and beyond.

The First Paper on AIDS: Michael Gottlieb

Imagine being a new assistant professor in the Department of Medicine at UCLA, authoring the first scientific paper on AIDS before it was labelled either "GRID" or "AIDS", as well as close to 50 more publications in high-end journals; getting an NIH grant worth 10.5 million dollars to start an AIDS research and treatment program at your institution; having Rock Hudson as your patient, and securing a $250,000 gift from him for AIDS research; co-founding the first major AIDS research foundation (amFAR—see earlier), and then not getting tenure because your Department Chair, Dean, and other colleagues just don't like you (Hillman 2017; Shilts 1987). That is the story of Michael Gottlieb, who published the first paper on AIDS on June 5, 1981, in *Morbidity and Mortality Weekly Report* (CDC 1981a) as well as the first major scientific paper on AIDS in the *New England Journal of Medicine* (Gottlieb et al. 1981). The CDC selected his first paper as one of their most important historical articles "of interest" in 1996 on the occasion of the journal's 50th anniversary (CDC 1996). In 2012, the *New England Journal of Medicine* selected his next paper as one of the most important publications in the journal's over 200-year history (Hillman 2017).

Michael Gottlieb found himself, like several other physicians at that time, treating the first known cases of AIDS in the United States at UCLA in November of 1980 (Shilts 1987; Gottlieb 1988). What distinguished Gottlieb as the "discoverer of AIDS" was his diagnostic prowess in identifying it as an entirely "new disease" based on the first five patients he saw who succumbed to pneumocystis carinii pneumonia. The first one he saw presented with out-of-control *Candida albicans* or thrush (Hillman 2017). According to Gottlieb: "The case smelled like an immune deficiency. You don't get a mouth full of *Candida* without being immune deficient" (Hillman 2017).

Gottlieb was born in 1947 and graduated from the University of Rochester medical school in 1973; he identified immunologist John Condemi as his mentor, and did a fellowship at Stanford University where he focused on the new field of cellular immunology and T-lymphocytes—his area of interest (Hillman 2017). He was hired in July 1980 at UCLA with a modest start-up lab package (Hillman 2017) to carry on with his burgeoning career in immunology research. When one of the UCLA Fellows came to him about a patient with unusual symptoms (weight loss and thrush, and later pneumonia). Gottlieb's work in immunology informed his idea to request a work-up of the patient by looking at his immune system because the patient's diseases appeared to be only known in patients with compromised immunity. Gottlieb's work-up allowed him to spot a CD4 T-lymphocyte abnormality, explaining the opportunistic infections ravaging the patient. Gottlieb began to note this same phenomenon in four other patients in a short time span—all homosexuals. Gottlieb felt this was a new phenomenon. He contacted the editor of the *New England Journal of Medicine* to discuss whether the journal would be interested in publishing on the new phenomenon, and the journal editor suggested that he first report it to the CDC through its weekly journal, *Morbidity and Mortality Weekly* Report (MMWR) and that his journal would be pleased to review his submission. So Gottlieb wrote the first known paper on AIDS, entitling it "*Pneumocystis* Pneumonia Among Homosexual Men in Los Angeles"; it was published as "*Pneumocystis* Pneumonia—Los Angeles" on June 5, 1981, forever marking that date as the official birthday of the AIDS epidemic. His biographer notes (Hillman 2017):

> Ultimately, Dr. Michael Gottlieb's discovery of AIDS and its underlying abnormality of cellular immunity was the product of happenstance and a mind willing to consider new possibilities. Unlike others who saw only what they expected to see or who tried to pigeonhole their findings into an ill-fitting, existing medical construct, Gottlieb made the necessary intellectual leap.

On the 30th anniversary of his first publication, Gottlieb recalled (SF AIDS Foundation 2011):

> It was in 1981 when I saw the first patients that we reported in the CDC's Morbidity and Mortality Weekly Report. I remember those four men more vividly than patients I saw yesterday. I saw the worry on their faces and witnessed their courage and patience as they realized that we had no idea of what their illness was or how to treat it. I knew right away that I would be in it for the long run….We should not forget that the Reagan administration refused to even acknowledge, much less prioritize AIDS, setting back the nation's response to the epidemic by a decade.

The June 5, 1981, paper in *MMRW* would be followed by a July 3, 1981, paper on Kaposi's sarcoma and pneumonia co-authored by Linda Laubenstein (see further), Gottlieb and many others entitled: "Kaposi's Sarcoma and *Pneumocystis* Pneumonia among Homosexual Men—New York City and California" which reported on both conditions showing up in gay male communities in San Francisco, Los Angeles and New York City." The July MMRW paper was picked up by the *New York Times* (Altman 1981) at which point the term "Gay-Related Immune Deficiency" (GRID) was first used in mainstream reports, as discussed earlier. The July 3rd article would be followed by the first CDC cluster study discussed further on, emphasizing the disease was spreading sexually among gay males. Gottlieb's *New England Journal of Medicine* article followed on December 10, 1981, entitled: "*Pneumocystis carinii* Pneumonia and Mucosal Candidiasis in Previously Healthy Homosexual Men—Evidence of a New Acquired Cellular Immunodeficiency" (Gottlieb et al. 1981). As the first medical journal publications emphasized the epidemic in the gay male patient population, the new disease would be called GRID within the medical community, which damned future heterosexual AIDS patients, too.

In 1981, UCLA was not interested in this new "gay-related" disease, and did not want to be known as an institution that treated it. Instead, it was vying for recognition as a major transplant center. Notwithstanding, as a result of his first publications, Gottlieb became the "go to" source for any journalist covering AIDS (including Shilts at the time) because he had essentially become the "physician of record" on this new disease. Gottlieb was also invited to major conferences all over the world, and was essentially living the type of life one would expect to see from a senior expert in a field—such as Robert Gallo (see further). But Gottlieb was living the senior expert's life prior to tenure, when he was only 33 years old. In 1983, Gottlieb was awarded half of a California state-funded grant of $2.5 million (1.25 million was provided each to UCSF and UCLA) for an "AIDS Clinical Research Center" for which he was named director. The California grant was perceived by UCLA's Department of Medicine brass as a "political grant" because it was a line item in its state budget; the state essentially asked Gottlieb to submit a proposal so they could give him the money, demonstrating at least some commitment to AIDS on the part of the state legislature, which was burdened with a significant proportion of cases. But as his notoriety in AIDS rose, it was rocking the boat in his department, as competitive colleagues began to resent him—and began to complain about him.

The running complaint was over authorship squabbles and being cut out of funded research.

The boat was about to get rockier; in July 1985, a month after the fourth anniversary of his *MMWR* paper, Gottlieb was called by the personal physician to actor Rock Hudson (see earlier), who asked him to come and see her patient because she suspected Hudson might have AIDS due to some unusual spots she thought might be Kaposi's sarcoma. Apparently, the spots were first noticed by First Lady Nancy Reagan who was hosting Hudson at an event at the White House (Hillman 2017). When Gottlieb confirmed that Hudson had AIDS, he became his physician of record and arguably, may have succumbed to "VIP Syndrome" (see under Healthcare Ethics Issues). As discussed earlier, once Rock Hudson revealed that he was dying of AIDS, it became one of the major media stories of the decade, and Gottlieb's fame and star power rose even higher, resulting in his befriending of Elizabeth Taylor with whom he co-founded NARF (see earlier), which eventually became amFAR (see earlier). But within Gottlieb's academic home institution, he was labelled "star struck" while his academic duties and priorities were questioned (Hillman 2017). One of the reasons cited for Gottlieb's institutional political problems was the Rock Hudson gift for "AIDS research". Gottlieb used it to start a foundation *outside* UCLA because he had concerns about whether Hudson's money would be properly allocated to AIDS research inside UCLA (Hillman 2017). However, Gottlieb was still the Golden Goose for his institution in that he secured a federally funded NIH grant for 10.3 million in 1986, which was an astronomical amount of extramural funding for an assistant professor at that time.

The NIH grant was indeed competitive and comprised an awarded contract by the National Institute of Allergy and Infectious Diseases (NIAD) to several institutions for an AIDS Clinical Trials Group (ACTG) to test potential therapies for HIV, which included Gottlieb's involvement in testing AZT. (The first grants awarded were for "AIDS Treatment and Evaluation Units"—or ATEUs, which is what Gottlieb started at UCLA.)

While it is predictable that one of these coveted ACTG awards would go to the man who "discovered" AIDS at UCLA, wrote the first papers, and authored numerous publications on it, it is also inconceivable that such a grant was *not supported* by Gottlieb's institution. But it wasn't—which reflects how work on AIDS was valued at the time—not very much. The stigma and disdain for AIDS research, which characterized most of the 1980s, is one of chief themes of *And the Band Played On*, and followed a host of early AIDS clinical researchers, including Linda Laubenstein (see further), whose colleagues thought she was foolish to waste her time on AIDS patients. A retrospective on Gottlieb's career in the *American Journal of Public Health* (Gottlieb et al. 2006) reported:

> Part of the problem, said Gottlieb, was that the UCLA Medical Center aspired to develop cardiac and liver transplant programs, and the physicians feared that if the hospital became too well known for AIDS, transplant patients might stay away. They also foresaw that there would ultimately be a lot of AIDS patients without good health care coverage. AIDS loomed as a threat to the well-being of the hospital and Gottlieb, so publicly and professionally identified with the disease, was becoming a nuisance.

Gottlieb's institution refused to allocate appropriate office or lab space for his NIH grant. Typically, a grant of that size requires the institution to guarantee that there will be sufficient space, personnel and resources devoted to the Principal Investigator (PI), in order to house the grant and benefit from the "indirect costs". In Gottlieb's case, however, UCLA did not offer him sufficient resources and forced him to set up a lab off-campus at the local Veteran's Association hospital with ancient facilities, making it even more difficult for him to fulfill his obligations on the grant as well as clinical care duties. Ultimately, Gottlieb's notoriety in HIV/AIDS publications and AIDS advocacy outside of his institution created political problems for him inside his institution. Despite clearly meeting the criteria for tenure based on his scholarly productivity and extramural funding, his Chair refused to support his candidacy for associate professor with tenure. Gottlieb did not attempt to pursue any appeals process that was undoubtedly available to him, nor were other L.A.-based institutions interested in having Gottlieb transfer his research and grant over to them (Hillman 2017) once they spoke to his Chair. So the man who discovered AIDS at UCLA in 1981 resigned in 1987 when he realized he would not get tenure. All of this played out the very same the year And the Band Played On was published, which prominently featured Gottlieb. Between 1981 and 1987 Gottlieb was author of over 50 scholarly publications on HIV/AIDS. One member of his tenure and promotion committee, who was interviewed over 35 years later and still at UCLA (Hillman 2017), recalls the tenure deliberations were split between two factions—some who thought Gottlieb met the criteria, and some who insisted he was not "academic material", acknowledging that in hindsight, fear and paranoia over AIDS may have been an underlying factor in a lack of support for Gottlieb (Hillman 2017). Indeed, Gottlieb's biographer was given fuzzy reasons from Gottlieb's former superiors for being denied tenure, such as not being a "serious enough" academic due to his celebrity/VIP patients, which also included pediatric AIDS activist Elisabeth Glaser (see further) Clearly, these reasons do not hold up to professional ethical standards and scrutiny today. Gottlieb's resignation was an ugly episode, covered by the Los Angeles Times. It read in part (Nelson 1987):

> Michael S. Gottlieb, the UCLA immunologist who in 1981 first reported cases of the myste-rious disease now known as AIDS has resigned his full-time university post because he said he could not gain tenure.
>
> "The climate at UCLA was not supportive of my academic advancement. It got too difficult to fight the disease and the system at the same time," Gottlieb, 39, said in an interview this week. "By 'the system,' I mean the system that was resistant to facilitating research."
>
> Dr. Roy Young, acting chairman of the department of medicine, disputed Gottlieb's account of why the celebrated researcher left. He said Gottlieb had "conflicts" with some colleagues and there were periods when he was not "research-wise." Young would not elaborate.
>
> He said Gottlieb resigned the middle of this month, before an academic committee had reached a decision on his promotion to tenured status, which confers a permanent appointment on the recipient.
>
> Gaining tenure on UCLA's medical faculty is a highly competitive hurdle for young physi-cians who must try for a limited number of new openings each year. Another potential

obstacle, say doctors who have had the experience, can come from already tenured faculty members who, because of personality conflicts or professional jealousy, may try to slow the advancement of a junior physician, especially one who has gained widespread recognition.

Last summer, Gottlieb became the principal investigator on a $10-million project awarded to UCLA's AIDS treatment evaluation unit by the National Institutes of Health to test new anti-AIDS drugs. The study, to continue for five years, will not be affected by Gottlieb's resignation.

Of note, Gottlieb eventually parted company with amFAR, too, when he and Krim did not see eye to eye. Gottlieb went into private practice (Gutierrez 2011) and remained an adjunct faculty member at UCLA in a clinical appointment.[2] Gottlieb's snubbing by his colleagues was a familiar narrative for early AIDS practitioners, which Linda Laubenstein experienced, too, when she was treating AIDS in her New York-based institution.

New York State of Mind: Linda Laubenstein

As Gottlieb was encountering pushback at UCLA over AIDS, hematologist/oncologist, Linda Laubenstein, was experiencing similar pushback—literally—at New York University Medical Center, where she was one of the first physicians in 1979 to begin treating the rare cancer, Kaposi's sarcoma, in the first cluster of New York City gay males with AIDS. She partnered with her colleague, Alvin Friedman-Kien, a dermatologist, to deal with the onslaught of cases; Friedman-Kien was a co-author on the July 3rd *MMRW* paper as well (see earlier). Her first patients would present with enlarged lymph nodes and generalized rashes, and Kaposi's sarcoma lesions. By May 1982, Laubenstein had treated 62 patients with Kaposi's sarcoma (Lambert 1992), which was significant, considering it was an otherwise rare cancer, but 1 in 4 known AIDS patients was a patient of Laubenstein's at that time (Larsen 2015).

Laubenstein was a polio survivor and was an outspoken paraplegic practitioner in a wheelchair, who fought for adequate resources to treat her AIDS patients, which included the famous "patient zero", Gaetan Dugas (see further). Her patients and colleagues who knew she was a fighter for her patients called her a "bitch on wheels" (Gordon 1993). Laubenstein is featured in Shilts' book; the character of "Dr. Emma Brookner"—the female paraplegic doctor character in Larry Kramer's "The Normal Heart"—is based on her. Julia Roberts was cast in the role for the HBO film version.

---

[2]Three years later, Gottlieb was also the subject of a 1990 reprimand from the California Medical Board for writing narcotic prescriptions for Elizabeth Taylor under a patient alias (a common practice for celebrity patients pre-HIPAA). But based on the historical record, it appears this episode had nothing to do with his tenure decision as it surely would have surfaced as a reason when the principals involved were interviewed 35 years later. The complaint was not filed until 1989, but does relate to prescriptions written between 1983 and 88—spanning much of the time Gottlieb was still at UCLA (Ellis 1990; Associated Press 1994).

Laubenstein was the first AIDS practitioner to call upon the gay community to practice abstinence, and advocated for closing the gay bath houses (Lambert 1992). Laubenstein was the doctor of Kramer's lover, who had succumbed to AIDS, and Kramer's portrayal of her in "The Normal Heart" remains a memorial to her dedication.

In 1983, well before amFAR, Laubenstein co-founded the Kaposi's Sarcoma Research Fund in 1983, and convened with Friedman-Kien, the first national medical conference on AIDS, which was held at New York University (Harvey and Ogilvie 2000). She and Friedman-Kien edited the conference proceedings, which were published in *JAMA* in 1984 as: "AIDS: The Epidemic of Kaposi's Sarcoma and Opportunistic Infections" (Smilack 1984).

By 1989, as the social stigma of AIDS began to affect housing and employment, she helped to found a non-profit organization called Multitasking, which was an employment service for AIDS patients, since so many lost their jobs—and also their health insurance. Laubenstein, essentially, was not just an AIDS practitioner, but tried to address the wider social disparities because so few practitioners were speaking out for their marginalized patients. Like Gottlieb, she eventually left NYU for private practice where she had more autonomy to exclusively treat AIDS patients when many physicians refused to see them. Unlike Gottlieb, Laubenstein was in a clinical appointment and so tenure was not the issue, but being able to see and treat AIDS patients the way she wanted to as a practitioner without pushback. She was known for making house calls in her wheelchair, or meeting patients at the emergency room at all hours of the night.

Laubenstein's own health was poor; because of post-polio syndrome, her lung capacity was compromised, and she eventually needed a ventilator and had difficulty weaning from it. Her death at age 45 in 1992 was due to a heart attack that was sudden and not foreseen. Ironically, she never lived to see the era of protease inhibitors, but some of her patients survived on AZT long enough to transition. At the time of her death, Kramer was close to a production agreement of "The Normal Heart" with Barbra Streisand, who would have played the Laubenstein-Brookner part, had the deal gone through. That particular deal never materialized; however, in the 1993 film, *Philadelphia* (discussed earlier), the female physician character who demonstrates compassion and fearlessness is likely a nod to Laubenstein, too. Laubenstein's colleague, Friedman-Kein was a medical advisor on the "The Normal Heart" and she was the inspiration for the character (Driscoll 2011). Currently, there is a named award for Laubenstein (New York State Health Department 2020), and she is also noted for establishing some of the first clinical practice guidelines on HIV/AIDS.

The CDC Cluster Study and Gaetan Dugas

The biggest controversy in And the Band Played On surrounds AIDS patient, Gaetan Dugas, who was labelled "Patient Zero" in an early cluster study. Several critics suggest that Shilts maligned Gaetan Dugas by (a) revealing his actual name in the book when Shilts was informed of it by the cluster study authors; and (b) suggesting Dugas maliciously and knowingly infected his lovers. This perception is unfortunate and misguided, but a 2016 paper in *Nature* (Worobey et al. 2016) vindicated Shilts' reporting.

In June 1982, the CDC published an epidemiological study of the first cluster of early AIDS patients seen in California (CDC June 18, 1982; CDC 1982). The 1982 paper referenced one "non-Californian" (who was Gaetan Dugas) with KS as part of the cluster. No mention of him as "Patient 0/Zero" was made in the 1982 CDC paper.

This was followed by a much more comprehensive write up of the cluster study in 1984 (Auerbach et al. 1984).

The 1984 analysis of that same cluster centered around an *index patient* in the paper's "Fig. 2.1"—a diagram of contacts (called a sociogram), placing "0" in the center of the sociogram (the Key reads: "0 = Index Patient") who was connected to 40 patients either through primary, secondary or tertiary connections, and then the sociogram labels all sexual contacts in accordance with the "sequence of onset" of symptoms. Of note, index patients are routinely called "Patient 0/[Zero]" in infectious disease contexts. The 40 patients in Auerbach et al.'s sociogram are labelled by geographic location as follows: NY 1 through NY 22; LA 1 through LA 9; FL 1 through FL 2; GA 1 through GA 2; PA 1; NJ 1; SF 1; and TX 1. NY 1, for example, was the first New York contact in New York City to get sick, while NY 21 was the 21st contact in New York City to get sick. A verbatim description of the sociogram reads (Auerbach et al. 1984):

> Sexual contacts among homosexual men with AIDS. Each circle represents an AIDS patient. Lines connecting the circles represent sexual exposures. Indicated city or state is place of residence of a patient at the time of diagnosis. "0" indicates Patient 0 (described in text)

In this analysis, the authors had interviewed 248 patients; contact-tracing revealed that 40 were all connected through sexual contact to the Index patient, who they describe as "Patient 0" in the text of the paper (This index patient was the same "non-Californian" referred to in the 1982 paper.). The Index patient was later revealed to Shilts to be Gaetan Dugas—three years after he was deceased in a pre-HIPAA age. In explaining the Auerbach 1984 study in his book, Shilts' writes: "At least 40 of the first 248 gay men diagnosed with GRID in the United States, as of April 12, 1982, either had had sex with Gaetan Dugas or had had sex with someone who had." (Shilts 1987: 147). This was a completely accurate reporting of the 1984 cluster study paper in 1987, when the book came out, and as of this writing, it remains true that the 40 patients in that cluster were connected through sexual contact.

Dugas was a Canadian flight attendant; he was a real study subject in the first cluster study that definitively linked HIV to sexual contact. He was correctly identified in 1982 by CDC investigators as a non-Californian. Dugas died from AIDS in March 1984, just when the 1984 Auerbach et al. paper was published. Shilts' descriptions of Dugas as Patient Zero and a frequent visitor to bath houses led to a misperception that Dugas brought AIDS to North America, which was never concluded in any empiric study, nor was this ever definitively stated by Shilts. Shilts does promote the idea that Dugas spread the disease carelessly, which was based on Shilts' interpretation of his interviews and notes with CDC investigators.

But years of questions about Dugas led to a reanalysis of data by a lead investigator at the University of Arizona (Worebey et al. 2016). The reanalysis found that the CDC's initial labelling of Dugas was "O" for "Outside-of-California", and he was patient number 057, for being the 57th case of AIDS reported to the CDC. When the cluster study was written up in 1984 to describe that 40 of the men were sexually connected, the O for "Outside" was misread as 0 for "Zero", and Dugas was improperly labelled the Index Patient in the paper's analysis of the cluster study. Accordingly (Worebey et al. 2016):

> Before publication, Patient 'O' was the abbreviation used to indicate that this patient with Kaposi's sarcoma resided 'Out(side)-of-California.' As investigators numbered the cluster cases by date of symptom onset, the letter 'O' was misinterpreted as the number '0,' and the non-Californian AIDS patient entered the literature with that title. Although the authors of the cluster study repeatedly maintained that Patient 0 was probably not the 'source' of AIDS for the cluster or the wider US epidemic, many people have subsequently employed the term 'patient zero' to denote an original or primary case, and many still believe the story today.

> According to Tara Smith (Smith 2016).

> When we think of Dugas's role in the epidemiology of HIV, we could possibly classify him as, at worst, a "super-spreader"–and individual who is responsible for a disproportionate amount of disease transmission. Dugas acknowledged sexual contact with hundreds of individuals between 1979 and 1981–but his numbers were similar to other gay men interviewed, averaging 227 per year (range 10-1560)…Dugas worked with researchers to identify as many of his partners as he could (~10% of his estimated 750), as the scientific and medical community struggled to figure out whether AIDS stemmed from a sexually-transmitted infection, as several lines of evidence suggested.…

> The media then extended Shilts's ideas, further solidifying the assertion that Dugas was the origin of the U.S. epidemic… *The New York Post* ran a huge headline declaring "The Man Who Gave Us AIDS." *Time* magazine jumped in with a story called 'The Appalling Saga Of Patient Zero." And *60 Minutes* aired a feature on [Dugas, describing him as] "One of the first cases of AIDS…"

It remains unclear whether Shilts distorted his notes about Dugas from CDC sources, or reported facts and concerns about Dugas contemporaneously, in a time-frame where facts were very difficult to validate. Worobey et al. note this about the 1984 paper (Worobey et al. 2016):

> Reports of one cluster of homosexual men with AIDS linked through sexual contact were important in suggesting the sexual transmission route of an infectious agent before the identification of HIV-1. Beginning in California, CDC investigators eventually connected 40 men in ten American cities to this sexual network [which included Dugas]. Thus, while [Dugas] did link AIDS cases in New York and Los Angeles through sexual contact, our results refute the widespread misinterpretation that he also infected them with HIV-1.

Thus, the 1984 sociogram is still accurate from the standpoint of connecting who had sex with whom, but Dugas was not the index case; the first case brought to the

CDC's attention is labelled on the sociogram as "NY 3", of which Dugas was a primary contact (Worebey et al. 2016).

As for Shilts' descriptions of Dugas' sexual activities, it was based on CDC sources, and their own interviews with Dugas. Dugas was very promiscuous, but exhibited typical behaviors for a sexually active gay male at the time who frequented the bathhouses. In 2013, video of Dugas surfaced that shows Dugas in March 1983 at an AIDS forum and panel hosted by AIDS Vancouver, asking questions that suggested he was indeed *not* practicing safe sex because of the absence of any clear HIV tests or guidelines for informing his lovers (AIDS Vancouver 2013). Dugas was not unique, and representative of hundreds of AIDS patients at the time. Moreover, contemporaneous minutes exist from June 20, 1983, surrounding AIDS Vancouver's concerns about Dugas' continued unsafe sexual activity (AIDS Vancouver 2013). Dugas' final primary care doctor in Vancouver, Brian Willoughby, also stated in 2013 that he met Harold Jaffe from the CDC (co-author on the 1984 paper), who specifically told him that the CDC had identified Dugas as "patient zero" for the first cluster of cases they investigated in the U.S (AIDS Vancouver 2013).

Thus, historical retribution surrounding Shilts' handling of the Dugas content does not consider: (a) validity limitations; (b) confidentiality limitations in infectious disease that had been long established; (c) the fact that there were *no confidentiality rules applied to deceased patients in 1984*; and (d) retrospective data or re-analyses of early epidemiological data that would not be clear until 29 years after the book was published (McNeil 2016).

In 2017, William Darrow, co-author of the 1984 paper and a source for Shilts, provided a retrospective analysis of Shilts' reporting for And the Band Played On . He states (Darrow 2017):

> The authors of at least a half-dozen books have addressed Randy's decision to identify Patient 0 as Gaetan Dugas, while the co-authors of a book published three years earlier [when Dugas was alive] chose to use a pseudonym. In Chap. 5 ("The Clusters") of The Truth About AIDS, Fettner and Check described a patient diagnosed with Kaposi's sarcoma known to Drs. Linda Laubenstein and Alvin Friedman- Kien "who traveled a lot for a Canadian company. Call him Erik" (p. 85). After giving a lecture to physicians in San Francisco, Dr. Friedman-Kien was able to speak with Erik and placed a telephone call to his office in New York City shortly afterwards. "Linda," he reportedly said, "I've located our Typhoid Mary."

Darrow clarified that the 1984 paper's sociogram reflects eight AIDS patients who slept with Patient Zero directly, and another eight patients who had sex with at least one of Patient Zero's partners, which came to 16 (Darrow 2017). The rest of the contacts shown in the paper's sociogram had tertiary contact, meaning they had sex with the partner of a partner of someone who had sex with Dugas. With respect to Shilts' sentence [emphasis mine]: "Whether Gaetan Dugas actually was the person who brought AIDS to North America **remains a question of debate and is ultimately unanswerable**." (Shilts 1987: 439), Darrow responded this way (Darrow 2017):

> The question was answered by Worobey and colleagues who showed that Patient 0 was not the person who brought AIDS to North America. The suggestion that Mr. Dugas was the first North American case did not emanate from any member of the CDC Task Force.

Darrow also clarified the following about the 40 sexually connected men (Darrow 2017):

> In the CDC cluster study, no assumptions were made about who infected whom… Some of the sexual partners of Patient 0 were healthy when initially contacted and subsequently developed AIDS, but CDC collected no evidence to prove that Patient 0 infected any of his sexual partners.

Ultimately, widespread excoriation of Shilts' reporting on the 1984 cluster study doesn't consider an historical lens: it would be decades until the etiology of HIV/AIDS was clear. In fact, as late as 2013, AIDS Vancouver was *still* reporting Dugas as the "first person diagnosed with AIDS in North America" (AIDS Vancouver 2013). Darrow points out in 2017 that efforts to trace the etiology in 1982 were primitive and not informed by what would become clear in the twenty-first century, and he clarifies how frustrated Shilts' was over failure to apply precautionary principles to mitigate presumed sexual spread of HIV/AIDS. Darrow concludes with a powerful vindication of Shilts' reporting on the cluster (Darrow 2017):

> Unfortunately, when AIDS was being recognized in June 1981, as many as 300,000 people may have already been infected with the virus on at least three continents. Only a few were aware of the disease and its potential consequences….
>
> Randy Shilts witnessed the exponential growth of the epidemic to over 50,000 AIDS cases reported in the United States by the time [he died in 1994]. He believed early on that a sexually transmitted agent was spreading rapidly among men who have sex with men (MSM) in San Francisco. He couldn't understand why city health authorities failed to close the bathhouses where he knew so many new infections were contracted. Subsequent studies have proved him right. The estimated incidence of HIV infection among MSM in the United States rose rapidly, from fewer than 100 men infected per year in 1977–1979, to 12,000 per year in 1980–1981, to 44,000 per year in 1982–1983, to 74,000 per year in 1984–1985. Sharp declines in HIV incidence followed as America finally began to respond to AIDS, but the irreversible damage of institutional failure and national neglect had been done.

From GRID to AIDS: Blood, Sweat and Tears

In covering the early days of the AIDS epidemic, it's important to make clear that the first label assigned to the disease by the medical community itself was "Gay-Related Immune Deficiency" (GRID). In fact, at UCLA, when Gottlieb first began to see gay males with the disease, the first patient that died from pneumocystic carini pneumonia (PCP) commented to his health team: "I guess I am one sick queen" (Hillman 2017). Thereafter, UCLA initially dubbed the new epidemic "SQUIDS"—for "Sick Queen Immune Deficiency Syndrome" until the term was banished from use (Hillman 2017), and the term "GRID"—Gay-Related Immune Deficiency—took hold. In the first published papers in *MMWR* (CDC 1981a, b) and the *New England Journal of Medicine* (Gottlieb et al. 1981), a causal association of "homosexual" and immune deficiency was made, reinforced by the *New York Times* article that mentioned the term GRID as well as "A.I.D." The terminology change to Aquired Immune Deficiency Syndrome, or "AIDS" was depicted in the film, *And the Band Played On*, which dramatized a July 27, 1982 meeting of stakeholders, comprising hemophiliac organizations, blood industry officials, members of the gay activist/political community, representatives from the CDC, NIH and the FDA. At this point, it was clear that the disease was not "gay related" but a virus transmitted through sexual contact as well as blood, because intravenous (IV) drug users and hemophiliacs were becoming infected as well. Additionally, Haitians of both sexes were becoming infected due to a cluster there traced to 1979. Between 1979 and 1982, 61 cases of AIDS were diagnosed in Haiti, where the infection initially spread in an urban area with high rates of prostitution, and there were questions as to whether Haiti brought HIV to the United States or was it the reverse (Koenig et al. 2010; Farmer 2006).

The historical July 1982 meeting wound up being very contentious with many groups opposed to implementing any guidelines for protecting the blood supply fearing social harms to certain groups through stigma. The CDC suggested that people in high-risk groups ought to refrain from donating blood: gay men, IV drug users and Haitians were considered "high risk" donors. The meeting was described this way (Nussbaum 1990):

> The meeting was a disaster. Hemophiliac groups didn't want their blood disorder to be associated with a gay disease. Gay community leaders were fearful that being prevented from donating blood was just the first step in quarantining all gay men. Indeed, right-wingers in Washington were already making noises about sending gays to "camps." The FDA and the CDC fought over turf. Regulation of the blood industry fell under traditional FDA authority. The involvement of the CDC was perceived as a threat. Many FDA doctors didn't even believe that a new disease existed. They thought the CDC was simply stitching together a number of unrelated diseases to boost their budget funding.

> No one was willing to agree to anything except to wait and see. There was one accomplishment, however. Different groups on different coasts were calling the new disease by many different names. Gay-Related Immune Deficiency was the most popular, but it was clearly untrue since IV drug users and Haitians were shown to be vulnerable. Gay cancer was used mostly in New York, but it focused on only one of the many opportunistic infections associated with the disease.

> Someone at the meeting suggested AIDS - Acquired Immune Deficiency Syndrome. It sounded good. It distinguished this disease from inherited or chemically induced immune deficiencies. It didn't mention the word gay or even suggest gender. AIDS. It stuck.

> July 27, 1982, the day the CDC adopted AIDS as the official name of the new disease, is the official date of the beginning of the AIDS epidemic. At that point, about five hundred cases of AIDS had already been reported to the CDC, of whom approximately two hundred had died. Cases had been diagnosed in twenty-four states, and the pace of new diagnoses was doubling every month. The CDC started calling the outbreak an epidemic.

Notes Shilts (1987:171):

> Somebody finally suggested the name that stuck: Acquired Immune Deficiency Syndrome. That gave the epidemic a snappy acronym, AIDS, and was sexually neutral. The word "acquired" separated the immune deficiency syndrome from congenital defects or chemically induced immune problems, indicating the syndrome was acquired from somewhere even though nobody knew from where.

Ultimately, the one notable good outcome of the meeting was the adoption of *Acquired* Immune Deficiency Syndrome (AIDS) as the formal label of the new epidemic. The one notable terrible outcome, however, was the refusal by decision-makers to take early steps to protect the blood supply. Although screening the blood supply could not occur prior to the discovery of HIV and an HIV test (see further), donor screening using questionnaires and data about high risk groups could have had a significant impact, and would have saved likely thousands of lives. At this meeting, three cases of hemophiliac deaths from PCP led to the conclusion that the virus was indeed in the public blood supply (IOM 1995), hence the line "how many dead hemophiliacs do you need?" It's worth noting that many students do not have a background or understanding of the disease, hemophilia, which is an X-linked disease afflicting mostly male children. It afflicted Richard Burton, who was married to Elizabeth Taylor (see earlier), as well as Roger Ailes, former CEO of Fox News, who died from it in 2017 after a fall. In covering the history of AIDS, a primer on hemophilia, the invention of clotting factor and Factor VIII is important, but I will not address it here. If hemophiliac infections via Factor VIII wasn't convincing enough, there was certainly strong evidence by early 1983 that AIDS was a blood borne virus that could be transmitted through blood transfusion, due to hundreds of transfusion-AIDS cases by then, including pediatric AIDS cases from maternal blood transfusions during childbirth. For example, Elizabeth Glaser (wife of actor Paul Michael Glaser) contracted AIDS as early as 1981, from a blood transfusion during childbirth, and passed it to her daughter through breastfeeding. Her second child then contracted the virus in utero. This became the story of thousands of women and children. By 1984, 12,000 cases of transfusion-HIV infections had occurred (IOM 1995). Yet despite the evidence that AIDS could be transmitted through transfusion, the blood industry declined to accept the CDC recommendations to implement any screening strategies (see under Healthcare Ethics Issues).

Astonishingly, between 1982 and 1985, absolutely no attempt to screen or protect the public blood supply was made. Even after the HIV virus was identified (see further), widespread screening of the blood supply did not occur until 1985, when the ELISA (and later Western Blot) tests were finally developed (see further). Even after 1985, untested blood continued to be used even after the ELISA testing was developed. The Institute of Medicine (IOM) concluded: "Inappropriate incentives inhibited reasonable decision-making by the responsible parties" (IOM 1995).

Some of the "inappropriate incentives" had to do with perception problems of the populations affected. Ultimately, although the transition from GRID to AIDS occurred early in the epidemic, the connotation that the disease was "gay related" would remain for at least 15 years, and likely affected delays with protecting the public blood supply. Ironically, although Rock Hudson's AIDS death was automatically associated with his being a homosexual and the presumption he got the virus through sex, it is also likely he got AIDS through a transfusion while undergoing heart surgery in 1982 (People 1982).

Discovery of the Virus that Causes AIDS: LAV, HTLV III and HIV

Not since the discovery of insulin in 1921, in which two research teams bitterly fought for the credit and shared recognition for the Nobel Prize (Bliss 1982) has there been such an ugly and public fight for credit as the discovery of the virus that causes AIDS. The book and the film, *And the Band Played On*, covers the Robert Gallo story and paints him as a highly problematic figure in the early years of AIDS. In fact, the film, more than the book, functions as a vehicle for Gallo's public shaming.

In a translational research race to discover the virus, researchers at the Pasteur Institute are the first to succeed in 1983, and published their discovery on May 20, 1983, in *Science* (vol. 220, p. 868). Dr. Luc Montagnier's research group found a new virus that causes AIDS, which they name lymphadenopathy virus (LAV). As the first article on a new virus, the paper made clear that more confirmatory research was required. A Scottish Inquiry that looked at this period in science history summarized it this way (Penrose 2015):

> The tentative conclusion of the article was that the virus belonged to a family of T-lymphotropic retroviruses that were horizontally transmitted in humans and might be involved in several pathological syndromes, including AIDS. The conclusion was uncommitted on the issue of whether the new virus was the etiological agent causing AIDS... The full significance of the French discovery was not widely acknowledged in 1983....

> Despite the hesitancy expressed by Montagnier, there was growing interest in the scientific community in the hypothesis that transmission of a virus caused AIDS, and for some specialists the Montagnier discovery was significant. After the publication of their paper in *Science*, however, the French scientists struggled to persuade some others in the field that the virus they had isolated was indeed the cause of AIDS....

When the retrovirus "King" Dr. Robert Gallo, a renowned researcher at the National Cancer Institute, learns of the news he clearly understood the significance, but plays down his French colleagues' discovery. As depicted in the public record, Shilts' book and the film, he "gaslights", suggesting LAV is merely a variation or relative of a retrovirus that he discovered years earlier, and part of the HTLV family of viruses linked to feline leukemia that he is working on. In April 1984 Gallo reports that his team has now found the definitive virus that causes AIDS, which he names the human T cell lymphotropic virus type III (HTLV-III)[3]; he publishes his finding on May 4, 1984 in *Science* (vol. 224, p. 500).

On April 23, 1984: "[Health and Human Services] Secretary Margaret Heckler announced that Dr. Robert Gallo of NCI had found the cause of AIDS, the retrovirus HTLV-III. She also announced the development of a diagnostic blood test to identify HTLV-III and expressed hope that a vaccine against AIDS could be produced within two years" (NIH 2019). According to the Scottish Inquiry (Penrose 2015):

> The tide only turned in France when Robert Gallo and his group in the United States made a similar discovery. In the spring of 1984, Gallo published more convincing evidence ...It

---

[3] Also reported as HTLV-IIIB, standing for human T-cell leukaemia/lymphoma virus type IIIB.

was announced that a retrovirus belonging to the HTLV family and designated HTLV-III had been isolated from a total of 48 subjects, some with AIDS, some with 'pre-AIDS' and some without symptoms but in risk groups. The authors concluded that HTLV-III might be the primary cause of AIDS.

Gallo's announcement was a turning point in developing knowledge worldwide. The evidence that people who had AIDS-like symptoms had antibodies against HTLV-III was more compelling circumstantial evidence that the virus was associated with the disease than finding the virus itself in somebody with the illness...Gallo's work was a major contribution to developing knowledge...However, the impression given overall is that, leaving aside all national prejudices, internationally there was cautious scepticism among many opinion leaders about the French research, until Gallo's announcement. The Montagnier/Barré-Sinoussi team did not have a long track record of discovering viruses. Their work did not have the international esteem required to spark the scientific research and development that followed the work of the Gallo group.

Montagnier stated in 2002 (Penrose 2015):

Our [1983 results] were still controversial … and we had difficulty in obtaining the funding needed to better characterize the virus and develop a blood test.

In June 1984, Drs. Robert Gallo and Luc Montagnier held a joint press conference to announce that Gallo's HTLV-III virus and Montagnier's LAV were virtually identical, which suggested foul play on Gallo's part, which is detailed in the film, exposing Gallo as potentially appropriating or manipulating Pasteur Institute materials he was sent for verification.

This led to delays in more robust translational research because there were disputes over whether LAV, HTLV-III or both, were the viruses that cause AIDS, which *would also affect patents*. Nonetheless, the NIH developed a test under Gallo's patent in 1985 that could detect exposure to so-called HTLV-III, while in France, an LAV test is also developed as well as experimental treatments for AIDS (see further). But in a "VCR versus Betamax" situation, the HTLV III test establishes itself as the mainstream lab test.

In 1985, the Pasteur Institute filed a lawsuit against the National Cancer Institute because the test for HTLV III was also a test detecting LAV, which they discovered, and was an identical virus. They wanted a share in the royalties from the HTLV-III test. Thus, early testing and treatment for AIDS was confusing; clinicians on the ground began referring to "LAV/HTLV-III" as the virus that causes AIDS, basically acknowledging they were the same virus found by different teams. While the United States took the lead in testing, raking in millions of dollars in profits from the test, the French took the lead in experimental treatments. Between 1984-1987, AIDS patients, including Rock Hudson, would go to Paris for experimental treatments with HPA-23 (see further).

As the LAV versus HTLV-III dispute persisted, the Pasteur Institute's lawsuit threatened to expose major scientific misconduct by Gallo, prompting an international committee of scientists in 1986 to intervene (Marx 1986). Called the "Varmus Committee", it was led by Dr. Harold Varmus, Chair of the Retrovirus Study Group within the Vertebrate Virus Sub-committee of the International Committee on Taxonomy of Viruses. The committee, which included Montagnier and Gallo, proposed new virus nomenclature, and that the virus be renamed to Human Immunodeficiency Virus (HIV). At the time, not only was there the "LAV" v. "HTLV- III" name recognition problem, but another AIDS researcher, Jay Levy, from the University of California San Francisco, had also isolated the AIDS virus in 1983 but didn't publish; he called it AIDS-Associated Retrovirus (ARV). Politically, whichever name was formally recognized would be associated with who was credited for the discovery. So, in May 1986, agreement was reached by all but two committee members (Gallo and Max Essex) to rename the virus HIV, and the test would become known thereafter as the "HIV antibody test" (aka HIV test). Despite this arrangement, the French pursued their lawsuit.

A memo from Don Francis at the CDC about what the lawsuit would uncover reads (Shilts 1987):

> If this litigation gets into open court, all the less-than-admirable aspects will become public and, I think, hurt science and the [U.S.] Public Health Service. The French clearly found the cause of AIDS first and Dr, Gallo clearly tried to upstage them one year later.

Don Francis is writing this memo as the Public Health Service is still licking its wounds from the Tuskegee study a decade before, which is now freshly in the news due to James Jones' book, Bad Blood, published in the early 1980s (see Chap. 1). Shilts further notes (Shilts 1987):

> On the most central issue of whether HTLV III was the product of viral pilfering Francis posed the hypothetical question: Could the prototype isolates of HTLV-III and LAV be identical merely by coincidence? And he answered "Probably not."

By 1987, Jonas Salk, who discovered the polio vaccine, became the mediator in the dispute with shuttle diplomacy, getting both parties to agree to be considered co-discoverers of HIV. The Pasteur Institute also demanded that the United States "turn over its half of the profits from the blood test—about $50 million since 1985" (Hilts 1992). In the end, the debacle became an international incident in which both the U.S. and French Presidents signed an agreement in 1987 between the two countries to share credit. As reported by the *Washington Post*, "the dispute over Dr. Gallo's claims became so linked to national scientific prestige that the Presidents of France and the United States attempted to end the conflict in 1987 when they agreed to a 50–50 split of credit and patent royalties from work with the AIDS virus and the blood test to detect it (Gladwell 1992).

1989: The Chicago Tribune Story and Federal Investigations

In 1989, an explosive 50,000-word investigative report by John Crewdson of the *Chicago Tribune* (Crewdson 1989), reviewed the Gallo matter again, and concluded that he had engaged in frank misconduct. The *Chicago Tribune* story led to a formal federal investigation—comparable to a science version of the "Mueller Report"— into Gallo's laboratory to see if he was indeed guilty of research misconduct. A 1991 summary of the report found Gallo and his laboratory guilty of scientific misbehavior and unprofessionalism but not overt misconduct. By 1992, a re-analysis by an independent panel concluded that Gallo indeed committed misconduct, and that he had "falsely reported" a key finding in his 1984 *Science* paper regarding isolating the AIDS virus (Michaelson 1993; Hilts 1992). According to the *New York Times* (Hilts 1992):

> The new report said Dr. Gallo had intentionally misled colleagues to gain credit for himself and diminish credit due his French competitors. The report also said that his false statement had "impeded potential AIDS research progress" by diverting scientists from potentially fruitful work with the French researchers.
>
> Most of his critics argued that Dr. Gallo had tried to take credit for work that French scientists had done and that he may even have taken the virus the French were studying and claimed it as his own. At the time, the virus was difficult to isolate and grow in sufficient quantity for research. In addition, the report found that Dr. Gallo warranted censure on …four other counts [including his] referring to his role as a referee for a different article submitted to a journal by his French competitors, in which he altered several lines to favor his own hypothesis about the AIDS virus [for reasons that were] "gratuitous, self-serving and improper."

The focus on Gallo in the film reflects the findings from both the *Chicago Tribune* article and the 1992 report. While the film was in production, Gallo's attorneys threatened HBO with a lawsuit (Cooper 2019), but HBO felt it was on solid legal ground. As the film was in production, the *New York Times* published a useful mini-history of the entire LAV-HTLV controversy for its readers, which included a chronology of events that took place over the same time line as Shilts' book and the film. Here it is (Hilts 1992b):

Chronology: "Years of Scientific Dispute"

1983: French scientists under Luc Montagnier at Pasteur Institute report discovery of a virus that might be the cause of AIDS.

1984: Ignoring French claim, U.S. scientists at National Institute of Health under Dr. Robert C. Gallo announce discovery of such a virus and proof that it causes AIDS.

1985: A blood test for antibodies to the AIDS virus is licensed. The French sue the U.S. Government over credit for the discovery of the virus.

1987: President Ronald Reagan and Prime Minister Jacques Chirac announce an agreement on sharing credit and divide royalties for the blood test.

1989: A Chicago Tribune article suggests that Dr. Gallo improperly took credit for the Montagnier discovery.

MARCH 1990: A [Chicago Tribune] report asserts that Dr. Gallo's virus was probably identical to the Pasteur Institute virus.

OCTOBER 1990: The National Institutes of Health says it will open a full-scale investigation of the matter by the Office of Scientific Integrity because a preliminary investigation suggested the possibility of misconduct.

MAY 1991: Dr. Gallo formally concedes that the viral cultures were probably contaminated by French samples but maintains that he is a co-discoverer.

SEPTEMBER 1991: Preliminary report by Office of Scientific Integrity finds evidence of misconduct by Dr. Gallo. Final report holds that he is not guilty of misconduct but deserves censure for permitting lapses and misrepresentations by those under him.

MARCH 1992: New investigation of charges of perjury and patent fraud announced by Inspector General of Department of Health and Human Services, the General Accounting Office and a House subcommittee.

DEC. 30, 1992: Report of Office of Research Integrity of Department of Health and Human Services says Dr. Gallo grew Dr. Montagnier's virus in his own laboratory and misled colleagues to gain credit for himself. (pg. A20)

A *Washington Post* review of the controversy in Spring 1992 fleshed it out this way (Gladwell 1992):

The Gallo controversy dates from 1982, when two laboratories – Gallo's at the National Cancer Institute and a French research team headed by Pasteur Institute researcher Luc Montagnier – began working simultaneously to track down the cause of the then-mysterious disease that was striking homosexual men.

First to publish was Montagnier, who described finding a virus, which he dubbed LAV, in an AIDS patient in May 1983. Montagnier, however, stopped short of claiming it was the cause of AIDS. In April 1984, Gallo went further. He announced that he too had identified a virus, and proposed that it was the cause of AIDS. Gallo also reported having developed a test to detect the virus in blood....

When Gallo's AIDS virus and Montagnier's AIDS virus were compared, they turned out to be the same. And because Montagnier had sent Gallo a sample of the French lab's virus early in the research process...Did Gallo steal the French virus and claim it as his own? Or did accidental contamination occur, as is common in laboratories working with highly infectious viruses like HIV?

The French, assuming the former, sued. That suit was settled in 1987 when the two parties agreed to share credit equally for having discovered the AIDS virus. There the matter rested until November 1989, when Chicago Tribune investigative reporter John Crewdson published a 50,000-word article implying that Gallo had deliberately stolen the virus and alleging a broader pattern of unethical behavior on Gallo's part.

The article prompted the NIH inquiry, which began in late 1989. The review entailed painstaking reconstruction of events of the period in question through numerous interviews and analysis of thousands of pages of documents.

By December 1992, a scathing conclusion was reached over Gallo's putative "explanation" for his 1984 *Science* paper's report that his lab had found the virus for AIDS (Crewdson 1992). In that same paper (Popovic et al. 1984), he essentially stated that LAV would not grow in his lab, so it couldn't possibly be labelled as the virus responsible for AIDS. Then, he backtracked in 1991, stating that his 1984 findings were accurate at the time, and he only recently deduced that his cultures must have become cross-contaminated with the Pasteur Institute's culture. The Office of Research Integrity didn't buy it (Hilts 1992):

[The Office of Research Integrity's] new report said, "Dr. Gallo falsely reported the status of L.A.V. research ...and this constitutes scientific misconduct...The explanations that Dr.

Gallo preferred for the statement are neither credible when the evidence is considered, nor do they vitiate the impropriety of falsely reporting the status of L.A.V. research.

In the end, Gallo is credited as a co-discoverer of HIV, but the treatment of the Gallo content in the 1993 film, *And the Band Played On*, forever branded his reputation and career. He left the NIH tarnished, and went to the University of Maryland, continuing to earn a profit from the HIV test of roughly $100,000.00 a year in royalties. In reality, Gallo's conduct stymied meaningful translational research into AIDS treatments. But Gallo continues to deny any wrongdoing, and now blames the entire controversy on cell line cross-contamination. Crewdson penned a book in 2002 about Gallo (Crewdson 2002; Horgan 2002) reviewing the entire controversy.

The Nobel Prize for HIV

In 2003, 20 years after of the discovery of HIV, Gallo and Montagnier published a joint retrospective on the discovery of HIV (Gallo and Montagnier 2003), in which they first give due credit to Gallo for having done the basic science work in retroviruses to begin with, which led multiple teams to search for such a virus as the cause of AIDS. The paper also implied that Gallo inadvertently mistook Montagnier's isolate for his own due to cell line contamination. Essentially, Gallo admits that he inadvertently "rediscovered" what Montagnier had first discovered from the exact same cell line. It is unknown if the 2003 paper by the "co-discoverers" of HIV was invited as a truce by the editor of the *New England Journal of Medicine*. But clearly it did not influence anyone on the Nobel committee.

By 2008, 25 years after the Pasteur Institute discovered LAV, the Nobel Prize in medicine was awarded to Luc Montagnier and Françoise Barré-Sinoussi for discovering the HIV virus in 1983; Robert Gallo was excluded. The rationale was straightforward: Montagnier's group was the first to publish in 1983. Through the years, it became clear that his group was indeed the first to discover the AIDS virus. By 1990, the seminal HIV discovery paper that was cited was Montagnier's *Science* paper (Barre-Sinoussi et al. 1983) and not Gallo's. Meanwhile, what Gallo published a year later in 1984 (Popovic et al. 1984) turned out to be from Montagnier's sample, but he renamed the virus and quickly filed a patent for a test that earned millions. Years of gaslighting, denial and finally, admission that it was all a terrible misunderstanding due to cell line contamination, left Gallo's reputation in tatters, with an explanation that was …well, prizeless.

To date, *And the Band Played On* remains the only film that documents one of the most notorious battles for credit in science history. But the Nobel decision also validates the role investigative journalism played (Miner 2008):

This [Nobel Prize decision] might be interpreted as the ultimate vindication of reporter John Crewdson, who in 1989, in a 50,000-word story in the Chicago Tribune, argued that Gallo – credited back then with codiscovering the virus – had merely rediscovered Montagnier's virus, which had been sent to Gallo as a professional courtesy

The HIV Test and Early Experimental Protocols

As discussed above, the first HIV test was patented by Gallo in 1985, and was an antibody assay designed to detect HTLV-III, and was manufactured by the pharmaceutical companies, Abbot and Electronucleonics, using HTLV-III isolates. At the same time, the Pasteur Institute had also developed a similar test to detect LAV, which was manufactured by the biotech company, Genetic Systems, using LAV isolates (Penrose 2015). The HTLV-III test is what was predominantly used. The first HTLV-III test was not designed to "diagnose AIDS" but to look for exposure to the virus that causes AIDS. The blood test was really designed as a tool to prevent transmission of the blood borne virus and to screen potential blood donors or banked blood. Ultimately, the first generation test became widely known in the medical literature as the "Anti-HTLV III/LAV assays" (Mortimer et al. 1985), but generally referred to the Gallo-patented test. The issue with the early tests is that they were unable to detect antigens, only antibodies. Despite the availability of all kinds of blood donor screening strategies (see under Healthcare Ethics Issues), it was not until the development of the first generation test in 1985, when widespread screening of the blood supply began. But untested blood continued to be used even after the test was developed. FDA approved the first enzyme linked immunosorbant assay (ELISA) test kit to screen for antibodies to HIV.

False Positives and Negatives

The following 1986 Letter to the Editor characterized early concerns with the HTLV-III test (Yomtovian 1986):

> During the last several months, much attention has been directed to the number of false-positive results of the test for the [HTLV-III] antibody. Concern has focused primarily on use of this test to screen blood donors and the worry that a large number of healthy donors would be falsely branded as HTLV-III viral carriers. While this is an important and valid concern, some of this worry has already been dispelled. What remains largely unaddressed, however, is the number of false-negative results of the HTLV-III antibody test. There are now published case reports of individuals who are HTLV-III viral culture positive, but HTLV-III antibody negative.

With the first generation of tests, testing negative was only reassuring if the test was done at least 12 weeks after potential exposure. The high sensitivity of the first antibody tests were designed primarily for protecting the blood supply, which was still slow-walked. But in many ultimately HIV-negative individuals, false-positive results were associated with other conditions such as infections, autoimmune disease or pregnancy (Alexander 2016). The testing was recommended for anyone who had unprotected sex since 1977, or had a blood transfusion since 1978, but this included many groups of low-risk individuals. Confirmatory testing was added for accuracy, which included the Western blot assay (Alexander 2016). By April 29, 1987, around the time Shilts' book is published, the FDA approved the first Western blot blood test kit—a more specific test, which began to be routine for certain hospital procedures (Banks and McFadden 1987). These second-generation tests could detect the virus within 4-6 weeks of exposure, and was the testing in place when the film came out in 1993.

By the late 1990s, HIV tests that combined antibody and antigen detection were developed, and could detect the virus within 2 weeks of exposure; this test gave one result but didn't distinguish whether the result was due to HIV antibodies or HIV antigens.

From a history of medicine standpoint, testing in the pre-treatment period was mainly for the benefit of third parties so that HIV-positive people could practice safe sex and other precautions; and that donor screening could prevent transfusion-AIDS. But pre-treatment testing also meant that testing positive could lead to enrollment in clinical trials (see further), which were few and far between. Frequently testing positive led to a social death where friends and employers abandoned the HIV-positive person, while depression and anxiety that came with any terminal illness diagnosis—even for those who were asymptomatic. A positive test did not predict when AIDS would materialize, given the high variability from HIV exposure to full blown AIDS. Testing positive for HIV prior to treatment raised very similar issues as genetic testing for diseases such as Huntington's disease.

The implication of HIV testing in the treatment era changed everything, as prophylactic therapy could be administered as soon as possible.

Ultimately, the main issues with HIV testing were the ethical issues: mandatory testing versus voluntary testing; access, privacy, insurance, and the duty to warn, which are all discussed further (see under Healthcare Ethics Issues). Many of these issues led to legal protections for HIV-positive patients, such as HIPAA, but access to treatment and insurance remained a problem even when treatment was available.

For current healthcare trainees, rapid HIV tests after exposure is the norm, while home HIV tests are now available on drugstore shelves (FDA 2018). For these reasons, it's important to emphasize the social stigma that was associated with earlier generations of testing, and the absence of legal protections, employment protections, healthcare access and social supports for those who tested positive.

Americans in Paris: HPA-23

HPA-23 was the first experimental antiretroviral used to treat HIV/AIDS patients; trials began in Paris in 1984, but were found to be ineffective. Montagnier and his team patented it, and began experimental trials almost as soon as HTLV-III/LAV were established as the AIDS virus. Preliminary research suggested that it slowed the progression of the virus. Because France had more permissive guidelines for experimental treatments, it went ahead with the HPA-23 trials in the absence of any other treatment or therapy.

The French team welcomed American patients as medical tourists to their HPA-23 trials, and treated over 100 Americans during this timeframe, which included Rock Hudson (see earlier), and San Francisco politician, Bill Kraus, featured as a major character in the film, *And the Band Played On*. American clinicians cautioned about pinning too much hope on HPA-23 (Raeburn 1985). Anthony Fauci noted at the time: "There really is no indication at all that [HPA 23] is any better, and it's considerably more toxic...HPA 23 causes a drop in the levels of blood cells called platelets, which in turn can lead to bleeding disorders" (Raeburn 1985). Eventually, the FDA succumbed to pressure from AIDS patient advocates to allow limited clinical trials in 1985. By 1986, it became clear that there was no therapeutic future with HPA-23, as the side-effects of the drug proved more problematic without any significant potential benefit (Moskovitz 1988).

From AZT to Cocktails

After the failed trials of HPA-23, a promising anti-retroviral therapy materialized with the drug, AZT (azidothymidine), a reverse-transcriptase inhibitor, also known by the name, Zidovudine. This drug was first discovered in 1964 as part of an NIH grant; it was investigated as a potential anticancer drug. It was found to inhibit the growth of the Friend leukemia virus in cell culture by inserting itself into the DNA of a cancer cell and interfering with its ability to replicate. It was abandoned when it didn't work well in mice and practical application seemed limited. After the Pasteur Institute discovered LAV in 1983 (see earlier), researchers at the NCI actively began a program to look for any drug or compound that had antiviral activity, and dug up the research results on the compound that became AZT, in addition to 11 other compounds that had potential. When tests were done in vitro, it blocked the HIV virus' activity.

By early 1985, Burroughs-Wellcome patented AZT, and a phase I clinical trial with started at Duke University. The results of that first trial demonstrated that AZT was safe and well-tolerated; increased CD4 counts and restored T cell immunity, and that AIDS patients were recovering on it. Then a large multi-center double-blinded randomized controlled trial with 300 patients began, which was halted after 16 weeks when it was clear that patients on the drug were getting better. In the placebo group, 19 patients had died, while the AZT group lost only one patient. Stopping the trial early was ethically defensible because the benefits clearly outweighed the risks of the placebo in the absence of any other treatment available for AIDS patients (Fischl et al. 1987).

AZT was approved by the FDA March 19, 1987, as an accelerated approval for both HIV-positive patients and AIDS patients, who had "AIDS-related Complex"—an antiquated term that just meant clinical symptoms of opportunistic infections. From the first clinical trial until FDA approval, 20 months had passed, but it had been, to date, the shortest approval process in the FDA's history at that time; in 1987, new drugs investigations took 8–10 years before they were approved.

One AIDS practitioner recalled (Sonnabend 2011b):

I'm not sure that it's even possible to adequately describe the terror and desperation felt in the early 1980s. By 1986 nothing of any use regarding treatments had come from the Public Health Service and very little from the academic medical community.

Then, after six years of inaction we were at last told that help was on the way. Dr. Samuel Broder who was head of the National Cancer Institute appeared on television shows trumpeting the benefits of a drug he called Compound S [later patented as AZT]. I well remember a TV show where he appeared with a person with AIDS who enthusiastically attested to the benefit he had received from the drug, presumably from 1.5G of it daily.

The clinical trial on which AZT's approval was based had produced a dramatic result. Before the planned 24 week duration of the study, after a mean period of participation of about 120 days, nineteen participants receiving placebo had died while there was only a single death among those receiving AZT. This appeared to be a momentous breakthrough and accordingly there was no restraint at all in reporting the result; prominent researchers triumphantly proclaimed the drug to be "a ray of hope" and "a light at the end of the tunnel". Because of this dramatic effect, the placebo arm of the study was discontinued and all participants offered 1500 mg of AZT daily.

But some clinicians were critical of AZT's early approval because (Sonnabend 2011a, b):

It took place in 12 centers across the country. There was no uniform approach to patient management during the trial; each of the 12 medical centers determined this independently. So the most important series of measures determining life or death in the short term was left unspecified. I suppose one has to conclude that government medical experts, unlike community doctors, must have felt that nothing could be done for people with AIDS, that the only hope to be found was in a new drug

AZT's approval coincided with the publication of Shilts' book; Shilts himself was on AZT until he died in 1994. AZT slowed the replication of the HIV virus, which prevented or delayed the progression of AIDS. Many HIV-positive patients were able to stave off full-blown AIDS for years on AZT until the protease inhibitor era arrived a decade later, in which AZT became part of a "cocktail" of drugs that typically included a second reverse-transcriptase inhibitor and an antiretroviral drug from another class of drugs (see further).

In June 1989, the first set of prevention guidelines for AIDS-related diseases and prevention of HIV transmission were published by the CDC, while the NIH endorsed access to experimental therapies for AIDS patients. Meanwhile, federal funding for hospice care was made available to community-based care programs in many states. While AZT was a breakthrough, it was not necessarily a sea-change as many patients either couldn't access it, or became resistant to it.

AZT Side-Effects, Efficacy, and Cost

AZT was initially prescribed at high doses (1500 mg), and had severe and sometimes toxic side-effects of anemia and malaise and liver damage. According to one critic of the initial dosing regimen (Sonnabend 2011a, b):

...[T]he drug was still approved at a dosage that proved to be so toxic that another trial
compared a similar dose with half that dose. This exercise resulted in excess deaths among
those taking the higher dose.

Indeed, a "randomized controlled trial of a reduced daily dose of zidovudine in
patients with the Acquired Immunodeficiency Syndrome" (Fischl et al. 1990) established that the initial dosing was too toxic, which also helped AZT gain approval in
1990 for pediatric patients.

But the biggest problem with AZT is that it stopped working after a time: the
virus eventually mutated and became AZT-resistant. The second biggest problem was
access and cost; at the time it was approved, AZT was considered the most expensive
drug in history. *Time* magazine described it this way on the 30th anniversary of AZT
(Park 2017):

But even after AZT's approval, activists and public health officials raised concerns about the
price of the drug. At about $8,000 a year (more than $17,000 in today's dollars) — it was
prohibitive to many uninsured patients and AIDS advocates accused Burroughs Wellcome
of exploiting an already vulnerable patient population.

As discussed earlier in this chapter, the cost of AZT led to the passage of the *Ryan
White CARE Act* (Comprehensive AIDS Resources Emergency) in 1990, which took
effect in 1991, and was designed to help vulnerable patients access HIV/AIDS treatments. Since then, millions of HIV/AIDS patients have used this law to get access
to treatment, including protease inhibitors (see further). Today, AZT is listed as an
Essential Medicine by the World Health Organization (WHO), and when the patent
expired on AZT in 2005, generic versions of the drug were FDA approved, and the
typical cost is now $2400 annually for U.S. patients, and between $5–25 per month
for patients in poorer countries.

The AIDS Cocktail: Protease Inhibitors

As drug resistance became a major side-effect of AZT, it was clear that a multi-
drug arsenal, known as the "cocktail therapy" would be the next generation of
HIV/AIDS treatment. An AIDS research team led by David Ho at Rockefeller University Hospital, noted that "within the body of a single infected person, many different
mutated versions of the virus exist" so it quickly evolves and becomes resistant with
a single drug, such as AZT. According to Rockefeller University, Ho's colleagues
found (Rockefeller University 2010):

[B]y giving patients three or four drugs at a time, [HIV] could not mutate rapidly enough
to evade all of them. By 1996 they had succeeded in reducing [HIV] levels to the point of
being undetectable in a group of patients treated with the new therapy. They discovered that
treating patients with a combination of three or more antiretroviral drugs could keep the
virus in check. The initial clinical trials of this therapy were carried out with patients at the
Rockefeller Hospital. Today, in the developed world, AIDS is a manageable chronic disease
thanks to the "AIDS cocktail" of combination antiretroviral therapy.

Ho began to publish his results on a class of drugs known as protease inhibitors, and by 1996, a protease inhibitor cocktail became the standard of care; 1996 was the year everything changed for HIV/AIDS patients.

Early drug research into the HIV protease led to specific protease inhibitors, now a component of anti-retroviral therapies for HIV/AIDS. When the HIV virus' genome became known, research into selective inhibitors and selective antagonists against the HIV protease began in 1987, and phase I trials of the first protease inhibitor, saquinavir, began in 1989. The results were published in 1995 (James 1995) and, and about four months later, two other protease inhibitors, ritonavir and indinavir, were approved.

In its December 30, 1996 issue, when the burden of HIV infection was now greatest in African Americans, *Time* magazine named David Ho, discoverer of the AIDS cocktail therapy, as its "Man of the Year", with several featured articles on Ho and AIDS. (In the same issue, there was an article on O. J. Simpson's post-acquittal life.). The main AIDS article by Philip Elmer-Dewitt called "Turning the Tide" started like this: (Elmer-Dewitt 1996)

> Some ages are defined by their epidemics. In 1347 rats and fleas stirred up by Tatar traders cutting caravan routes through Central Asia brought bubonic plague to Sicily. In the space of four years, the Black Death killed up to 30 million people. In 1520, Cortes' army carried smallpox to Mexico, wiping out half the native population. In 1918 a particularly virulent strain of flu swept through troops in the trenches of France. By the time it had worked its way through the civilian population, 21 million men, women and children around the world had perished…Today we live in the shadow of AIDS–the terrifyingly modern epidemic that travels by jet and zeros in on the body's own disease-fighting immune system."

CNN noted about the *Time* pick for Man of the Year on December 21 1996 (CNN 1996):

> Time magazine picked AIDS researcher Dr. David Ho, who has pioneered the use of drug "cocktails" to fight HIV, as its 1996 Man of the Year.
>
> "Some people make headlines while others make history," Time said.

By 1998, Stanford University began the "HIV Drug Resistance Database" which helped to track the resistance mutations to refine cocktail therapies (Stanford University). In 2014, when *Time* named Ebola researchers Persons of the Year, the magazine reflected on the last time it had named any medical researcher as "Person of the Year", and it was Ho. *Time* recalled (Rothman 2014):

> [In 1996] AIDS was a death sentence — but Ho, by successfully lowering the virus count in patients who received a combination of new and powerful drugs when they'd only just been infected, helped change the way the medical community looked at HIV and AIDS.

In 2017, 30 years after AZT had been discovered, and 21 years after David Ho pioneered the cocktail therapy, *Time* summarized treatment like this (Park 2017):

> Today, if someone is diagnosed with HIV, he or she can choose among 41 drugs that can treat the disease. And there's a good chance that with the right combination, given

at the right time, the drugs can keep HIV levels so low that the person never gets sick...-
Today, there are several classes of HIV drugs, each designed to block the virus at specific
points in its life cycle. Used in combination, they have the best chance of keeping HIV at bay,
lowering the virus's ability to reproduce and infect, and ultimately, to cause death. These
so-called antiretroviral drugs have made it possible for people diagnosed with HIV to live
long and relatively healthy lives, as long they continue to take the medications...And for
most of these people, their therapy often still includes AZT.

Treatment with the "AIDS cocktail", or protease inhibitors—now known as
antiretroviral therapy (ARV), led to early prophylactic therapy. With rapid testing
(see above), and starting early ARV before any symptoms begin, HIV/AIDS is not
only treated like a chronic disease, but recent drug trials reveal that HIV-positive
persons with undetectable viral loads may not even be infectious anymore. Over the
years, efforts to improve compliance with ARV led to the "one pill fits all" modality,
in which combination therapy is delivered through one pill, which comprises several
antiretroviral drugs.

To deal with high-risk groups, a new treatment modality evolved, known as "Pre-
exposure prophylaxis" (PrEP), which is when, according to hiv.gov:

...people at very high risk for HIV take HIV medicines daily to lower their chances of getting
infected. Daily PrEP reduces the risk of getting HIV from sex by more than 90%. Among
people who inject drugs, it reduces the risk by more than 70%. PrEP involves a combination
of two anti-retroviral drugs, and is currently sold as Truvada®.

One of the "Truvada for PrEP" commercials shows representatives from high-risk
groups (e.g. young African Americans of both sexes and white males) announcing:
"I'm on the Pill", explaining that Truvada will protect them from getting HIV while
having sex.

While an HIV vaccine remains elusive as of this writing, treatment has changed
everything—to the point where the history and social location of *And the Band Played
On* could even be forgotten if it is not taught. While writing this chapter, commercials
for the drug, Biktarvy, were running, which is an ARV cocktail in one pill, or as the ad
notes. "3 different medicines to help you get to undetectable". The ad shows happy
HIV-positive people: young African Americans doing an array of activities, along
with a presumably married gay male couple back from gourmet food shopping. The
ad looks indistinguishable now from an allergy drug commercial. As the ad explains
to potential patients: nothing in their lives needs to change, they should "keep on
loving" themselves for who they are, and the drug, combined with others, would help
them get to "undetectable". One wonders what Randy Shilts would say if he saw that
ad today (which actually aired while Pete Buttigeig was declared the winner of the
2020 Iowa Caucus).

Healthcare Ethics Issues

As mentioned at the start of this chapter, *And the Band Played On* is what I call a
"Maleficence blockbuster" because of the laundry list of ethical violations housed

under the Principle of Non-Maleficence, which obligates healthcare providers not to knowingly or intentionally cause harm, knowingly neglect, discriminate against, or abandon a patient, or knowingly withhold or not disclose information about material risks or harms, which are necessary for informed consent. As Shilts' makes clear in his book, the chaotic early years of AIDS was "allowed to happen" through poor decision-making in every critical theater of healthcare from policy to bedside. For these reasons, it's useful to cover the most egregious ethical violations shown in the film from the standpoint of each healthcare theater: The Bedside Theater, the Public Health Theater and the Research Theater. There are also a range of ethical violations in the HIV/AIDS story that are beyond the scope of the film, as the film ends before the testing and treatment era, which I will summarize briefly at the end of this section.

The Bedside Theater: Clinical Ethics Violations

The original sin that occurred at the bedside was provider discrimination against AIDS patients based upon their status as homosexuals. Next, labelling the newly recognized syndrome "gay-related immune deficiency" (GRID—see under Medical History) had unintentional consequences and became highly problematic once the syndrome was understood to be a blood-borne virus in 1982. Although the nomenclature was changed to AIDS in July 1982, the disease by then was inexorably linked to being "gay-related" forever after, no matter what it was called. The association of "gay" with "AIDS" seriously interfered with public policy, funding and resources that could have helped contain/manage the virus earlier on. For example, at UCLA, the idea of a major grant focused on AIDS was so undesirable to institutional stakeholders, the institution denied tenure to a faculty member with a multi-million dollar grant to start an AIDS treatment center to avoid the "stigma" of servicing so many AIDS patients (see under History of Medicine with respect to the Michael Gottlieb story). In other words, UCLA did not want to be a center of excellence in the early 1980s for gay males and IV drug users. In the pre-testing era, this level of discrimination against AIDS patients also made non-gay AIDS patients a target of much more discrimination from healthcare providers, their employers and social networks. In this timeframe, many healthcare providers, in fact, refused to treat, examine or care for AIDS patients, which led to the creation of multiple gay community-based caregiving programs, such as the Gay Men's Health Crisis (GMHC) in New York City, and similar programs elsewhere. Although healthcare provider discrimination is a theme that is far more expansive in the *Normal Heart* (2014), *And the Band Played On* features scenes in which the "buddy system" is discussed—where laypeople in the gay community took up the caregiving burden in the absence of healthcare providers willing to treat, which frequently included caregiving for in-patients at hospitals when healthcare providers refused to enter the rooms. In fact, the rise of articles on Professionalism and Humanism in medicine (see Chap. 8) coincided with the AIDS crisis.

Moral Distress and 1112 and Counting

As discussed earlier in this chapter, the first clinical ethics treatise on the AIDS crisis was written by Larry Kramer with a powerful "moral distress" piece that laid bare the discrimination and abandonment at the bedside, as well as the failure of adequate resources to deal with the epidemic. It is also a clarion call for safe sex in his community in the context of gay liberation, closeted homosexuals, and promiscuity (For the entire article see link at the reference, Kramer 1983[4]):

> There are now 1,112 cases of serious Acquired Immune Deficiency Syndrome. When we first became worried, there were only 41. In only twenty-eight days, from January 13th to February 9th [1983], there were 164 new cases - and 73 more dead. The total death tally is now 418. Twenty percent of all cases were registered this January alone. There have been 195 dead in New York City from among 526 victims. Of all serious AIDS cases, 47.3 percent are in the New York metropolitan area….All it seems to take is the one wrong fuck. That's not promiscuity – that's bad luck.

> There is no question that if this epidemic was happening to the straight, white, non-intravenous-drug-using middle class, money would have been put into use almost two years ago, when the first alarming signs of this epidemic were noticed by Dr. Alvin Friedman-Kien and Dr. Linda Laubenstein at New York University Hospital.

> During the first two weeks of the Tylenol scare, the United States Government spent $10 million to find out what was happening….

> I am sick of closeted gays. It's 1983 already, guys, when are you going to come out? By 1984 you could be dead. I am sick of guys who moan that giving up careless sex until this blows over is worse than death. How can they value life so little and cocks and asses so much? Come with me, guys, while I visit a few of our friends in Intensive Care at NYU. Notice the looks in their eyes, guys. They'd give up sex forever if you could promise them life. …I am sick of guys who think that all being gay means is sex in the first place. I am sick of guys who can only think with their cocks.

The second "original sin"—likely stemming from the first one—was failure to disclose the risks of blood transfusions to unsuspecting surgical patients, especially when undergoing elective procedures, or non-emergent procedures. The courts had established in 1976—five years before the first article on AIDS—that healthcare providers had a duty to warn (see further), which was also violated. All healthcare providers were aware that AIDS was a blood borne virus before it was isolated as LAV (Barre-Sinoussi et al. 1983) or HTLV III (Popovic et al. 1984), and knew that there was no screening being implemented. Thus, the first line of defense for reducing transfusion-AIDS was informed consent and disclosure of transfusion-AIDS as a risk, and providing the option of autologous transfusion. This did not become widespread practice until it was too late. Moreover, patients who had been infected and developed AIDS, as depicted in the film, were routinely not told they had been transfused with unsafe blood, which meant they could infect others unknowingly.

Similarly, hemophiliac patients were not warned about the risks of taking Factor VIII until it was too late (IOM 1995); many hemophiliacs even questioned their

---

[4]See also Clews, Colin (July 6, 2015) Discussion and Reprint of article by Kramer, Larry (March 1983) https://www.gayinthe80s.com/2015/07/1983-hivaids-1112-and-counting/.

personal physicians, or patient foundations about risks, and were encouraged that the risks were overblown when directed plasma donor programs could have been an alternative (IOM 1995). Risk disclosure of procedures or therapy, in the absence of any defined guideline or policy, is up to medical judgment. Disclosure of the risks of transfusion-AIDS ultimately became complicated in a risk averse, for-profit medical model, and in the setting of public hysteria over HIV/AIDS.

As discussed in the History of Medicine section, early AIDS practitioners were ostracized by their peers when they wanted to devote their interests to AIDS; this led to a clinical care vacuum for early AIDS patients, and also influenced institutional budgets with respect to resource allocation for AIDS care.

The failure to warn about risks about transfusion-AIDS was summarized this way by the Institute of Medicine (IOM) in its scathing report on its investigation into the epidemic of transfusion-AIDS cases, HIV and the Blood Supply (IOM 1995):

> The introduction of HIV into the blood supply posed powerful dangers to those individuals with hemophilia and recipients of blood transfusions during the early years of the epidemic. During the period from 1982 to 1984, before the AIDS virus was finally identified and a test developed to determine its presence, there was considerable speculation about whether the blood supply could be a vector for this new, fatal infection. As evidence about risk developed, consumers of blood and blood products—as well as their physicians—found themselves in a complex dilemma. Continued use of blood and blood products might heighten the risk of acquiring a new disease. Reducing or discouraging the use of blood products might increase the morbidity and mortality… Approximately half of the 16,000 hemophiliacs and over 12,000 recipients of blood transfusions became infected with HIV during this period. More effective communication of the risks associated with blood and blood products and the opportunity to choose from a wide spectrum of clinical options might have averted some of these infections….
>
> Several social and cultural impediments in the relationships between patients and physicians interfered with the communication of information about the risks associated with using blood and blood products. These included the tendency of physicians to not discuss, or to downplay and deny, the risk of AIDS;…the difficulties of communicating dire news to patients; and the problems associated with communicating uncertainty….[P]hysicians often responded to the initial questions of patients with reassurances that the risk was not serious, that the patient was overreacting…Or, the physician conveyed the impression that the risk was a problem associated with homosexual behavior [only].

Invalid Consent, Nondisclosure and Withholding of Information

Obviously, no patient who required blood products or a potential blood transfusion in the period before 1985, when the blood industry finally began screening for HIV/AIDS, had informed consent about the risk of HIV/AIDS through the blood supply. Unfortunately, informed consent served as both the first, and last, line of defense in the early years of the epidemic. Unlike the physicians involved in the Tuskegee study (see Chap. 1), who practiced medicine in the pre-informed consent era, there was no excuse by 1981, when AIDS was first described. By then, informed consent in medicine was legally and ethically required (Katz 1986; Faden and Beauchamp 1985), and even the judicial council of the American Medical

Association stated in 1981 that "Informed consent is a basic social policy... Social policy does not accept the paternalistic view that the physicians may remain silent because divulgence might prompt the patient to forego needed therapy." But the IOM discovered that (IOM 1995):

> [I]n some special medical practices, such as hemophilia and transfusion medicine, [disclosure of risks] was not fully adhered to until the early 1980s... [and] the practice of hemophilia and transfusion medicine was somewhat removed from recognized medical norms....The Committee also found that some physicians were reluctant to discuss bad news, including a prognosis with dire implications, once symptoms of AIDS began to occur in their patients. Even when confronted with initial symptoms of AIDS, the physician's message to his patient sometimes was to not worry....The appearance of AIDS in a previously healthy individual with hemophilia became a frightening experience for physicians [and once] physicians realized that the majority of individuals with severe hemophilia were infected with HIV, they became uncomfortable with discussing the implications of the widespread infection with their patients....

The IOM concluded (IOM 1995):

> [T]here were serious shortcomings in effective communication of risks associated with the use of blood and blood products. The Committee's analysis of physician–patient communications at the beginning of the AIDS era illustrates the tragedies that can accompany silence about risks.... One powerful lesson of the AIDS crisis is the importance of telling patients about the potential harms of the treatments that they are about to receive.

With respect to hemophilia, there was also a failure of the National Hemophilia Foundation's (NHF) medical advisory council to warn patients about risks. Its clinical practice guidelines were compromised by the organization's "financial connections to the plasma fractionation industry...The NHF provided treatment advice, not the information on risks and alternatives that would enable physicians and patients to decide for themselves on a course of treatment. Hemophilia patients did not have the basis for informed choice about a difficult treatment decision" (IOM 1995).

The Public Health Theater

Public health ethical violations at the health policy level led to widespread contamination of the public blood supply and no warning to the public that the blood supply was unsafe. The IOM's findings were alarming and damning. There was enough evidence by 1982, and irrefutably, by early 1983, that AIDS was a blood borne virus that could be transmitted through blood transfusion. By 1984, 12,000 cases of transfusion-HIV infections had occurred, which meant that infected patients had also likely spread the virus to untold thousands through unsafe sex or other means.

Despite the evidence that AIDS could be transmitted through transfusion, the blood industry, which comprised of multiple agencies, declined to accept the CDC recommendations to implement a number of wise strategies that would have significantly reduced transfusion-AIDS cases. Each recommended strategy is outlined here.

Donor Screening and Deferral

Donor screening and deferral involved screening interviews, questionnaires, or medical exams, that would identify high-risk donors so that they could be coded as either "not for transfusion" or rejected as donors. Lab tests on donor blood that had another infectious disease could also have been used as a "flag" for blood that was "not for transfusion". The history of donor pools and blood donation is vast, and will not be discussed here. However, it was clear for years that paid blood donors (prisoners, mental health patients, IV drug users) were typically higher risk than unpaid volunteers, and so could have been easily flagged as higher risk. Before 1973, blood was routinely screened for syphilis (see Chap. 1), and hepatitis and was not used if it was infected, but after this timeframe, it was labelled merely as paid donor or unpaid donor, and typically not screened; most Factor VIII was derived from the paid donor pool (IOM 1995).

On January 4, 1983, there was an historic contentious meeting of stakeholders surrounding the blood supply, which was a disastrous meeting in which no one could agree to any standards or guidelines. The meeting is re-enacted in the film, and was the scene in which Don Francis yells in frustration: "How many dead hemophiliacs do you need?" On January 6, 1983, Francis suggested deferring any blood donors who were high-risk (IV drug use; unprotected sex/sexually promiscuous; have Hepatitis B, or have lived in Haiti), and estimated that this would have eliminated 75% of infected blood donors (IOM 1995).

In 1976, in a landmark "duty to warn" case involving the duty to breach confidentiality in order to warn a third party of a known danger (see further), the courts made clear that "privacy ends where the public's peril begins" (Tarasoff 1976). Nevertheless, the IOM found in 1995 that "special interest politics" interfered with sound public health policy (IOM 1995):

> Some people viewed direct questioning about sexual behavior and drug use as a violation of an individual's right to privacy. Public health officials countered by saying that the individuals' rights were less important than the collective public health. Many in the gay community objected to direct questioning and donor deferral procedures as discriminatory and persecutory. Many in the blood bank community questioned the appropriateness of asking direct or indirect questions about a donor's sexual preference. Other individuals and organizations were concerned about providing the public with information about AIDS that might scare them away from donating blood....

The blood banks were worried about optics and losing a motivated donor pool they depended on, while the National Gay Task Force, a strong political lobby, were against banning blood donation from homosexual men. Blood banks were worried that donor screening/deferral strategies might also "out" gay men who were closeted, which had serious social repercussions (IOM 1995). But as the courts had ruled by that time, there were limits to autonomy, privacy and confidentiality when the public's health was at risk in a public health emergency or epidemic, which this was. Proposals for "donor deferral" were also rejected, in which donors themselves read a questionnaire that surveys if they are high-risk and opt out themselves (IOM 1995).

Surrogate Testing Using the Hepatitis B Test

If the blood industry did not like the idea of donor questionnaires and interviews, the CDC had another idea that would protect privacy and confidentiality, which was called "surrogate testing". At the same disastrous January 4, 1983, meeting, the CDC proposed this idea, too. By merely screening donors for the Hepatitis B virus, it would also catch a large percentage infected with the AIDS virus. Donors who tested positive for Hepatitis B could reasonably be considered too high risk to donate blood. But, alas, surrogate testing was not considered cost-effective, and on January 28, 1983, "the American Blood Resources Association (ABRA), recommended against large-scale surrogate testing of donated blood until ABRA had evaluated its feasibility" (IOM 1995). By June 22, 1983, the American Association of Blood Banks, the Council of Community Blood Centers, and the American Red Cross jointly stated that the risk of AIDS was too low to justify screening strategies that would "disrupt the nation's blood donor system" (IOM 1995), while gay activists were against screening measures that could increase discrimination. The only blood bank that began to screen using a surrogate test was the Stanford University Blood Bank, which began screening blood July 1, 1983. Nonetheless, public anxiety over AIDS was mounting (Daly 1983).

Warning the Public About Transfusion-AIDS

Incredibly, even the suggestion of direct warnings to the public about transfusion-AIDS was summarily rejected by the blood industry. Warning the public about the possibility of AIDS through transfusion and educating them about directed or autologous donation was the minimal requirement, and not doing so clearly violated the Duty to Warn (see further). According to the IOM report (IOM 1995):

> The blood industry was concerned about providing information on AIDS to the public lest donors take fright and stop donating blood. In January 1983, The [American Association of Blood Banks' (AABB)'s] Committee on Transfusion Transmitted Diseases [stated] "we do not want anything we do now to be interpreted by society (or by legal authorities) as agreeing with the concept—as yet unproven—that AIDS can be spread by blood".

The IOM also found that politics, competition for resources in the parsimonious Reagan administration, and hidden agendas interfered with sound public health policy. For example, a January 26, 1983, interoffice memo between two American Red Cross officials stated (IOM 1995):

> CDC is likely to continue to play up AIDS – it has long been noted that CDC increasingly needs a major epidemic to justify its existence...especially in light of Federal funding cuts.

Even after the HTLV-III virus was isolated by Gallo in 1984 (see earlier), no interim strategies to safeguard the blood supply were implemented, including heat-treating Factor VIII to kill the AIDS virus, which was suggested in 1983, confirmed to work well by 1984 (IOM 1995), *but not required by the FDA until 1989,* when it finally recalled all untreated Factor VIII.

Ultimately, widespread screening of the blood supply did not occur until 1985, when testing was finally developed (see earlier). Even so, untested blood continued to be used even after the ELISA testing was developed (IOM 1995). However, with the first generation of HIV tests (see earlier), HIV-positive donors could still squeak by if they donated too early after exposure, given the several weeks' window it could take for antibodies to develop.

Failure to Notify At-Risk Blood Recipients

Even with the delays in screening the blood supply, there was no program in place to notify recipients of infected blood that they had been infected, a situation conveyed in a powerful scene in *And the Band Played On*, in which a patient (played by Swoosie Kurtz) became infected when she had cardiac surgery, but was never told why she had opportunistic infections. When this scene aired in 1993, many recipients of tainted blood had not yet been made aware of the risk of AIDS, the risks of transmission to their partners through sex, maternal transmission at birth, or postnatal transmission through breastfeeding. These critical "lookbacks" had not begun until 1991, which was egregious. Yet by 1985, it was clear that transfusion-AIDS was a big problem. It was also clear that infected recipients of blood products could infect intimate partners, and could lead to maternal AIDS transmissions. In 1988 the Reagan administration asked the Department of Health and Human Services (DHHS) to find a way to trace all recipients of possibly infected blood products to inform them that they are at-risk so they can could get tested, and practice safe sex. But for reasons not clear, it was not until September 1991 that any type of "lookback" program was put into place. The FDA did, however, contact recipients of any "repeat blood donor" who had newly tested HIV-positive.

The IOM stated (1995):

> Individuals with hemophilia, transfusion recipients, and their families have raised serious concerns about why there were not better safeguards and warning systems to protect them from viruses transmitted by blood products…Were consumers of blood and blood products appropriately informed of the possible risks associated with blood therapy and were alternatives clearly communicated?…Perhaps no other public health crisis has given rise to more lasting anger and concern than the contamination of the nation's blood supply with HIV…. The safety of the blood supply is a shared responsibility of many organizations—the plasma fractionation industry, community blood banks, the federal government, and others….

> When confronted with a range of options … to reduce the probability of spreading HIV through the blood supply, blood bank officials and federal authorities consistently chose the least aggressive option that was justifiable…Interagency squabbling, lack of coordination, and miscommunication …[was the] most important organizational factor at work in explaining the cautious choices of public health authorities with regard to donor screening and deferral….The FDA instituted the "lookback" recommendation in 1991, at least six years after it was clear that AIDS had a long incubation period during which the patient could transmit HIV through sexual contact or contact with blood.

Catastrophically (IOM 1995):

> Blood banks, government agencies, and manufacturers were unable to reach a consensus on how extensively to screen for high-risk donors in order to substantially reduce the risk of HIV transmission through the blood supply. There was no consensus at the January 4, 1983, meeting, and it appears that no individual or organization took the lead…Lack of agreement on the interpretation of scientific data, pressure by special interest groups, organizational inertia…resulted in a delay of more than one year in implementing strategies to screen donors for risk factors associated with AIDS…[A] failure of leadership…during the period from 1982 to 1984… led to incomplete donor screening policies, weak regulatory actions, and insufficient communication to patients about the risks of AIDS.

The Duty to Warn is discussed in Chap. 1 and elsewhere.

The Research Ethics Theater

And the Band Played On devotes considerable time to the Montagnier-Gallo controversy, which is fully discussed earlier. Research misconduct and violations of professional ethics in the research setting ultimately affect patient care because it leads to delays in necessary translational research. Gallo had been sent samples of LAV to verify, which he did not do. Instead, he confused and delayed progress in AIDS research by naming a different virus (HTLV III) which wound up being LAV after all. Gallo's mischaracterization of LAV (which he currently admits was unintentionally) and manipulation of his own data (which he admits occurred unintentionally) was fueled by his desire to usurp credit for the discovery of HTLV-III, and be the first to patent the HTLV-III (later HIV) test. But in the process, may have delayed the development of an LAV test by at least a year, which could have screened blood donors earlier, and may have also sped up development of a treatment. It is not possible to quantify what Gallo's actions ultimately cost in lives from the standpoint of avoidable infections or earlier treatment. Professionalism and collegiality are discussed later in this book (Chaps. 8 and 9).

Most of all, as soon as AIDS was labelled a "gay disease", it became less desirable as a research priority due to stigma and discrimination. The example of the Michael Gottlieb story (see under History of Medicine) is testament to how major academic research centers viewed AIDS research—they did not want to fund it or champion it. Nor did funding agencies.

And the Harms Go On: Post-Band Ethical Issues in HIV/AIDS

The film, *And the Band Played On* ends around 1984, and does not cover the eras of HIV testing or AZT. Testing raised many issues surrounding privacy and confidentiality as a result of the social stigma related to a positive test, which eventually led to the passage of the HIPAA privacy rules that took effect in 2003. Throughout the mid-1980s and 1990s, many ethical debates raged over anonymous versus mandatory testing because of tensions between protecting autonomy and confidentiality and the duty to warn.

Meanwhile, access to treatment raised many issues surrounding health disparities, insurance coverage for AIDS patients, end of life care and palliative care, but these were much broader systemic issues that had to do with the discriminatory practices upon which the entire U.S. healthcare system was based (see Chap. 4). Conservative opposition to sex education and contraception also led to a ban on federal funding for *safe sex* education, which resulted in more HIV transmissions in teens and young adults. Similarly, although needle exchange programs flourished throughout Europe, there was resistance from all levels of government to initiate such programs in the United States. By the 1990s, AIDS "treatment activism" begins to emerge, as AIDS patients fought for access to experimental therapies and clinical trials as a means to accessing treatment.

Drug pricing for AZT (see earlier) also led to huge debates; AZT cost about $8000 annually in 1987, comparable to about $20,000 annually today. The cost of AZT treatment, in fact, drove researchers dealing with the global AIDS epidemic in poor countries to seek a cheaper maternal AIDS protocol, which resulted in the resignation of the editor of the *New England Journal of Medicine* at the time because AZT protocols designed to "undertreat" to save money was compared to a repeat of the Tuskegee study (see Chap. 1). Today, similar questions are being raised about the treatment costs of PrEP (see earlier), which is not reaching the populations the regimen was designed to protect.

From the time *And the Band Played On* aired in 1993 until 2008, healthcare reform in the United States had stalled. By the time healthcare reform became an election issue again in 2008 (thanks to the 2007 film, *Sicko*, discussed in Chap. 4), roughly one million people were living with HIV in the United States, and 21% were unaware of their serostatus. In 2010, the CDC estimated that 56,300 Americans were becoming infected with HIV each year, with approximately 18,000 deaths from AIDS annually—equal to six times the casualties on 9/11. Between 1981 and 2009, when the *Affordable Care Act* was passed, 60 million people globally had contracted HIV; 33 million were living with AIDS (2 million under age 15) and 25 million had died of AIDS, equal to the entire respective populations of several countries.

Lingering issues surrounding stigma and the duty to warn remained for years, but anonymous testing was abandoned in 37 U.S. states by the early 2000s when it was clear that the duty to warn and partner notification was a higher priority. Mandatory HIV screening in newborns mitigated the spread of pediatric AIDS while "opt out" testing policies in prenatal populations were successful in reducing maternal AIDS transmission. Finally, the AIDS crisis led to changes in access to experimental treatments and clinical trials that ultimately led to a sea change in expedited clinical research overall.

Conclusions

When planning a healthcare ethics curriculum that covers the history of harms to vulnerable populations, teaching the early history of HIV/AIDS is as critical now as teaching about Nazi medical experiments or the Tuskegee study (see Chap. 1). Within the HIV/AIDS story lies a significant chapter in the "history of healthcare ethics". When the first bioethics courses began to be taught in the 1970s, we were about 30 years away from the liberation of the Nazi death camps and the Nuremberg trials. When the Tuskegee study began to be taught to bioethics students in the mid-1980s, we were only about a decade out from the exposure of that notorious trial. We are now further away from the early days of AIDS than we were from either the Nuremberg trials or Tuskegee study when they became a fixed menu item on the healthcare ethics curriculum. When the story of HIV/AIDS is taught to healthcare trainees today, it leads to many epiphanies because this history is just not taught—even to medical students studying HIV/AIDS as part of their infectious disease units. Until recently, experts in HIV/AIDS in charge of medical school curriculums had very clear memories of the 1980s and 1990s, but many did not want to "relive" it, sticking to solely the treatment advances narrative. It's likely that healthcare professionals on the front lines in those years suffered moral injury and are unable to revisit the ethical violations they witnessed during this period. However, the discriminatory attitudes toward the LBGT community have not disappeared, and are still playing out. Fortunately, HBO's decision to keep the film, *And the Band Played On* faithful to Shilts' book, and not dilute it with needless subplots and melodrama, makes it an enduring docudrama that has immense pedagogic utility for medical educators and should become a fixed menu item, too.

Theatrical Poster

And the Band Played On (1993)

> **Based on**: And the Band Played On: People, Politics and the AIDS Epidemic by Randy Shilts
> **Screenplay by**: Arnold Schulman
> **Directed by**: Roger Spottiswoode
> **Starring**: Matthew Modine, Alan Alda, Ian McKellen, Lilly Tomlin, Richard Gere
> **Theme music composer**: Carter Burwell
> **Country of Origin**: United States
> **Producers**: Sarah Pillsbury, Midge Sanford
> **Running time**: 141 min
> **Distributor**: HBO
> **Original Network**: HBO
> **Original Release**: September 11, 1993

References

Associated Press. (1994). Three doctors reprimanded for falsifying actress' patient records, August 11, 1994. https://www.apnews.com/2ba2cc562cc3b6a6e71060d4b1f7571b.

Alexander, T. S. (2016). Human immunodeficiency virus diagnostic testing: 30 years of evolution. *Clinical and Vaccine Immunology.* https://doi.org/10.1128/cvi.00053-16, https://cvi.asm.org/content/23/4/249.

Altman, L. K. (1982). New homosexual disorder worries health officials. *New York Times,* May 11, 1982. https://www.nytimes.com/1982/05/11/science/new-homosexual-disorder-worries-health-officials.html?pagewanted=all.

Ashe, A. (1992). Secondary assault of AIDS spells the public end to a private agenda. *Washington Post,* April 12, 1992. https://www.washingtonpost.com/archive/sports/1992/04/12/secondary-assault-of-aids-spells-the-public-end-to-a-private-agenda/9ce701de-b39c-4465-827d-9e8c3d392b7c/?utm_term=.1ec620148ea9.

Ashe, A. (1993). *A hard road to glory: A history of the African American athlete (Volumes 1–3).* New York: Amistad.

Ashe, A. (1979). Ashe heart attack: Why me? *Washington Post,* August 29, 1979. https://www.washingtonpost.com/archive/politics/1979/08/29/ashe-heart-attack-why-me/b7be2d21-98b7-4e06-8f6a-f743b46e3a11/?utm_term=.0fa8f1e611c6.

Auerbach, D. M., Darrow, W. W., Jaffe, H. W., & Curran, J. W. (1984). Cluster of cases of the acquired immune deficiency syndrome: Patients linked by sexual contact. *The American Journal of Medicine, 76,* 487–492.

Avert (2018). *History of AIDS.* https://www.avert.org/professionals/history-hiv-aids/overview#footnote7_t5oqmaq.

Banks, T. L., & McFadden, R. R. (1987). Rush to judgement: HIV test reliability and screeing. *Tulsa Law Review, 23,* 1–26. https://digitalcommons.law.utulsa.edu/cgi/viewcontent.cgi?referer, https://www.google.com/&httpsredir=1&article=1790&context=tlr.

Barker, K. (1985). Arthur Ashe jailed in Apartheid protest. *Washington Post,* January 12, 1985. https://www.washingtonpost.com/archive/local/1985/01/12/arthur-ashe-jailed-in-apartheid-protest/8a138f41-107c-475d-a8c8-9bd669fa3655/?utm_term=.2137e6a41086.

Barre-Sinoussi, F., Chermann, J. C., Rey, F., Nugeyre, M. T., Chamaret, S., Gruest, J. et al. (1983, May 20). Isolation of a T-lymphotropic retrovirus from a patient at risk for acquired immune deficiency syndrome (AIDS). *Science, 220*(4599), 868–871.

Bloom, A. (2018). *Divide and conquer: The story of Roger Ailes.* A&E Indie Films.

Bull, C. (1998, July 7). *Goldwater's legacy.* The Advocate.

Casey, C. (1993). 'Conduct Unbecoming': In defense of gays on the front lines. *The Los Angeles times,* March 29, 1993. http://articles.latimes.com/1993-03-29/news/vw-16447_1_conduct-unbecoming.

CBS. (1994). Randy Shilts interview. *60 Minutes.* Season 26 Episode 23. Airdate: February 20, 1994.

CDC. (1981a, June 2). Pneumocystis pneumonia—Los Angeles. *MMWR, 30* (21), 250–252. https://www.ncbi.nlm.nih.gov/pubmed/6265753.

CDC. (1981b, July 3). Kaposi's sarcoma and pneumocystis pneumonia among homosexual men—New York City and California. *MMWR, 3.* https://history.nih.gov/nihinownwords/assets/media/pdf/publications/MMWRJuly31981.pdf.

CDC. (1982, July 16). Epidemiologic notes and reports pneumocystis carinii pneumonia among persons with hemophilia A. *MMRW, 31*(27), 365–367. https://www.cdc.gov/mmwr/preview/mmwrhtml/00001126.htm.

CDC. (1982, June 18). A cluster of Kaposi's sarcoma and pneumocystis carinii pneumonia among homosexual male residents of Los Angeles and range counties, California. *MMWR, 31*(23), 305–307. https://www.cdc.gov/mmwr/preview/mmwrhtml/00001114.htm.

CDC. (1996, August 30). Pneumocystis pneumonia—Los Angeles. *MMWR, 45*(34), 729–733. https://www.cdc.gov/mmwr/preview/mmwrhtml/00043494.htm.

CDC. (2013, Nov). HIV infection—United States, 2008 and 2010. *MMWR, 62*(03), 112–119. https:// www.cdc.gov/mmwr/preview/mmwrhtml/su6203a19.htm.

Collins, N. (1992). Liz's AIDS Odyssey. *Vanity Fair*, November 1992. https://www.vanityfair.com/ news/1992/11/elizabeth-taylor-activism-aids.

Crewdson, J. (1989). Science under the microscope. *Chicago Tribune*, November 19, 1989. https:// www.chicagotribune.com/news/ct-xpm-1989-11-19-8903130823-story.html.

Crewdson, J. (1992). Inquiry concludes data in AIDS article falsified. *Chicago Tribune*, February 9, 1992. https://www.chicagotribune.com/news/ct-xpm-1992-02-09-9201120824-story.html.

Dafoe, D. (1772). *A Journal of the Plague Year*. https://www.gutenberg.org/files/376/376-h/376-h.htm.

Daly, M. (1983). AIDS anxiety. *New York Magazine*, June 20, 1983. http://nymag.com/nymag/features/47175/.

Darrow, W. W. (2017). And the band played on: Before and after. *AIDS and behavior, 21*, 2799–2806.

De La Garza, A. (2018). 'Don't Ask, Don't Tell' was a complicated turning point for gay rights, July 19, 2018. http://time.com/5339634/dont-ask-dont-tell-25-year-anniversary/.

Driscoll, K. S. (2011). Doctor inspiration for 'Normal Heart' character. *Cape Cod Times*, April 23, 2011. https://www.capecodtimes.com/article/20110423/LIFE/104230304.

Dulin, D. (2015). Aileen Getty, December 1, 2015. https://aumag.org/2015/12/01/aileen-getty/.

Ehrenstein, D. (1993). Movies: AIDS and Death. *Los Angeles Times*, March 14, 1993. http://articles.latimes.com/1993-03-14/entertainment/ca-541_1_silverlake-life-tom-joslin-mark-massi.

Ellis, V. (1990). Taylor Doctors accused. Los Angeles times, September 8, 1990. https://www.latimes.com/archives/la-xpm-1990-09-08-me-466-story.html.

Elmer-Dewitt, P. (1996). Turning the tide. *Time*, December 30, 1996. http://content.time.com/time/ magazine/0,9263,7601961230,00.html.

Faden, R., & Beauchamp, T. (1985). *A history and theory of informed consent*. New York: Oxford University Press.

Farmer, Paul. (2006). *AIDS and accusation: Haiti and the geography of blame*. Berkeley: University of California Press.

FDA. (2018). Information regarding the OraQuick in-home HIV test. February 2, 2018. http:// www.fda.gov/BiologicsBloodVaccines/BloodBloodProducts/ApprovedProducts/PremarketApprovalsPMAs/ucm311895.htm.

Finn, R. (1993). Arthur Ashe: Tennis star is dead at 49. *New York Times*, February 8, 1993. https:// www.nytimes.com/1993/02/08/obituaries/arthur-ashe-tennis-star-is-dead-at-49.html.

Fischl, M. A., Richman, D. D., Grieco, M. H., Gottlieb, M. S., Volberding, P. A., & Laskin, O. L et al. (1987). The efficacy of azidothymidine (AZT) in the treatment of patients with AIDS and AIDS-related complex. *The New England Journal of Medicine, 317*, 185–191. https://www.nejm.org/doi/full/10.1056/NEJM198707233170401.

Fischl, M. A., Parker, C. B., Pettinelli, C., Wulfsohn, M., Hirsch, M. S., & Collier, A. C. et al. (1990). A randomized controlled trial of a reduced daily dose of zidovudine in patients with acquired immunodeficiency syndrome. *The New England Journal of Medicine, 323*, 1009–1014. https:// www.nejm.org/doi/full/10.1056/NEJM199001113231501.

Freeman, M. (1992). Arthur Ashe announces he has AIDS. *Washington Post*, April 9, 1992. https://www.washingtonpost.com/archive/politics/1992/04/09/arthur-ashe-announces-he-has-aids/eeb305b9-e36e-4e7c-8fba-c7912a2d5368/?utm_term=.8649373d8ac5.

Friend, T. (2001). Still stunning the world ten years later. *ESPN*, November 7, 2001. https://www.espn.com/espn/news/story?id=1273720.

Gallo, R. C., & Montagnier, L. (2003). The discovery of HIV as the cause of AIDS. *New England Journal of Medicine, 349*, 2283–2285. https://doi.org/10.1056/nejmp038194, https://www.nejm.org/doi/full/10.1056/nejmp038194.

Gans, A. (2011). Letter from Larry Kramer distributed following 'Normal Heart' performances. *Playbill*, April 25, 2011. http://www.playbill.com/article/letter-from-larry-kramer-distributed-following-normal-heart-performances-com-178544.

Geiger, J. (1987). Plenty of blame to go around. *New York Times*, November 8, 1987. https://www.nytimes.com/1987/11/08/books/plenty-of-blame-to-go-around.html.

Gladwell, M. (1992). NIH vindicates researcher Gallo in AIDS virus dispute. *Washington Post*, April 4, 1992. https://www.washingtonpost.com/archive/politics/1992/04/26/nih-vindicates-res earcher-gallo-in-aids-virus-dispute/af132586-d9c9-40ee-82f6-4d4a50f1cb0d/?utm_term=.cb9 a9c371806.

GMHC website. (2018). *HIV/AIDS timeline*. http://www.gmhc.org/about-us/gmhchivaids-timeline.

Gordon, M. (1993). The woman who "discovered" AIDS. *McCalls*, May 1993. pp. 110–119.

Gottlieb, M. S., et al. (1981). Pneumoycystic carinii pneumonia and mucosal candidiasis in previ-ously healthy homosexual men: Evidence of new acquired cellular immunodeficiency. *NEJM*, *305*, 1425–31.

Gottlieb, M. S., Fee, E. & Brown, T. M. (2006). Michael Gottlieb and the identification of AIDS. *American Journal of Public Health*, *96*, 982–983. https://www.ncbi.nlm.nih.gov/pmc/articles/ PMC1470620/.

Grimes, W. (1994). Randy Shilts: Author dies at 42, one of first to write about AIDS. *New York Times*, February 18, 1994.https://www.nytimes.com/1994/02/18/obituaries/randy-shilts-author-dies-at-42-one-of-first-to-write-about-aids.html.

Gutierrez, O. R. (2011). Frontline physician. *Poz*, May 2, 2011. https://www.poz.com/article/Frontl ine-Physician-HIV-20329-1953.

Harvey, J., & Ogilvie, M. B. (2000) Laubenstein, Linda (1947–1992). *The biographical dictionary of women in science*. New York: Taylor and Francis.

Higgins, B. (2015). 30 Years ago, Rock Hudson revealed he had AIDS. *Hollywood Reporter*, July 30, 2015. https://www.hollywoodreporter.com/news/throwback-thursday-30-years-rock-811346.

Hillman, B. J. (2017). *A plague on all our houses*. Lebanon, New Hampshire: University Press of New England.

Hilts, P. J. (1992). Magic Johnson quits panel on AIDS. *New York Times*, September 26, 1992. https://www.nytimes.com/1992/09/26/us/magic-johnson-quits-panel-on-aids.html.

Hilts, P. (1992). Federal inquiry finds misconduct by a discoverer of the AIDS virus. *New York Times*, December 31, 1992. https://www.nytimes.com/1992/12/31/us/federal-inquiry-finds-mis conduct-by-a-discoverer-of-the-aids-virus.html.

Horgan, J. (2002). Autopsy of a medical breakthrough. *New York Times*, March 3, 2002. https:// www.nytimes.com/2002/03/03/books/autopsy-of-a-medical-breakthrough.html.

Ifil, G. (1993). Clinton accepts delay in lifting military gay ban. *New York Times*, January 30, 1993. https://www.nytimes.com/1993/01/30/us/the-gay-troop-issue-clinton-accepts-delay-in-lif ting-military-gay-ban.html.

Institute of Medicine (IOM). (1995). *HIV and the blood supply: An analysis of crisis decision-making*. Washington, D.C.: National Academy Press.

Jones, J. (1993). *Bad blood* (Revised Edition). New York: Free Press.

Katz, J. (1986). Informed consent: A fairy tale? Law's vision. *University of Pittsburgh Law Review*, *39* (1977).

Koenig, S., Ivers, L. C., Pace, S., Destine, R., Leandre, F., Grandpierre, R., Mukherjee, J., Farmer, P. E., & Pape, J. W. (2010, March 1). Successes and challenges of HIV treatment programs in Haiti: aftermath of the earthquake. *HIV Therapy*, *4*(2), 145–160.

Kopstein. (2013). The valiant muscians. *World Military Bands website*. https://archive.is/201301 05160913/, http://www.worldmilitarybands.com/the-valiant-musicians/.

KQED. (2009). *The history of the Castro*. https://www.kqed.org/w/hood/castro/castroHistory.html.

Kramer. (1983). 1112 and counting. *New York Native*, (59), March 14–27, 1983. https://www. indymedia.org.uk/en/2003/05/66488.html and http://bilerico.lgbtqnation.com/2011/06/larry_kra mers_historic_essay_aids_at_30.php.

Lambert, B. (1992). Linda Laubenstein. *New York Times*, August 17, 1992. https://www.nytimes. com/1992/08/17/nyregion/linda-laubenstein-45-physician-and-leader-in-detection-of-aids.html.

Larsen, A. E. (2015). An historian goes to the movies. *A. E. Larson blog*, June 1, 2015. https://ael arsen.wordpress.com/tag/linda-laubenstein/.

Marx, J. L. (1986). AIDS virus has a new name—Perhaps. *Science*, *232*, 699–700. https://science. sciencemag.org/content/232/4751/699.

McNeil, D. G. (2016). HIV arrived in the U.S. long before "Patient Zero". *New York Times*, October 26, 2016. https://www.nytimes.com/2016/10/27/health/hiv-patient-zero-genetic-analysis.html.

Michaelson, J. (1991, September 21). HBO Delays 'Band Played On' Project. *Los Angeles Times*. http://articles.latimes.com/1991-09-21/entertainment/ca-2476_1_band-played-on.

Michaelson, J. (March 21, 1993). Television: Finally, the band will play: Despite countless hurdles, journalist Randy Shilts' book about the first five years of AIDS in America will make it to TV screens. *Los Angeles Times*. http://articles.latimes.com/1993-03-21/entertainment/ca-13531_1_journalist-randy-shilts-book.

Miner, M. (2008). Tribune lays off John Crewdson, others. *Chicago Tribune*, November 12, 2008. https://www.chicagoreader.com/Bleader/archives/2008/11/12/tribune-lays-off-john-crewdson-others.

Mortimer, P. P, Parry, J. V, & Mortimer, J. Y. (1985). Which anti-HTLV III/LAV assays for screening and confirmation testing? *The Lancet, 326*, 873–876. https://www.sciencedirect.com/science/article/pii/S0140673685901369.

Moskovitz, B. L. (1988). Clinical trial of tolerance of HPA-23 in patients with acquired immune deficiency syndrome. *Antimicrobial Agents and Chemotherapy, 32*, 1300–1303. https://www.ncbi.nlm.nih.gov/pmc/articles/PMC175855/.

Navarro, A. (1994). Vindicating a lawyer with AIDS years too late. *New York Times*, January 21, 1994. https://www.nytimes.com/1994/01/21/nyregion/vindicating-lawyer-with-aids-years-too-late-bias-battle-over-dismissal-proves.html.

Nelson, H. (1987, April 30). Doctor who reported first AIDS victims resigns UCLA post. *Los Angeles Times*, April 30, 1987. https://www.latimes.com/archives/la-xpm-1987-04-30-me-2910-story.html.

New York State Dept of Health AIDS Institute. (2020). *Laubenstein Award*. https://www.hivguidelines.org/quality-of-care/laubenstein-award/.

NIH. (2019). History of AIDS. *NIH website*. https://history.nih.gov/NIHInOwnWords/index.html.

Nussbaum, Bruce. (1990). *Good intentions: How big business and the medical establishment are corrupting the fight against AIDS*. New York: Atlantic Monthly Press.

Nuwer, R. (2011). Please know: Larry Kramer and 'The Normal Heart'. *Scienceline*, April 27, 2011. https://scienceline.org/2011/04/please-know-larry-kramer-and-the-normal-heart/.

O'Conner, J. J. (1989). AIDS and hemophelia. *New York Times*, January 16, 1989. https://www.nytimes.com/1989/01/16/arts/review-television-aids-and-hemophilia.html.

Oates, M. (1985). Hollywood gala [for AIDS]. *Los Angeles Times*, September 20, 1985. http://articles.latimes.com/1985-09-20/news/mn-6330_1_aids-project.

Park, A. (2017). The story behind the first AIDS drug. *Time*, March 19, 2017. https://time.com/4705809/first-aids-drug-azt/.

Parsons, D. (1992). Sickest part of the story: Arthur Ashe loses his privacy. *Los Angeles Times*, April 12, 1992. http://articles.latimes.com/1992-04-12/local/me-432_1_arthur-ashe.

Penrose Inquiry Final Report. March 25, 2015. http://www.penroseinquiry.org.uk.

Popovic, M., Sarngadharan, M. G., Read, E., & Gallo, R. C. (1984, May 4). Detection, isolation, and continuous production of cytopathic retroviruses (HTLV-III) from patients with AIDS and pre-AIDS. *Science, 224*(4648), 497–500 https://doi.org/10.1126/science.6200935, https://science.sciencemag.org/content/224/4648/497.

Raeburn, P. (1985). French AIDS drug no better than American drug, American doctors say with PM-Rock Hudson Bjt. *Associated Press*, July 26, 1985. https://www.apnews.com/6ed12830f4c62010e1e5e113e50bcfd3.

Reimer, S. (1992). Newspaper defends its inquiry into Arthur Ashe's health. *Baltimore Sun*, April 9, 1992. https://www.baltimoresun.com/news/bs-xpm-1992-04-09-1992100129-story.html.

Reverby, S. M. (2009). *Examining Tuskegee*. University of North Carolina Press.

Rich, F. (1985). Theater: 'The Normal Heart,' by Larry Kramer. *New York Times*, April 22, 1985. https://www.nytimes.com/1985/04/22/theater/theater-the-normal-heart-by-larry-kramer.html.

Rockefeller University. (2010). *Combination antiretroviral therapy: The turning point in the AIDS pandemic*. http://centennial.rucares.org/index.php?page=Combination_Antiretroviral_Thera.

Rothman, L. (2014, December 10). Until 2014, this man was Time's only medical person of the year. *Time*, December 10, 2014. https://time.com/3627996/david-ho-person-of-the-year/.

Rostker, B., Harris S. A., Kahan, J. P., Frinking E. J., Fulcher, C. N., Hanser, L. M. et al. (1993). *Sexual orientation and U.S. military personnel policy: Options and assessment.* Santa Monica, CA: RAND Corporation. https://www.rand.org/pubs/monograph_reports/MR323.html. Also available in print form.

San Francisco AIDS Foundation. (2011). Hot topics from the experts. *San Francisco AIDS Foundation website,* May 6, 2011. http://sfaf.org/hiv-info/hot-topics/from-the-experts/the-view-from-here-michael-gottlieb-paul-volberding.html.

Scheer, R. (2006). *Playing president.* New York: Akashic Books.

Schmalz, J. (1993). Randy Shilts, Writing valiantly against time. *New York Times,* April 22, 1993. https://www.nytimes.com/1993/04/22/garden/at-home-with-randy-shilts-writing-against-time-valiantly.html?sec=health.

Schmitt, E. (1992). Marine corps chaplain says homosexuals threaten military. *New York Times,* August 26, 1992. https://www.nytimes.com/1992/08/26/us/marine-corps-chaplain-says-homosexuals-threaten-military.html.

Schmitt, E. (1993). U.S. military cites wide range of reasons for its gay ban. *New York Times,* January 27, 1993. https://www.nytimes.com/1993/01/27/us/military-cites-wide-range-of-reasons-for-its-gay-ban.html.

Shilts, R. (1987). *And the band played on: Politics, people and the AIDS epidemic.* New York: St. Martin's Press.

Shilts, (1982). *The Mayor of Castro street: The life and times of Harvey Milk.* New York: St. Martin's Press.

Shilts, (1993). *Conduct unbecoming.* New York: St. Martin's Press.

Smilack, J. D. (1984, November 16). AIDS: The epidemic of Kaposi's Sarcoma and opportunistic infections. *JAMA, 252*(19), 2771–2772. https://doi.org/10.1001/jama.1984.03350190063028.

Smith, T. (2016). HIV's patient zero was exonerated long ago. *Science Blogs,* October 27, 2016. https://scienceblogs.com/aetiology/2016/10/27/patient-zero-was-exonerated-long-ago.

Sonnabend, J. (2011a). AZT: The clinical trial that led to its approval. *AIDS Perspective,* January 28, 2011. http://aidsperspective.net/blog/?p=749.

Sonnabend, J. (2011b). Remembering the original AZT trial. *Poz,* January 29, 2011. https://www.poz.com/blog/-v-behaviorurldefau.

Stanford University. *HIV drug resistance database.* http://hivdb.stanford.edu.

Sullivan, D. (1985). Stage review: Normal heart. *Los Angeles Times,* December 13, 1985. http://articles.latimes.com/1985-12-13/entertainment/ca-16850_1_aids-crisis.

Thrasher, S. W. (2018). It's a disgrace to celebrate George H. W. Bush on World AIDS Day. *The Nation,* December 1, 2018. https://www.thenation.com/article/george-hw-bush-world-aids-day-obit/.

Walton, C. R. (2017). Remembering Arthur Ashe's HIV announcement 25 years later. *The Atlanta Journal Constitution,* April 12, 2017. https://www.ajc.com/sports/remembering-arthur-ashe-hiv-announcement-years-later/x6pVAE7ESvmSTrAuiDkmEK/.

Weiss, M. (2004). Randy Shilts was gutsy, brash and unforgettable. He died 10 years ago, fighting for the rights of gay American society. *San Francisco Chronical,* February 17, 2004. https://www.sfgate.com/health/article/Randy-Shilts-was-gutsy-brash-and-unforgettable-2794975.php.

Worobey, M. et al. (2016). 1970s and 'Patient Zero' HIV1 genomes illuminate early HIV history in the United States. *Nature, 539,* pp. 98–101. https://www.nature.com/articles/nature19827.

Wright, L. (2006). *The looming tower.* New York: Vintage Books.

Yomtovian, R. (1986). HTLV-III antibody testing: The false-negative rate. *JAMA, 255*(5), 609. https://jamanetwork.com/journals/jama/article-abstract/402751.

Chapter 3
The Kevorkian Chapter: Physician-Assisted Death in *You Don't Know Jack* (2010)

If you're teaching about end of life issues in any healthcare ethics class, you need to discuss physician-assisted suicide (PAS), more recently known as physician-assisted death (PAD). That includes the story of Dr. Jack Kevorkian (1928–2011), who played a pivotal role in the American societal debate over euthanasia during the Clinton Administration, which failed to pass universal healthcare. The film, *You Don't Know Jack*, was released just as the *Affordable Care Act* (ACA) was becoming the law of the land during the Obama Administration's first term. The film particularly resonated during this time frame because opponents to the ACA falsely decried that the law created "death panels"—merely because early drafts of the legislation provided reimbursement for end of life discussions with patients.

This accurate film tracks the troubled history of Dr. Jack Kevorkian, a retired, quirky pathologist who became notorious for challenging state and federal laws surrounding physician-assisted suicide throughout the 1990s. Actor Al Pacino's embodiment of Kevorkian is so exact, it's difficult to tell the difference between Kevorkian and Pacino at times. The film intersperses Pacino into actual video footage of Kevorkian's discussions with patients seeking his help, and it is difficult to see the difference.

Between 1990 and 1998, Kevorkian apparently assisted 130 patients to die, engaging in passive euthanasia in 129 of these cases. His first patient, Janet Adkins, had Alzheimer's disease, and requested his assistance in 1990. His last patient was Thomas Youk, who had amyotrophic lateral sclerosis (ALS), and Kevorkian taped himself injecting him with lethal drugs. The tape aired on *60 Minutes* on November 22, 1998—coinciding with the 35th anniversary of the JFK assassination. In April 1999, Kevorkian was ultimately convicted of second-degree murder for actively euthanizing Thomas Youk (Johnson 1999). The judge stated to Kevorkian about his last patient: "When you purposely inject another human being with what you know is a lethal dose, that, sir, is murder." (Johnson 1999). And yet, by 2015, that action became legal in Canada through its Medical Assistance in Dying (MAID) law, which is a groundbreaking active euthanasia law (see further).

© Springer Nature Switzerland AG 2020
M. S. Rosenthal, *Healthcare Ethics on Film*,
https://doi.org/10.1007/978-3-030-48818-5_3

As the third film discussed in this section that looks at medical harms in the context of the Principle of Non-Maleficence, it is important to literally examine the Kevorkian Chapter of the 1990s. Kevorkian emerges as the wrong messenger at the right time. Kevorkian both advanced and delayed the evolution of physician-assisted death in the United States, a discussion that also led to debates over conscientious objection. The film's title also makes a sad statement: although a prominent medical figure who bookended the 1990s, by the time of the film's release a decade later, few medical students had ever heard of Kevorkian. This begs the question as to whether Kevorkian's notoriety was really that impactful from a history of medicine context in the final analysis. This chapter will explore the film, *You Don't Know Jack*, Kevorkian's life within his social and historical location, the movement of physician-assisted death from a history of medicine context; and finally, it will review the healthcare ethics themes in the debate over physician-assisted death.

From Biography to Biopic: Origins of the Film

You Don't Know Jack is essentially a filmed version of a 2006 biography of Kevorkian titled, <u>Between the Dying and the Dead</u>: <u>Dr. Jack Kevorkian's Life and the Battle to Legalize Euthanasia</u> by Neal Nicol and Harry Wylie, who knew Kevorkian well. Neal Nicol (played by John Goodman in the film) had been a friend and coworker of Kevorkian's since 1961, and participated in Kevorkian's early research studies that involved transfusing cadaver blood (see further). Nicol was a corpsman and laboratory technician who worked with Kevorkian, and had a laboratory supply company that provided Kevorkian with materials while he was active. Nicol regularly visited Kevorkian when he was in prison. Harry Wylie had a long friendship with Kevorkian, who was his next-door neighbor, and also visited him in prison frequently. The book is intended as an in-depth biography on Kevorkian but is written by clear advocates, rather than objective historians of his work. The biography was penned while Kevorkian was in prison, and Kevorkian fully endorsed the book as accurate. What the biography reveals inadvertently is that it seems clear that Jack Kevorkian, like other eccentric individuals in the STEM fields (Science, Technology, Engineering and Medicine) probably had undiagnosed/unrecognized Asperger's syndrome. To any reader familiar with the traits of Asperger's syndrome (now simply known as being on the autism spectrum), the diagnosis leaps out from the book's opening chapter (Nicol and Wylie 2006) (bold emphasis mine):

> When his fight to legalize euthanasia was making headlines in the 1990s, the public saw only the macabre 'Dr. Death' – **the often abrasive, always outspoken** proponent of the right of the terminally ill to end their suffering on their own terms. But behind that persona lies a complicated man with a compelling story. **He was a former child prodigy**, the son of Armenian refugees who came to America to escape the Turkish genocide...**His early talents ranged from woodwork to linguistics to science experiments in the basement**. Later, he became a brilliant pathologist, devoting his life to the unusual pursuit of extracting social benefit from death. ... Jack Kevorkian is a complex individual, full of fascinating contradictions...**He is also a shy, eccentric man who lived a monastic, ethical life, buying**

his clothes at the Salvation Army and subsisting on the plainest food, particularly white bread! He lacks the capacity to lie so much that when he played poker with his friends, he never bluffed and if he bet, everyone folded…Jack was spectacularly unsuccessful at cultivating a long-term relationship with a member of the opposite sex …[and when asked about his dating life, said) 'I just haven't found the right woman yet'…

Understood within the context of a man with Asperger's syndrome, everything about Kevorkian's work and personality can be easily explained, especially his singular obsession and focus on proving his point. His extremely poor decision to defend himself in his final trial was likely related to his over-appreciation of his intellectual grasp of the facts and under-appreciation of the social skills and emotional intelligence required in a proper defense before a jury. Notwithstanding, Kevorkian did not have a clear grasp of the legal issues in his last trial, either. His biographers ultimately tell the story of a man who had an unusual mix of interests in very specific topics; strange obsessions in his research endeavors; had difficulty getting along with people; bounced around in his employment; and had no social life to speak of other than a few individuals who appreciated his complexity and knew him well. His life consisted of his work and hobbies, and he was socially isolated by his own design. He was known for a strange, quirky sense of humor, which included heavy usage of puns and plays on words in labelling of his machines and practices, such as "medicide"—what he called compassionate active euthanasia; names for his death machines—the "Thanatron" (lethal injection machine) and "Mercytron" (gas mask method).

However, the authors falsely attribute living wills and advance directives to Kevorkian's crusade, when they state: "Regardless of how one feels about his politics, Dr. Kevorkian changed the way most of us think about dying. Because of him, we now have living wills and the right to refuse resuscitation." This is just not so, and an enlargement of his influence; living wills and Advance Directives are the result of the Nancy Cruzan case (see further), which was decided in 1990, before Kevorkian made major headlines.

The biography also tracks Kevorkian's professional reprimands: he lost his medical license in 1991 in Michigan and then in California in 1993. He was denounced by the American Medical Association as a "reckless instrument of death" (Nicol and Wylie 2006).

The authors discuss Kevorkian's own writings, including his early research projects (see further). But they gloss over one of his most notorious ideas: the promotion that we should offer prisoners on death row the option to *die as medical research subjects*—an idea that would likely have been in violation of the Nuremberg Code (1947) when he proposed it in 1958, and then wrote a book on it in 1960, entitled Medical Research and the Death Penalty (Kevorkian 1960). It begins with an eerie prescient passage, in which Kevorkian puts himself on trial with the reader, complaining that when he proposed this great idea to his peers, he was ridiculed.

> You are to be more than a mere reader of this book. You are to be judge and jury…no matter what the motivation, it is hardly conceivable that you will be able to read them without arriving at a tentative or final decision regarding the central theme – as every good judge and jury should.

…A little over a year ago I presented a formal scientific paper before the American Society of Criminology at the annual meeting of the American Association for the Advancement of Science in Washington, D.C., in which I proposed that condemned men be allowed the choice of submitting to anesthesia and medical experimentation, as a form of execution in lieu of conventional methods now prescribed by law…but to my dismay, discussion has been difficult, in fact, impossible, to stimulate…opposition has been open and vociferous… And with this one final request – that you judge sincerely within the limits of your innate wisdom – we enter your chambers and take our stands…

Kevorkian was quite obsessed with the death penalty and methods of execution; he was an early proponent of lethal injection for capital punishment, which the authors of his biography note was a standard execution method by 1982. Undoubtedly Kevorkian would have been influenced by the well-publicized executions of Julius and Ethel Rosenberg by electric chair, which occurred June 19, 1953. At that time, Kevorkian was already a physician, having completed medical school in 1952 at the University of Michigan, and just starting his internship in 1953, which he interrupted to serve in the Korean war for 15 months as an Army medical officer. The Rosenbergs' trial and convictions were a huge story that dominated the headlines, and their executions were described in media reports in elaborate detail, including the fact that Ethel was required to undergo two cycles of electrocution because she survived the first attempt. The legal and ethical issues regarding the Rosenbergs' executions were a subject of speculation for years. Roy Cohen, a prominent lawyer in the 1950s, was a lead prosecutor in the case, and admitted in his autobiography (Cohn 1988) that he personally convinced the judge to sentence Ethel to her death. Cohen was featured as a major character dying of AIDS and haunted by the ghost of Ethel Rosenberg in the play "Angels in America" and was played by Al Pacino in the HBO 2003 miniseries version of the play. It is all the more ironic that Pacino would go on to play Kevorkian seven years later, given Kevorkian's writings and obsessions about the death penalty.

Nicol and Wylie also discuss Kevorkian's 1991 book, Prescription: Medicide: The Goodness of Planned Death (Kevorkian 1991), which served to be his operations manual during his time in the limelight, explaining his worldview and argument for physician-assisted death.

Nicol and Wylie discuss Kevorkian's early observations of dying patients during his medical training, in which he would try to photograph the first moments of death (this is what led to his nickname "Dr. Death"); they quoted the following from Prescription: Medicide : "…I was sure that doctor-assisted euthanasia and suicide are and always were ethical, no matter what anyone says or thinks."

But Kevorkian's 1991 book also circles back to his obsession with research on death row prisoners. In his 1991 iteration of the 1960 argument to give the option of "death by medical experimentation" to death row prisoners, he also proposed that the conditions for "opting in" for this terminal medical research would require organ donation after death. A 1991 review of the book (Kirkus Reviews 1991) stated the following, also inadvertently noting potentially Asperger's traits (bold emphasis mine) with respect to problematic social skills, and obsessiveness over a topic.

Its title notwithstanding, this is not primarily a discussion of euthanasia–or "medicide," the author's term for euthanasia performed by professional medical personnel–but, rather, largely a defense of his position that death-row inmates should be given the option of execution by general anaesthesia, thus permitting use of their bodies for experimentation and harvesting of their organs. Since his days as a medical resident, Kevorkian has attempted to convince legislators, prison officials, and physicians of the value of this approach. **However, the art of persuasion is not Kevorkian's forte; indeed, he seems unable to resist attacking and insulting those who disagree with him, referring to his medical colleagues as "hypocritical oafs" with a "slipshod, knee-jerk" approach to ethics. Those seeking a thoughtful discussion of euthanasia will not find it here, but Kevorkian does offer a revealing look at gruesome methods of execution.** …Kevorkian concludes with a recounting of his development of the "Mercitron" (as he has named his suicide machine), his reasons for creating it, and his difficulties in promoting its use. A model bioethical code for medical exploitation of humans facing imminent and unavoidable death is included in the appendix. **An angry doctor's rambling and repetitious harangue, certain to arouse the ire of the medical establishment**.

Ultimately, one of the purposes of the Nicol and Wylie biography was to offer the public a more empathetic view of Kevorkian, his arguments and worldview as his advocates. Their book is not, however, a critical presentation of his work; that would be offered more in the screenplay based on the book (see further).

From Video Tape to Book to Film

When it came to Kevorkian, most Americans became acquainted with him on television; he made more than one appearance on *60 Minutes*, which was the quintessential television news magazine in the 1990s. The CBS iconic journalism institution had particular skill for protecting whistleblowers, having been responsible for taking down Big Tobacco, for example, by interviewing tobacco company scientist, Jeffery Wigant on February 5, 1996, who told the public that tobacco companies were in the business of nicotine addiction. In fact, in the 1999 film, *The Insider*, which was based on the *60 Minutes* negotiation with Wigand, Al Pacino plays *60 Minutes* producer, Lowell Bergman. That film debuted exactly a year after Kevorkian's own fateful appearance on *60 Minutes*. According to Mike Wallace, "of all the interviews he conducted for *60 Minutes*, none had a greater impact" than Kevorkian's 1998 interview in which he showed the videotape of his euthanizing Thomas Youk (CBS 1998; Nicol and Wylie 2006). Prior to that, Kevorkian had given hundreds of media interviews throughout the 1990s, a time frame in which the country was just beginning to examine the power of citizen journalism, in which citizens videotaped events that would then become big news in a pre-social media/YouTube culture. By the time Kevorkian's 1998 *60 Minutes* interview aired, he was a well-known U.S. figure and had become part of the American zeitgeist. He had carved out a significant media space for himself in the 1990s amidst other dominant social justice/legal stories such as the Rodney King beatings (1991), subsequent L. A. Riots (1992) and the O. J. Simpson murder trial (1994–5). But by November 1998, when Kevorkian made his last appearance on *60*

Minutes, he was competing with a far more compelling story for the country: the Clinton-Lewinsky affair, the Kenneth Starr Report (September 11, 1998), and the looming impeachment of President Bill Clinton that dominated the news throughout much of that fall and spring of 1999, when Kevorkian was finally sentenced.

By the fall of 1999, nobody was talking about Kevorkian anymore, and his star had faded. The country's attention span had moved into the coming Millennium, concerns over "Y2K" and then the start of a presidential election year, which became the most embattled electoral college tie in U.S. history, leading to *Bush v. Gore*. In science, the story of the year was the completed first draft of the Human Genome Project.

By January 2001, as George W. Bush was being sworn in as the 43rd President of the United States, Kevorkian had completely faded from the public's attention as he sat in prison. He remained in prison on September 11, 2001, which was also the seventh anniversary of his sister, Margo's death (see further). He stayed in prison throughout the early post-9/11 years, as the country transformed into a completely different place, fighting two wars—one in Afghanistan and an ill-conceived war in Iraq. The Nicol and Wylie biography was published in 2006 in the middle of the second term of Bush 43, after the 2005 Terri Schiavo case. The Schiavo case had dominated headlines for weeks, and embroiled the President's brother, Jeb Bush, then-governor of Florida in a case that revisited the Cruzan decision (see further). Terri Schiavo was not a case about physician-assisted death, but about a surrogate's decision to withdraw/withhold nutrition and hydration from a patient in a persistent vegetative state. Nonetheless, it put the topic of end of life into daily headlines. The Kevorkian biography essentially re-introduced readers to a figure who now seemed out of place and out of his time. But a year after its publication in 2007, the Kevorkian story would start to find another audience as another feature film would put the topic of healthcare back into news and make healthcare reform a central issue in the 2008 election. Michael Moore's *Sicko* (see Chap. 4), which was about the insanity of the pre-ACA U.S. healthcare system, re-invigorated debates about healthcare access for people with chronic illnesses, terminal illnesses, and healthcare spending at the end of life. Meanwhile, several states had begun to legalize physician-assisted death following Oregon (see further). Coinciding with *Sicko*'s debut, the documentary *The Suicide Tourist* (2007) about an American who goes to Switzerland to end his life, puts physician-assisted death back in the public square.

Kevorkian's Re-entry

On June 1, 2007, 79-year-old Kevorkian was released from prison and put on probation. His release was covered by major news organizations, and he did a few prominent interviews, but he struggled to find the same relevance. His probation was conditional upon his promise not to participate in any more assisted deaths, but "[that] promise does not preclude Mr. Kevorkian from speaking in support of physician-assisted suicide" (Peligian and Davey 2007). Kevorkian's re-entry timing was put this way (Peligian and Davey 2007):

> As he emerged today, assisted suicide and death choices — spurred in part by the case of Terri Schiavo in Florida — were still being fought over; in California, legislators are expected to vote on a "Compassionate Choices Act" next week. Oregon remains the only state in the country with a law that allows a terminally ill patient to ask a doctor to prescribe a lethal amount of medication under certain circumstances. Other states, including Vermont, have rejected such proposals.

Kevorkian was allowed to leave prison early as part of a compassionate release due to failing health, but he lived long enough to endorse his friends' biography of him and see his own biopic before he died in 2011 of natural causes. Kevorkian was eager for a "reset" and actually launched a bid for Congress in 2008 (see further). He vouched for the accuracy of both the biography and biopic.

The Film Project: 2007–2010

Steve Lee Jones, one of the executive producers of *You Don't Know Jack*, had been trying to get the film on Kevorkian made since he read the Nicol and Wylie biography when it was first published. Jones obtained the film rights to the Nicol and Wylie book and then began shopping the project around as a feature film. HBO expressed interest in 2007, when Al Pacino and director, Barry Levinson said they were interested in coming on-board. Barry Levinson burst onto the Hollywood directorial scene in the 1980s with films such as *Diner* (1982); *Good Morning Vietnam* (1987), starring Robin Williams; and *Rain Man* (1988), about an autistic adult played by Dustin Hoffman, which arguably rivaled Hoffman's 1969 performance in *Midnight Cowboy*, which also starred Brenda Vaccaro, (see further). Levinson began as a screenwriter and co-wrote one of the most notable Pacino films, *And Justice for All* (1979) with its iconic "You're out of order!" scene.

Pacino's career is vast; He studied at the Actors Studio with its founder, Lee Strasberg, and by 2010 was President of the Actors Studio himself. Pacino had a long list of enduring characters he brought to life in the decades that preceded his portrayal of Kevorkian, including Michael Corleone of the *Godfather* films and Frank Serpico. Pacino had won his first Oscar for the role of Colonel Frank Slade in *Scent of a Woman* (1992), which was about a suicidal veteran who had retired from the military in disgrace. His most recent HBO project prior to *You Don't Know Jack* was as a terminally ill and dying Roy Cohen in *Angels in America* (2003). The topics of suicide and terminal illness were not foreign to Pacino's acting palette.

A young screenwriter, Adam Mazer, wrote the screenplay for *You Don't Know Jack,* while Levinson, Jones, Lydia Dean Pilcher and Glenn Rigberg were the executive producers. Jones had concomitantly been involved in a documentary on Kevorkian's life post-prison, and his run for a Congressional House of Representatives seat in 2008 (Kim 2008). The Jones documentary, *Kevorkian* (2010), explains his intense interest in the biopic project. With respect to *You Don't Know Jack*, Len Amato, president of HBO Films noted: "We were not just taking on Jack's story or the surrounding issues, but the thematic idea of mortality. It's a challenge to try to make a film that deals with those aspects and still draws an audience in" (Associated Press 2010). Mazer noted: "I don't think it's about sympathizing or empathizing with Jack. I think it's about understanding him, the choices he made, who he was. I think we show a very honest portrayal of the man, his foibles, his strengths, his weaknesses and his flaws" (Canyon News 2010).

Unlike typical biopic projects, this film was being made while Kevorkian was still alive, which the filmmakers saw as an opportunity to present a balanced perspective. But Pacino decided not to meet with Kevorkian prior to filming. The following was noted about his performance (Stanley 2010):

> A credible biography of Dr. Kevorkian has to focus on the self-serving zealotry beneath the martyr's guise, but Mr. Pacino has a subversive gift for tapping into the endearing underside of the most despicable villains…So it is a credit to Mr. Pacino that while he burrows deep into the role, he never lets Dr. Kevorkian's crackpot charm overtake the character's egomaniacal drive.

The film also wove Pacino into existing video records of Kevorkian's patients, with permission obtained from their family members. The video footage was the legal property of Kevorkian, who granted consent to its use as well.

However, two other major characters were no longer alive: Janet Good (played by Susan Sarandon) and Kevorkian's sister, Margo Kevorkian Janus (played by Brenda Vaccarro). Janet Good had founded a Michigan chapter of the Hemlock Society and later worked with Kevorkian to find him patients, and assisted him at times. She herself, had been diagnosed with pancreatic cancer, and asked Kevorkian to assist with her death, which he did, and which is covered in the film (Stout 1997). Margo, however, died in 1994 of natural causes from a heart attack on a future day of infamy—September 11th. According to her obituary in the *New York Times* (1994):

> [S]he had kept all the patient records involving the assisted suicides, and videotaped sessions between her brother and the 20 patients he helped commit suicide since 1990. She was present at the first 15 of the suicides, and later helped organize meetings of the survivors of Dr. Kevorkian's patients. She also worked in Dr. Kevorkian's campaign for a statewide referendum on doctor-assisted suicide…"She was my record-keeper, my videographer and my chronicler," Dr. Kevorkian said. "She was also my supporter when I had no other supporters."

Kevorkian later named a fledgling death clinic after his sister: the Margo Janus Mercy Clinic, which did not last long. When Levinson asked Brenda Vaccarro to read for the part of Margo, she had not worked in years due to ageism in Hollywood, which did not affect Pacino (born 1940), who was only a year younger than Vaccarro (born 1939). In fact, they had both started their careers around the same time. A *Los Angeles Times* article made note of how important the film was to her at that juncture, reporting their interview with Vaccarro as follows (McNamara 2010):

> "Oh, I wanted to work, Toots," she says, over lunch on Ventura Boulevard. "But I couldn't get roles. Hell, I couldn't get representation. These agents, they would say, 'Let's face it, it's the age' or 'You don't make enough money,' or they already had old cows on their plate and couldn't take another one because there are too many old cows out there already….Barry Levinson, he saved me," she says, turning serious. "He gave me back my vocation. It had been taken from me."

As for landing the role of Margo (McNamara 2010):

> Out of the blue, she got a call from casting director Ellen Chenoweth asking if she'd like to
> read for the part of Margo. Vaccaro thought it was a joke, but off she went to HBO and made
> a tape. Having watched her own beloved mother die all too slowly, she knew precisely what
> Margo and Jack were fighting for. And what she herself was fighting for as well.
>
> "I was so excited just to get up and act again," she says. "I had such a good time, but when I got
> back in the car I said to [my husband], 'Honey, forget about it. It's Al Pacino.'"...Levinson
> liked the tape, Pacino liked Vaccaro and, suddenly, she was back in business, working nonstop
> and having the time of her life.

Susan Sarandon, whose larger than life career was noted for films such as *Bull Durham* (1988); *Thelma and Louise* (1991); and *Lorenzo's Oil* (1992—see Clinical Ethics on Film). Her "Janet" characters spanned from *Rocky Horror Picture Show* (1975) to an easy morphing into Janet Good in *You Don't Know Jack*. Ultimately, as the *New York Times* stated: "it is a credit to the filmmakers that a movie dedicated to a fearless, stubborn man's campaign against the medical establishment and the criminal justice system doesn't overly romanticize his struggle or exonerate him from blame" (Stanley 2010). Kevorkian noted to the *New York Daily News* (2010) that he thought the film was "superb", a testament to the accuracy of the supporting characters in the film, which also include Nicol himself, played by John Goodman, as well as Kevorkian's lawyer, Jeff Feiger. When the film premiered on April 14 at the Ziegfeld Theater in New York City, *The Affordable Care Act* had just been signed into law (March 23, 2010) by President Obama, but only passed with the removal of the provision that doctors could get reimbursed for "end of life" conversations with patients. That provision led to a disinformation campaign by the Republican Party that the law would lead to "death panels" (see Chap. 4 for a much more detailed discussion of the ACA). Kevorkian, who would die the following Spring (on June 3, 2011) walked the red carpet alongside Pacino. Kevorkian stated that both the film and Pacino's performance "brings tears to my eyes—and I lived through it".

Synopsis

This 2010 biographical portrait of Jack Kevorkian's work in the 1990s should be treated as a History of Medicine film with respect to the study of physician-assisted death. The *New York Times* summarized *You Don't Know Jack* this way:

> The film, directed by Barry Levinson, looks at times like a documentary, inserting Mr.
> Pacino into real archival footage, including the infamous 1998 Mike Wallace segment on
> "60 Minutes" that showed Dr. Kevorkian administering a lethal injection to Thomas Youk
> and daring the authorities to stop him...Dr. Kevorkian taped his patients, creating a video
> record of their concerns and consent, and a few of those poignant, real-life interviews are also
> worked into the film. Others are skillful re-creations, including scenes with his first patient,
> Janet Adkins, who at 54 was only beginning to show symptoms of Alzheimer's disease when
> she asked Dr. Kevorkian to help her die....The film captures his zeal, his self-righteousness

and also the creepy tawdriness of his right-to-die practice: the macabre, ghastly art works he painted himself, sometimes with his own blood; his shabby apartment; his rickety DIY death contraptions; and the battered van he used as a death chamber. Even [his sister and assistant] Margo is shocked by how makeshift and crude the process is, exclaiming after the first assisted suicide in his van, "I guess somehow I just thought the whole thing would be nicer."

Ultimately, *You Don't Know Jack* teaches students about Kevorkian's role in the physician-assisted suicide/death debate, which may have even set back the debate by years due to his methods. What is important to point out to students, however, is that the film is not only accurate, but was judged to be accurate by Kevorkian himself. Considering that the film does not present him as a moral hero, and that Pacino never met him prior to the film, Kevorkian's thoughts on the film help with its validity because he is indeed presented as a controversial figure with controversial methods. Most students will probably not be familiar with Kevorkian prior to the film, and you may want to pair it with the documentaries, *How to Die in Oregon* (2011) and the aforementioned *Suicide Tourist* (2007).

Death Becomes Him: The Social Production of "Dr. Death"

Murad (Jack) Kevorkian was born May 26, 1928, and was of the same generation as many famous figures from the Beat Generation, which was an "earlier adopter" of experimentation and off-the-beaten track ideas within their fields, including choreographer Bob Fosse (see Clinical Ethics on Film), Diane Arbus (1923–71), Lenny Bruce (1922–66), Truman Capote (1924–84), and Andy Warhol (1928–87). Kevorkian was the child of Holocaust survivors—not the Nazi genocide, but the Armenian genocide of 1915, which involved the systematic extermination of 1.5 million Armenians living in what is currently Istanbul (then-Constantinople). Kevorkian grew up within the Armenian diaspora community, which fled persecution and sought asylum in other countries. His father got out of Turkey in 1912 with the help of missionaries, and settled in Pontiac, Michigan and worked in the auto industry. His mother was able to flee and lived in Paris with relatives and then reunited with her brother in Pontiac, which had an Armenian community (Nicol and Wylie 2006). It was there that Kevorkian's parents met and married, and had Margo in 1926 (played in the film by Brenda Vaccaro), followed by Jack in 1928, and then Jack's younger sister, Flora (Flora Holzheimer who settled in Germany).

In one interview about their family history, Kevorkian's sister, Margo Janus stated: "Only our parents, barely teenagers were able to escape [the genocide]. Growing up we heard stories of bloodshed and it frightened Jack" (Aghajanian 2010).

Kevorkian was a gifted student who skipped a few grades forward, and during the war, taught himself German and Japanese. Consistent with Asperger's syndrome, he had problems making friends, was labelled a "bookworm" and he didn't date. He graduated high school in 1945 when he was 17, a time frame when news of Nazi concentration camps had started to become public.

Extrapolating from research done on children of Nazi Holocaust survivors, Jack Kevorkian, as a second-generation survivor of an earlier genocide, was likely trying "to make sense of [his] background" in a time frame when there was no research into "the growing field of epigenetics and the intergenerational effects of trauma…which had previously established that survivors of the Holocaust have altered levels of circulating stress hormones [and] lower levels of cortisol, a hormone that helps the body return to normal after trauma…" (Rodriguez 2015).

Given Kevorkian's likely unrecognized spectrum disorder of Asperger's syndrome, his obsession with death in his research, and later clinical practice, may have been linked to his parents' tales of escaping genocide, which he had difficulty processing emotionally. In fact, he consistently claimed that the Armenian genocide was far more gruesome than the Final Solution of the Nazi genocide, which would upset his colleagues (Hosenball 1993):

> Kevorkian argued in a 1986 article that Nazi experiments, such as dropping prisoners into icy water to see how quickly they froze to death, were "not absolutely negative. Those who can subordinate feelings of outrage and revulsion to more objective scrutiny must admit that a tiny bit of practical value for mankind did result (which, of course, did not and never could exonerate the criminal doctors)." In a 1991 interview with the Detroit Free Press Magazine, Kevorkian went so far as to minimize the Holocaust, insisting that his Armenian ancestors had it even worse. "The Jews were gassed. Armenians were killed in every conceivable way…So the Holocaust doesn't interest me, see? They've had a lot of publicity, but they didn't suffer as much."….
>
> [Kevorkian] maintained that the Nazi experiments had some validity [but when provided with] scholarly works attacking the experiments, Kevorkian toned down references to them in his 1991 book.

Kevorkian graduated from the University of Michigan's medical school in 1952 and then began a residency in pathology. Kevorkian led a number of failed crusades around death research. When he was a pathology resident at the University of Michigan hospital, he would observe dying patients and photograph their eyes as they died to see what changes occurred, based on a strange nineteenth century French study he read (Nicol and Wylie 2006). He did this to apparently pinpoint the precise moment death occurred, which is why he was nicknamed "Dr. Death". This resulted in a 1956 journal article, "The Fundus Oculi and the Determination of Death" (Kevorkian 1956). Although Kevorkian suggested this information could help to distinguish death from fainting, shock or coma, and also when to do cardiopulmonary resuscitation (CPR), he said later that he did the research because "it was interesting"(Biography.com 2019), a "taboo subject" (Biography.com 2019), and that it "had no practical application. I was curious, that's all" (Hopkins 2011).

In reading more about his family's background, Kevorkian discovered that Armenians had once performed experiments on men who were condemned to death— likely intellectuals and professionals who had been imprisoned in the first wave of the Armenian genocide. He also would have read this in the wake of concentration camp experiments and the 1953 Rosenberg executions (see earlier). Oblivious to the optics and timing, however, Kevorkian would visit prisons and began to promote research on death-row prisoners as well as harvesting their organs after death. In 1958, he presented a paper (Kevorkian 1959) at an American Academy of Arts and Sciences (AAAS) meeting in Washington, D.C., in which he advocated the Armenian-inspired idea of first killing death-row inmates through medical experimentation, and then using their organs for transplantation. But of course, in a freshly post-Holocaust environment in which the details of Nazi medical experiments were still being absorbed as war crimes, the paper was a disaster. In fact, the paper was so controversial and dismally received, the University of Michigan told Kevorkian to cease the research or to leave the institution in the middle of his residency (Nicol and Wylie 2006), which today would probably be in clear violation of his academic freedom. Kevorkian left. He next went to Pontiac Hospital to practice. This episode is what led to his 1960 book (see earlier) and his quest for a "retrial" (Kevorkian 1960). Here, he advocates that death row prisoners could provide a service by donating their bodies for science in "terminal human experimentation" that would be "painless". He also published a 1960 pamphlet entitled "Medical Research and the Death Penalty" which stated that "experiments on humans could provide far greater benefits than those on animals, and that killing condemned convicts without experimenting on them was a waste of healthy bodies… [and] if the experiment was not excessively mutilating…the subject could conceivably be revived if evidence of innocence was uncovered" (Hosenball 1993).

While at Pontiac Hospital, he next became obsessed with transfusing cadaver blood into live people, an ill-conceived practice that had a long history and was actually detailed in a 1962 *Archives of Surgery* article (Moore et al. 1962). Kevorkian had come across much of this history on his own before the *Archives of Surgery* paper was out, and wanted to pursue it. It was at this juncture that he met Neal Nicol, who was a medical technologist at the same hospital who assisted him. Kevorkian was under the impression that the military would be interested in his cadaver research, but his pursuit of this wound up alienating his colleagues who thought the idea was insane (Hopkins 2011). Kevorkian actually performed cadaver transfusions on a number of volunteers, including himself, which resulted in his acquiring Hepatitis C, and eventually led to his own death from liver cancer. He published his cadaver transfusion research in *The American Journal of Clinical Pathology* (Kevorkian and Bylsma 1961), a year before the 1962 *Archives of Surgery* paper (Moore et al. 1962), and it is unknown whether the authors were aware of it. By 1963, the field of pathology and cadavers would suddenly become of intense interest to the public in the wake of the 1963 Kennedy assassination and The Warren Commission Report (1964), which would lead to decades of speculation about the handling of a President's cadaver during autopsy (Altman 1992). Kevorkian's cadaver transfusion idea died when he couldn't get funding for it, while further alienating and alarming his colleagues.

Kevorkian eventually left Pontiac Hospital for Saratoga General Hospital in Detroit, and by 1970 became the chief pathologist at the institution (PBS 2014). By then, there was more interest in death as new criteria for brain death had been published (JAMA 1968), as well as a groundbreaking book by a contemporary of Kevorkian's titled On Death and Dying (Kubler-Ross 1969), which I discuss further on. By the late 1960s, there was apparently a "Summer of Love" for Kevorkian. Although he was unsuccessful in dating women (Nicol and Wylie 2006), he reportedly had a fiancée in the late 1960s but broke off the relationship in 1970 (Hopkins 2011).

In the late 1970s, in one of his most bizarre moves, he retired early and travelled to California to invest all his money in directing and producing a feature film based on the composer, George Frideric Handel, and Handel's symphony "Messiah", a project that completely failed[1] since Kevorkian had absolutely no experience in the film industry (Hopkins 2011; PBS 2014). Kevorkian lost all of his money on this failed film project, and by 1982 was unemployed and was sometimes living out of his car. He spent some time in Long Beach, California around this period working as a part-time pathologist for two hospitals, but wound up quitting when he got into an argument with his pathology chief (Nicol and Wylie 2006). He returned to Michigan in 1985, and also returned to the topic of experimenting on death row prisoners in a series of articles in low impact medical journals (Kevorkian 1985a, b, 1987). Here's a sample of what he proposed (Kevorkian 1985a; Hosenball 1993):

> Some audacious and highly motivated criminals may agree to remain conscious at the start of certain valuable experiments…That would make possible probes like those now carried out during surgery on brains of conscious patients…" In the same article, he seemed to suggest that experiments could go beyond death-row inmates to almost anyone facing "imminent death," including: "(a) all brain-dead, comatose, mentally incompetent or otherwise completely uncommunicative individuals; (b) all neonates, infants and children less than (–) years old (age must be arbitrarily set by consensus). (c) all living intrauterine and aborted or delivered fetuses.

Thus, his nickname "Dr. Death" was well-earned prior to Kevorkian's last crusade into euthanasia, a project that would finally have some public support despite the criticism of his peers.

The Final Death Spiral

In 1986, Kevorkian learned that lethal injection was decriminalized in the Netherlands under certain conditions endorsed by the courts and the Dutch Medical Society (Nicol and Wylie 2006). He then became obsessed with promoting physician-assisted death and euthanasia as his new crusade. At this point, he began to publish numerous articles on the topic in the German journal *Medicine and Law* (Nicol and Wylie 2006). He then began to build his death machines and look for potential patients/clients through the Hemlock Society, where he met Janet Good. His "Thanatron" (Greek for "Instrument of Death"), which cost him only $45.00 in materials, consisted of three bottles that delivered successive doses of a saline solution, followed by a painkiller and, finally,

---

[1] Kevorkian embarked on this project years before a quintessential Handel biography was published (Keate's 1985, Handel: The Man and his Music) which eventually did lead to a film after Kevorkian's death.

a fatal dose of potassium chloride. As a passive euthanasia aid, patients would self-administer the fluids. Kevorkian also proposed "obitoriums,"—a type of death clinic in which patients would go to be euthanized. In his 1991 book, he writes (Kevorkian 1991):

> Euthanasia wasn't of much interest to me until my internship year, when I saw first-hand how cancer can ravage the body...The patient was a helplessly immobile woman of middle age, her entire body jaundiced to an intense yellow-brown, skin stretched paper thin over a fluid-filled abdomen swollen to four or five times normal size.

As a pathologist, Kevorkian was actually *not* experienced in treating patients at the end of life, nor did he really have any training in palliative care, end of life care, or dealing with death and dying. In fact, his interest in championing physician-assisted death was almost *more* outside the competencies of a pathologist of his limited clinical experience than the practice of transfusing cadaver blood into live people. And yet, because he had little else to do professionally, and was obsessed with death, he became the unlikely "patron saint" of physician-assisted death.

After Kevorkian helped his first patient to die with his Thanatron (see further), the Michigan court issued an injunction barring his use of it, and also suspended his medical license. He next assembled the "Mercitron", which delivered a lethal amount of carbon monoxide through a gas mask. But with a suspended license, his role when using the Mercitron was as *a civilian*, not a doctor (see under Healthcare Ethics Issues).

By 1992, the Michigan Legislature also passed a bill outlawing physician-assisted suicide, but at the same time, upheld the Principle of Double-Effect (see under Healthcare Ethics Issues) as Kevorkian's lawyer argued that if the physician's "intent to relieve pain and suffering," was met, it could not be considered a crime, even it if did increase the risk of death.

Kevorkian's limitations and crossing of practice boundaries were well-recognized. A 1993 *Newsweek* (Hosenball 1993) article asked:

> Is Kevorkian a prophet or a pariah? Supporters say that just because some of the ideas he advances are extreme, that doesn't mean all of his proposals are worthless or crazy. Still, his writings raise questions about whether "Dr. Death" is the best person to counsel the sick and the despondent in a decision that would be their last.

When Kevorkian first began to advance his ideas around 1989–1990, death was very much in the news: AIDS deaths were soaring (see Chap. 2); cancer diagnoses were on the rise as people were living longer, and dietary/environmental causes of cancer were increasing along with more screening; and cruel autonomy-robbing illnesses were becoming more publicized, such as Alzheimer's disease. Kevorkian did create some of his own Guidelines and Screening Questionnaires (Kevorkian 1992). As Kevorkian became more known, popular books such as Tuesdays with Morrie (Albom 1997)—about a man dying of ALS, and documentaries such as *Complaints of a Dutiful Daughter* (PBS 1996)—about the ravages of Alzheimer's disease— increased awareness of neurological diseases. Meanwhile a major Canadian Supreme

Court Case involving Sue Rodriguez, an ALS patient who was fighting for the right to a physician-assisted death in 1993–4 became international news (Farnsworth 1994).

At the same time, healthcare access in the United States was limited, as was palliative care (see Clinical Ethics on Film). Kevorkian was able to advance the idea of physician-assisted death alongside a movement that was gaining ground on its own without any "physician assistance"—the death with dignity movement, an outgrowth of the rise of the patient autonomy movement in which patient-led organizations were forming around "do-it-yourself" deaths. Patients were already organizing around the "right to die" well before Kevorkian was a household name; what Kevorkian brought to the forefront was a physician willing to participate with tools, and was building a clientele. Kevorkian's first patient, 54-year-old Janet Adkins came through the Hemlock Society. Adkins was already a member of the Hemlock Society before her diagnosis. Kevorkian assisted her in his Volkswagen van, and became well known after her death. The State of Michigan charged Kevorkian with her murder, but the case was later dismissed, because Michigan did not have a specific law preventing physician-assisted suicide. By 1993, it did. The *Washington Post* noted (Bhanoo 2011):

> His goal was to make it legal for a doctor to actively help a patient commit suicide. But to date, no state has made this legal and only three states, Washington, Oregon and Montana, have legalized any form of physician-assisted suicide. To the contrary, the state of Michigan, where Dr. Kevorkian did much of his work, explicitly banned physician-assisted suicide in 1993 in direct response to his efforts.

1993: Death and the Health Security Act

Kevorkian's *Time* had come in 1993 when he made the magazine's cover as "Dr. Death" May 31, 1993. By this point, Kevorkian's infamy grew and was also the subject of the cover of *Newsweek* (Hosenball 1993). At this juncture in U.S. history, the Clinton Administration was just getting started on its plans for universal healthcare through what was being called the "Health Security Act"—designed to become as integral to American life as the *Social Security Act* of 1935. President Clinton had put then First Lady, Hillary Rodham Clinton in charge of designing the new health legislation, and by 1993, she was heralded for her flawless appearances before Congress to discuss the rationale for the new legislation and what it would entail. One Republican critic, Senator Dick Armey from Texas, had referred to the planned legislation as "a Dr. Kevorkian prescription" that would apparently "kill American jobs" (Priest 1993). This led to one of Hillary Clinton's most famous exchanges when being grilled by Congress, described by the *Washington Post* like this (Priest 1993):

> One exchange during [Hillary Clinton's] testimony yesterday before the House Education and Labor Committee left Rep. Richard K. Armey (R-Tex.), a critic of the plan, red-faced and flustered.
>
> "While I don't share the chairman's joy at our holding hearings on a government-run health care system," Armey began, "I do share his intention to make the debate as exciting as possible."

"I'm sure you will do that, you and Dr. Kevorkian," Clinton shot back, in a sharp reference to Armey's recent comment comparing the administration plan to "a Dr. Kevorkian prescription" that would kill American jobs.

"I have been told about your charm and wit," Armey said. "The reports of your charm are overstated, and the reports on your wit are understated." His face bright red, Armey laughed and shook his head. Then he left the room.

As discussed in Chap. 4, the 1993 Health Security Act legislation also experienced a "physician-assisted death" when a Conservative group of physicians—American Association of Physicians and Surgeons (AAPS) sued to gain access to the list of members of the task force Hillary had put together to draft the new legislation. Ultimately, the Clinton Administration effort died and led to a Republican-controlled Congress in 1994 while debates over healthcare access would simmer and eventually lead to the passage of the *Affordable Care Act* (see Chap. 4), just as the film, *You Don't Know Jack* was being released. The two debates were symbiotic: without any health security or affordable healthcare access, the right to die movement in the United States would continue to gain momentum until the first U.S. states began to establish Death with Dignity laws (see under History of Medicine).

Ultimately, before he was finally convicted in 1999, Michigan prosecuted Kevorkian four times for assisted suicides of patients.[2] He was acquitted three times, with a mistrial on the fourth case. By 1998, the Michigan legislature enacted a law, making physician-assisted suicide a felony punishable by a maximum five-year prison sentence or a $10,000 fine. As a direct challenge, Kevorkian actively euthanized Thomas Youk via lethal injection, which was even more legally fraught because Kevorkian's medical license had been revoked. On his fifth indictment, he was charged with second-degree murder, and foolishly served as his own legal counsel. He was convicted March 26, 1999, of second-degree murder and in April 1999, he was sentenced to 25 years in prison with the possibility of parole. As the country's focus turned to the Clinton Impeachment, the 2000 *Bush v. Gore* election and 9/11, Kevorkian tried to appeal his conviction and wanted his case to go to the Supreme Court, but was not successful. He was released from prison on June 1, 2007, and in another strange move, actually ran for a House seat in Michigan's 9th Congressional District and got 2.6% of the vote.

Death as a Civil Rights Battle

From the perspective of patients and the public, Kevorkian's crusade was ultimately seen as a civil rights battle. Overall, the "Death with Dignity" movement refers to a broader "right to die" movement in which several groups lobbied for greater control over the dying experience, which includes groups such as the Hemlock Society (founded in 1980), now known as the Final Exit Network, titled after the "guidebook"

[2]For a comprehensive list of all of Kevorkian's legal battles and activities, see Nicol and Wylie, pp. 185–7.

Final Exit (Humphries 1991). This became a patients' rights lobby and movement championing physician-assisted suicide/physician aid in dying laws. Other popular groups are SOARS (Society for Old Age Rational Suicide) and Compassionate Choices.

Ultimately, this movement was an outgrowth of civil rights; the patient autonomy movement evolved along similar historical timelines as the civil rights movement, and can be traced to several important societal shifts where autonomy and healthcare frequently intersected.

First, major technological breakthroughs in life prolonging treatments, combined with the groundbreaking Harvard Criteria for Brain Death in 1968, literally changed the definition of death so that technology could prolong the "cardiovascular" lives of patients, regardless of their neurological realities (JAMA 1968).

Second, several racially or politically marginalized groups fought for equal rights, legal protections and equal access to societal goods, including healthcare. The battle over healthcare access for several groups became magnified by the AIDS crisis, but had begun much earlier when the details of the Tuskegee study (see Chap. 1) became known in 1972.

Third, the women's autonomy movement was a major contributor in the death with dignity battle as women fought for the right to make decisions about their own bodies, reproductive lives, as well as gender equality. In addition to *Roe v. Wade* (1973), in both Canada and the United States, landmark end of life cases revolved around women's bodies (Re: Quinlan, 1976; Cruzan v. Director, 1990; Nancy B. v. Hotel, 1992; Rodriguez v. B.C. 1993) which I have argued elsewhere, is not a coincidence (See Clinical Ethics on Film). Kevorkian's first patient was a woman, whose death coincided with the Cruzan case (see further).

The Death with Dignity movement was informed by many of the growing debates within the patients' rights battles. The discussion over death and dying was advanced by a major qualitative study by Elisabeth Kubler Ross (Kubler-Ross 1969), in which patients overwhelmingly expressed the desire to have more control over their dying experiences. Kubler-Ross' work (see further), published only a year after the Harvard Brain Death criteria, informed future debates in patient autonomy and end of life, including hospice and palliative care. In the United Kingdom, as early as 1969, there was a fierce debate over euthanasia, for example in the British House of Lords (House of Lords 1969; Voluntary Euthanasia Society 1971) as patients requests to die began to raise ethical questions in a paternalistic medical model.

The Disability Rights movement, which began to peak in the late 1970s/early 1980s, consequent to prenatal technology debates raising concerns about eugenics, and growing awareness of disabled veterans and spinal cord trauma. This movement was validated by the United Nations (UN) in 1975 with the UN Declaration on the Rights of Disabled Persons (UN 2019). Ultimately, this movement culminated in the United States into the passage of the *American with Disabilities Act* (1990), which had far-reaching consequences for patient autonomy issues, and became a major driver in right-to-die arguments. (See Clinical Ethics on Film).

The psychiatry abuse movement was another factor in patients advancing bodily autonomy arguments. Broad practices such as involuntary hospitalization of competent but oppressed persons, such as women, minorities and the LGBT community; harmful therapies with absolutely no ethical oversight, such as lobotomy, induced diabetic coma and early ECT; forced sterilization (*Buck v. Bell* 1927) and research ethics abuses such as the Willowbrook trial (Beecher 1966) led to dramatic reform and current laws that protect mental health patients as vulnerable persons. Major and influential works by Frieden (1963); Kesey (1962); and Chesler (1972) called attention to how rigid social constructs and unfair social arrangements contributed to poor mental health outcomes. However, by the 1990s, patients suffering from mental health issues would also argue that they had a right to end their suffering through PAD.

By the 1980s, the patient autonomy movement grew stronger as the HIV/AIDS pandemic and cancer clinical trials dominated medical news. New concerns arose over access to end of life treatments and innovative therapies, and privacy and confidentiality as some diseases (e.g. AIDS) led to the patient's "social death"—whereby they were abandoned by employers, friends and colleagues, which was more consequential to overall wellbeing (see Chap. 2). Ultimately, all of these civil rights movements contributed to a sea change in the doctor-patient relationship and healthcare organizational culture. Patients in all democratized countries identified with these movements as they recognized problems with paternalism and patriarchy, voicing preferences, obtaining informed consent, capacity to consent, and self-determination. By 1990, the U.S. Congress passed the *Patient Self-Determination Act* (see discussion of the Nancy Cruzan case further), which legally bound hospitals to ask about Advance Directives (see further). Similarly, throughout the 1990s a myriad of advance directive/living will/and proxy decision maker laws passed throughout the democratized world. It was at this time that Kevorkian assisted his first patient, Janet Adkins, to die. Adkins, as an Alzheimer's patient embodied all of the patient autonomy battles. Ultimately, constitutional scholars looked at Kevorkian's cause as a Ninth Amendment issue, which meant that because the right to die was not enumerated, or listed, in the Constitution, it is still a right that people have. Alan Dershowitz, in an interview noted (Bhanoo 2011):

> [Kevorkian] was part of the civil rights movement — although he did it in his own way…He didn't lead marches, he didn't get other people to follow him, instead he put his own body in the line of fire, and there are not many people who would do that. In the years that come, his views may become more mainstream.

Death with Dignity: History of Medicine Context

The history of the medical community's debate over death with dignity, or physician-assisted death, did not start or end with Kevorkian, but he entered the debate at the right time—just as patient Advance Directives were taking off with the passage of the *Patient Self-Determination Act* (1990), on the heels of the Nancy Cruzan case discussed further (also see Clinical Ethics on Film). The history of medicine context to Kevorkian actually starts in Switzerland and ends in Oregon. It was a Swiss-born physician and contemporary of Kevorkian's who completely changed the paradigm around death and dying. And just as Kevorkian was flying too close to the sun in 1997, the state of Oregon passed the first Death with Dignity legislation in the United States, which went on to become model legislation for dozens of other states and countries. By the time Kevorkian was released from prison, the management of end of life in most American hospitals had transformed as palliative care, hospice, "Allow Natural Death" and physician-assisted death in several states had become the norm. The question remains whether Kevorkian sped up or slowed down the progress.

Life *and Death … and Dying*

It is impossible to teach the Kevorkian period in medical history without first discussing the contributions of Elisabeth Kubler-Ross to death and dying discourse. Kubler-Ross' book, On Death and Dying, was published in 1969 and then was featured in *Life* magazine that same year. Her book famously introduced the Five Stages of Grief/Death. Kubler-Ross' work was a culmination of a large qualitative study she had undertaken during the turbulent 1960s in which she interviewed over 200 dying patients (and later, over 500). Kubler-Ross was doing this work after Dame Cicely Saunders had first pioneered the concept of hospice in 1963, and after the passage of Medicaid and Medicare in 1965.

Kubler-Ross was a contemporary of Kevorkian's, born in 1929 in Switzerland. She had a very different life experience, having come of age in post-war Europe. She had left home at 16 to volunteer during World War II in hospitals and caring for refugees, and then in 1945, had joined the International Voluntary Service for Peace to help in war-torn France, Poland, and Italy. This included going to the Majdanek concentration camp in July 1945 just after its liberation, which inspired her interest in death and dying. According to one interviewer, "She told me that it was hitchhiking through war-ravaged Europe as a teenager and visiting liberated concentration camps that transformed her. 'I would never have gone into the work on death and dying without Majdanek … All the hospices—there are 2100 in America alone—came out of that concentration camp' (Rosen 2004).

Post-war Europe was a hellish place to spend one's formative years, and was described like this (NPR 2013):

> Imagine a world without institutions. No governments. No school or universities. No access to any information. No banks. Money no longer has any worth. There are no shops, because no one has anything to sell. Law and order are virtually non-existent because there is no police force and no judiciary. Men with weapons roam the streets taking what they want. Women of all classes and ages prostitute themselves for food and protection…This is not the beginning to a futuristic thriller, but a history of Europe in the years directly following World War II, when many European cities were in ruins, millions of people were displaced, and vengeance killings were common, as was rape.

Kubler-Ross' views were shaped by what she saw in this time frame. After her post-war volunteer years, she returned home and went to university and medical school around the same time as Kevorkian but at the University of Zurich between 1951 and 1957. In 1958, she married a U.S. medical student, Emanuel Robert Ross, and moved to the United States, where she did a residency in psychiatry in the 1960s (around the time Ken Kesey wrote One Flew Over the Cuckoo's Nest, discussed in Clinical Ethics on Film), and eventually worked at the University of Chicago where she began a course in death and dying with theology students. She noted (Lattin 1997):

> When I came to this country in 1958, to be a dying patient in a medical hospital was a nightmare…You were put in the last room, furthest away from the nurses' station. You were full of pain, but they wouldn't give you morphine. Nobody told you that you were full of cancer and that it was understandable that you had pain and needed medication. But the doctors were afraid of making their patients drug addicts. It was so stupid.

She published some of her work in 1966 (Kubler-Ross 1966) before completing her now-famous book in 1969, which was on the heels of a country that was going through intense cycles of grief between 1963 and 1968, after assassinations of President Kennedy, Martin Luther King, Robert F. Kennedy, as well as thousands of casualties and injuries from the Vietnam War. Kubler-Ross noted the following in the Introduction of her book, which was a sign of the times in which she lived (Kubler-Ross 1969):

> Physicians have more people in their waiting rooms with emotional problems than they ever had before….The more we are making advancements in science, the more we seem to fear and deny reality of death….Dying becomes lonely and impersonal because the patient is often taken out of his familiar environment…. [I]f we come home from Vietnam, we must indeed feel immune to death….Wars, riots and…murders may be indicative of our decreasing ability to face death with acceptance and dignity."

Kubler-Ross' famous Five Stages of Grief and Death were as follows, but not necessarily linear, as she would repeat throughout her career: Denial (and Isolation); Anger (and rage); Bargaining; Depression; and Acceptance (not to be mistaken for "happy"). "Elisabeth always said that the most misunderstood thing in her life is her stages. They are as unique as we are. Not everyone goes through them in the same way. If you go back and read her initial papers on it, they're not the same for everyone" (Oransky 2004).

Her book had resonated in many contexts, but Kubler-Ross and her work became a household name when *Life* magazine did an expose on her in its November 21, 1969 issue—just at the 6th anniversary of the JFK assassination, and just over a year since the Martin Luther King and Robert F. Kennedy assassinations in the Spring of 1968 on April 4th and June 6th, respectively. Kubler-Ross then went on to provide testimony to Congress in August 1972, just a couple of weeks after the Tuskegee Study had been "outed" on July 25, 1972. As Kubler-Ross testified before Congress about death and how end of life management needed to change, the Tuskegee Federal Ad Hoc Panel was not yet chartered; it would be on August 28th of that year (see Chap. 1).

As a direct of result of Kubler-Ross' work and her Congressional testimony, both the hospice movement as well as the death with dignity movements began to flourish. By September 13, 1982, President Reagan announced "National Hospice Week" (Federal Register 1982) in "Proclamation 4966—National Hospice Week, 1982" in which he states: "The hospice concept is rapidly becoming a part of the Nation's health care system…" Soon after, Medicare added hospice services to its coverage. By 1984, accreditation standards emerged for hospice care. Kubler-Ross' influence on death and dying discourse made Kevorkian's work possible. According to Byok (2014):

> In a period in which medical professionals spoke of advanced illness only in euphemisms or oblique whispered comments, here was a doctor who actually talked with people about their illness…[This] book captured the nation's attention and reverberated through the medical and general cultures…. During the socially tumultuous mid-twentieth century, one diminutive Swiss-American psychiatrist had the temerity to give voice to people facing the end of life.

The Battle for Withdrawal: Quinlan and Cruzan

Physician-assisted death and the "right to die" are not the same arguments, which the Quinlan and Cruzan decisions would make clear. The Quinlan case (re: Quinlan, 1976) was about a 21-year old woman who collapsed at a party after combining alcohol with valium; she had an anoxic brain injury, and remained in a persistent vegetative state. When it was clear that there was no hope of improvement, the parents requested her ventilator be withdrawn. The hospital refused to do this because at the time, it was believed to be a criminal act. In a landmark decision, the courts ruled in favor of the parents, and allowed withdrawal of life sustaining treatment in this circumstance. The case led to a sea-change in clinical ethics, and paved the way for patients/families to refuse therapies that did not improve their quality of life, or offered no medical benefit other than prolonging suffering.

Seven years after the Quinlan decision, 30-year-old Nancy Cruzan would be thrown through her car windshield in 1983 and suffer a head injury; Cruzan's 1963 car had not been equipped with a seatbelt, as they had not been mandatory at that time. She lingered in a persistent vegetative state breathing on her own until 1990, when a Missouri court decision established the right of competent patients to refuse life-sustaining treatments that included nutrition and hydration. In her case, although she did not need a ventilator, her parents agreed to a feeding tube (gastrostomy tube) in the belief that she had a chance to recover. However, when it became clear there was no chance of improvement, they requested withdrawal of the feeding tube, and the hospital refused. Eventually, the courts ruled in the parents' favor. The case also established that such refusals could be honored by a substituted judgment made by a surrogate decision-maker, based on *known or previously stated preferences* of the patient. In the Cruzan case, the courts recognized that the parents provided "clear and convincing evidence" about what their daughter would have wanted if she were competent because she had prior conversations about not wanting to live in a compromised state (she indicated she wanted a "half normal life"); it is likely these wishes were triggered by a discussion of the Quinlan case, which had been highly publicized in the media). The Cruzan case also established that nutrition and hydration count as medical treatment, and can be withheld. The case paved the way for the *Patient Self-Determination Act* (1990), which requires all hospitals to discuss Advance Directives with patients.

But Kevorkian did not champion withdrawal/withholding of life-sustaining treatment and actually argued it was "torture"—claiming hospitals were "starving patients to death" (CBS 1996), which is not an accurate description of withdrawal of either life support or nutrition and hydration in patients who request it, or are candidates for it. Kevorkian argued that physician-assisted death was a far more humane way for patients to die rather than merely removing the barriers to a natural death, which could prolong suffering. Many of Kevorkian's patients would have been good candidates for palliative and hospice care, which can include making patients comfortable if they refuse nutrition and hydration, known as Voluntary Stopping of Eating and Drinking (VSED). That said, in the 1990s, palliative care was not a popular service in the United States. Ultimately, the Quinlan and Cruzan cases led to the establishment of Advance Directives/Living Wills as part of the patient autonomy movement, in which patients and their surrogates could refuse medical interventions that interfered with a natural death, and request comfort care only. However, the advance directives did not have an option for physician-assisted death or euthanasia. It is important to separate the "Allow Natural Death" movements, for which Advance Directives are vehicles, from the physician-assisted death movement, which is what Kevorkian championed.

By 1993, 48 states with two pending had passed Advance directive laws, also prominently promoted by President Clinton and Hillary Rodham Clinton who had signed their own living wills in the wake of Hillary's father's death. In the spring of 1994, former President Richard Nixon and former First Lady, Jacqueline Kennedy Onassis, died almost a month apart, on April 22nd and May 19th, respectively. It was reported that both had signed advance directives surrounding end of life options.

The Road to Oregon

Between 1994 and 1997, during the Kevorkian chapter in American medical history, the first right-to-die legislation was introduced and finally passed in the state of Oregon, known as the *Death with Dignity Act* (ORS 127.800-995). In fact, most of the world's current right-to-die legislation is modelled after the Oregon law. Janet Adkins' husband, Ron, was among the Oregonians who lobbied to get the law passed (Nicol and Wylie 2006).

But the road to the Oregon law started in 1906, when the first failed euthanasia bill was drafted in Ohio. In 1967, the first "right-to-die" bill was introduced by Dr. Walter W. Sackett in Florida's State legislature, which also failed to pass (Death with Dignity 2019); this was followed by a failed Idaho bill in 1969 for voluntary euthanasia.

In 1980, Derek Humphry, a British journalist and right-to-die advocate moved to the United States and founded the Hemlock Society in Santa Monica, California. Humphry was born in 1930 and roughly the same age as Kevorkian and Kubler-Ross. The role of his organization was to distribute "do it yourself" guides to successful and painless suicides. Through the organization, the book, Let Me Die Before I Wake was published and was the first publicly available book for sale on suicide methods. In 1984—the same year the first Apple Macintosh computer was introduced—voluntary euthanasia under certain conditions is legalized in The Netherlands. By 1988, a famous article in the *Journal of the American Medical Association* (Anonymous 1988) is published by an anonymous resident doctor who admits to performing active euthanasia by lethal injection to an ovarian cancer patient, which invites scrutiny from a public prosecutor but the resident is never identified (Wilkerson 1988). Kevorkian then adds to this history when he helps his first patient to die in 1990. In 1991, another physician is also in the news by publicizing assisted suicide. Timothy Quill publishes an article in the *New England Journal of Medicine*, which describes his assistance to help a leukemia patient, "Diane" die at home by prescribing her lethal drugs she could take at home (Quill 1991). That same year, Derek Humphry publishes a new book, Final Exit, on how to properly commit suicide, which becomes a best-seller and is highly publicized.

1994–7: Measure 16 and Measure 51

The Oregon *Death with Dignity* law took three years to pass. It began with a ballot measure in the 1994 midterm elections held November 8 that year. Measure 16 asked the voters to decide whether physician-assisted suicide should be legalized by allowing physicians to prescribe lethal doses of medications for patients to take themselves, and 51.3% of the voters approved of the measure. But once the voters decided, the National Right to Life organization filed a motion for injunction against Measure 16 in a federal district court (Lee v. State of Oregon) arguing that the law would violate the Constitution's First and Fourteenth Amendments as well as several federal statutes. The same year as Measure 16, Timothy Quill filed a lawsuit with New York State (Quill et al. v. Koppell) to challenge the New York law prohibiting assisted suicide, but Quill loses and files and appeal.

In 1995, *the Journal of the American Medical Association* published the results of the SUPPORT study, a groundbreaking multi-site clinical research project that looked at end-of-life care and how to reduce "mechanically supported, painful and prolonged process of dying" (Connors et al. 1995). The study looked at 9105 adults hospitalized with one or more life-threatening diagnoses, and found poor communication surrounding code status discussions, and goals of care. The study results exposed that hospitals handled end of life decisions poorly, and led to the expansion of palliative care (Childress 2012). The study concluded that its goals were "[to] improve the experience of seriously ill and dying patients, greater individual and societal commitment and more proactive and forceful measures may be needed" (Connors et al. 1995).

Two years after the SUPPORT study was published, the injunction to the Oregon bill was lifted October 27, 1997, which was followed by a repeal effort as another ballot measure—Measure 51, which was rejected by 60% of voters (Death with Dignity 2019) (The *Death with Dignity Act* was challenged again by the Bush Administration in *Gonzales v. Oregon* in 2006, but was upheld.)

In 1996, a Gallup poll survey demonstrated that 75% of respondents favored physician-assisted death legislation, and so the rejection of Measure 51 was consistent with shifting attitudes, which is why Kevorkian's timing in challenging the law was reflecting the cultural shifts on this issue (Death with Dignity 2019).

By 1997, the U.S. Supreme Court rules on two physician-assisted death laws upholding the New York State ban on the practice (*Vacco v. Quill*), and clarifying in another case (*Washington v. Glucksberg*) that there is no liberty interest or "right" to assisted death, but the court makes clear it's a matter for individual states to decide. At the same time, the court validates that the Principle of Double Effect (see under Healthcare Ethics Issues) is legal and consistent with beneficent medical care. The opinion states: "Throughout the nation, Americans are engaged in an earnest and profound debate about the morality, legality and practicality of physician-assisted suicide. Our holding permits this debate to continue, as it should in a democratic society."

The Oregon law paved the way for other states to follow suit, and in 1998, the state also legislated greater access to the law through state funds to lessen health disparities. In its first year of legalization, only 16 people make use of the law (Childress 2012).

Ultimately, all of the physician-assisted death laws in the United States revolve around the Oregonian model of passive euthanasia, harkening back to legal versions of Kevorkian's "Thanatron" in which physicians assist with prescribing the best methods that allow patients to end their lives, but the patients have to actually take the medications prescribed themselves. For example, in the documentary *How to Die in Oregon* (2011), the physician goes to the patient's home, mixes up a concoction of medications that will terminate the patient's life, but must hand the concoction to the patient to drink without any further interference. There are also several conditions the patients must meet in order to access the Oregonian law, which includes being diagnosed with a terminal condition—a criterion that Janet Adkins in 1990 did not meet.

Patients' rights organizations have argued that passive euthanasia is too limiting, and have lobbied for legal versions of the Thomas Youk case—for the physician to be enabled to actively inject lethal doses of drugs to end a life, which in his case consisted of a barbiturate and muscle relaxant, followed by potassium chloride (Nicole and Wylie 2006). As discussed further on, such "active euthanasia" laws present many ethical dilemmas and have not passed in the United States. Canada, which in 1994 refused to allow any physician to help an ALS patient to die, has reversed course: it has now passed an *active euthanasia* law, which is important to outline when teaching this content to students.

The Canadian Exception: Medical Assistance in Dying (MAID)

Canada's *Medical Assistance in Dying* (MAID) law, known as Bill C-14, allows patients to access active euthanasia if they meet certain criteria, which include that the patient must be diagnosed with a "grievous and irremediable" condition; (b) the patient's condition is "incurable" and that natural death is "reasonably foreseeable"; and (c) the illness or condition "causes enduring suffering that is intolerable to the individual in the circumstances of his or her condition." The provision that "natural death is reasonably foreseeable" is an important societal "check". Otherwise, patients who opt to live with devastating diagnoses for which there are no good treatments, such as early onset dementia, spinal cord injury, Locked In Syndrome; aggressive cancers, or any range of congenital disorders that cause life-long disability, may be coerced into requesting MAID against their wishes to save resources in a universal healthcare context. Without that provision, healthcare providers could begin to suggest MAID prematurely to patients who prefer aggressive measures and *want to live as long as possible with these conditions.* Consider cases where an adult has no family and is unbefriended: social isolation and/or poverty could magnify the intolerability of chronic illness. Based on a recent report by the Canadian Council Academies (2018) the authors noted (CCA 2018):

> Under Canada's current legislation, patients can initiate a MAID request if *they* decide that their condition is causing them intolerable suffering. Although healthcare practitioners need to confirm that the intolerable suffering criterion has been met, the legislation does not require an independent judgment of the patient's level of suffering by a third party.

Healthcare Ethics Issues

When using this film to discuss the broader ethical and societal implications of physician-assisted death, there are a number of considerations and perspectives to review. This section looks at the competing bioethics principles involved, including the concept of the "rational suicide"; healthcare provider obligations and duties including conscientious objection; slippery slope arguments and societal implications; and finally, other end of life options to physician-assisted death that can be specified in advance directives or advance care planning discussions, including withdrawal and withholding of life support, intubation, and code status preferences such as Allow Natural Death (AND) and Do Not Resuscitate (DNR).

Competing Bioethics Principles in Physician-Assisted Death

When looking at physician-assisted death from the standpoint of the four core bioethics principles (Beauchamp and Childress 2008), it is housed as a "non-maleficence" dilemma because healthcare providers indeed have an ethical obligation not to intentionally or knowingly cause harm without any intended medical benefit, as for many, how can "death" be seen as a medical benefit when you have a duty to heal? Regardless of what the laws may allow in some states, many healthcare providers view either passive or active euthanasia as a clear violation of the Principle of Non-Maleficence because it indeed causes "intentional harm." In several of the Kevorkian cases, there are ethical questions as to whether his patients would even meet the current Oregonian criteria for passive euthanasia or Canadian criteria for active euthanasia, for example. However, physician-assisted death may also be viewed as a patient autonomy issue. The ethical dilemma results when a physician feels morally compelled to refuse a patient's request to die as a necessary autonomy-limiting act in order to uphold the Principle of Non-Maleficence. Then there are concerns regarding the Justice implications and whether more permissive euthanasia laws would lead to the slippery slope of weaponizing physician-assisted death for economic benefits in the hands of morally bankrupt political leaders, which I discuss further on.

In considering the two most relevant competing bioethics principles (non-maleficence and autonomy), there are several clinical ethics concepts in the lived experiences of patients, and their decision to end their lives: quality of life; the "rational suicide"; the role of palliative care and comfort care; patient refusal of treatment, nutrition and hydration; VSED (voluntary stopping of eating and drinking); "Do Not Intubate" (DNI); and "Allow Natural Death" (AND). However, there are also specific guiding and obligatory ethical principles that healthcare providers must follow, which may take priority or limit patient requests for certain treatments or procedures.

A basic tenet in patient autonomy is the right to be "left alone" and to refuse medical intervention. Another basic tenet in patient autonomy is to allow the patient to determine his/her own quality of life, rather than imposing our own norms. Quality of life is thus subjective and not normative. Although the Principle of Respect for Autonomy obligates healthcare providers to honor patients' wishes, values and goals, which may include respecting their right to refuse medical treatments or interventions that result in earlier deaths, the Principles of Respect for Persons, Beneficence and Non-Maleficence obligate healthcare providers to *protect* patients who do not demonstrate decision-making capacity, or do not meet the U-ARE criteria, an acronym for Understanding, Appreciation, Rationality, Expression of a choice, and who therefore are not autonomous (see <u>Clinical Ethics on Film</u>). When discussing physician-assisted death in the context of an autonomous patient's request to die, in those whose deaths are foreseeable, advance care planning, DNR/DNI orders and aggressive palliative care will typically hasten death and remove the need for such requests. The dilemma generally presents in patients *whose death is <u>not</u> imminent or* *reasonably foreseeable*, notwithstanding that we all have a "terminal condition" as part of being mortal. These are the "Kevorkian cases" such as Janet Adkins, his first patient with early symptoms of dementia, and Thomas Youk, with long-suffering ALS. In such cases, there may be compelling arguments for granting the patient's request to die; but there are also recent cases of healthy people who simply do not want to live to become too old to enjoy life (Donnelly 2015).

Quality of Life and Existential Suffering

Autonomous persons may reach a conclusion that they are at the end of life, or that continuing their lives is tantamount to living in a prison of either psychological or physical misery and suffering. The patient's conclusions are based on the subjective, lived experience. However, the healthcare provider has a beneficence-based obligation to consider the patient's request by objective, or evidence-based criteria, such as whether all potentially beneficial therapeutic options have been reviewed, and whether the patient's medical/mental status would significantly improve with therapy. For these reasons, an ethical dilemma may exist when the patient and healthcare provider do not agree that the patient is at the end of life. In the vast majority of such cases, the patient's conclusions may be based on a diminished quality of life due to financial constraints, lack of access to quality care, and caregiving issues within the family system (Kerr 2000; Mehta and Cohen 2009). It's been well-established that primary care providers are not well trained in the social determinants of health (Lauren et al. 2019), while studies looking at doctors in universal healthcare countries still found numerous problems with capacity/resources to address such issues (Pinto and Bloch 2017), and cultural barriers (Girgis et al. 2018). In other words, with the right social, financial, palliative and other supports, patients may come to different conclusions about their quality of life. The healthcare provider must consider what factors are contributing to the patient's request, and consider all the options available to the patient. There is also an unavoidable intersection between depression and chronic illness; the bioethics literature is clear that depression and autonomy may coexist, and that situational depression may, in fact, be clear evidence that the patient has profound understanding and appreciation of his/her circumstances (McCue and Balasubramaniam 2017). Although clinical ethicists do not consider situational depression necessarily to be a barrier to capacity, when depression is diagnosed, psychiatric evaluation and treatment for depression may be among the options considered.

The Rational Suicide

Many cases of early requests to die can be deemed to be the "rational suicide". For example, there are several cases of elderly persons requesting to die while they are still "well" and not wait for disability, dementia or a terminal diagnosis (McCue and Balasubramniam 2017). Similarly, many early-stage dementia patients request to "opt out" of living longer in a diminished state, as was the case with Janet Adkins. For these reasons, the typical definition of end of life or terminal condition may not be applicable when a patient determines s/he has reached the end of his/her quality of life. Rational suicides, also called "remedial suicides" (Nelson and Ramirez 2017), are frequently acted upon without involving medical professionals when individuals exercise their autonomy to end their lives by committing suicide. Rational suicides are complicated when healthcare providers are asked to assist, and there is no moral obligation for healthcare providers to assist if, in their medical judgment, the decision is not based on medical evidence. For example, in the Robin Williams case (Tohid 2016), he had a history of psychiatric co-morbidities when he was diagnosed with a form of Parkinsonian dementia, and hung himself. However, had he made the request for his doctor to assist him, such might have been denied on beneficence-based grounds, as Williams likely had many more years of life ahead, and most healthcare providers would posit that clinical intervention, or even palliative care, pastoral/spiritual care may have eased his transition and at least facilitated better closure with his loved ones and friends. Notwithstanding, this case may still be considered a "rational suicide" in that Williams did not want to live in a diminished state, and many would argue that his diagnosis was a "bridge too far" given his psychiatric history. Yet healthcare providers do not have an ethical obligation to actively assist patients in a "rational suicide" if there is no medical evidence that they are at the end of life, as defined by medical/clinical criteria. What, then, is the role of a healthcare provider when a patient wants to die prematurely?

Post-Kevorkian: Healthcare Provider Duties and Obligations in Death Requests

When reviewing the Kevorkian story, it's important to make clear to students that although he attempted to set some sort of inclusion/exclusion criteria, he was not qualified to assess a patient's request to die because he had no experience whatsoever in end of life care, psychiatry, palliative care, or even in treating patients who were dying in either a hospice or intensive care setting. As a pathologist, he had very limited experience with patients, and by his own admission, even a "checkered" employment history within his own subspecialty field. For these reasons, some may argue that Kevorkian actually engaged in egregious behaviors because he crossed practice boundaries by seeing dying patients when he was not properly trained as a clinician, and then by 1991, when his medical license was revoked, some could argue that he actually *practiced medicine without a license*. But others argue that he indeed offered a service for dying patients desperate to end their suffering in a time frame when there was no other way to access such a service, and by doing so, he made physician-assisted death a household discussion.

Regardless of how we might judge Kevorkian's actions now, from a medico-legal standpoint, in his final trial and conviction, he indeed met the legal criteria for second-degree murder and not "physician"-assisted death because he was no longer licensed as a physician by that point. Bioethicist Arthur Caplan gave Kevorkian a pass when Caplan was interviewed in 2011, when Kevorkian died (Bhanoo 2011):

> He was involved in this because he thought it was right, and whatever anyone wants to say about him, I think that's the truth… He didn't do it for the money, he didn't do it for the publicity, he wasn't living a luxurious life – he wanted change.

Kevorkian's obituary in the *Washington Post* ended like this (Bhanoo 2011):

> Despite his best efforts, Dr. Kevorkian was, for the most part, a lone soldier who had an abrasive personality. Although he was the best-known figure in fighting for euthanasia's legalization, the legislative results of his efforts were largely unsuccessful, if not counterproductive.

Therefore, in any classroom debate about whether Kevorkian was right or wrong, it's paramount to point out that there is a difference between Kevorkian as the messenger and the ethical argument for physician-assisted death. Kevorkian's lack of training as a clinician in treating patients at the end of life, self-made "criteria" for patient candidacy for physician-assisted death, and his inability to recognize he was crossing practice and professional boundaries, made him a more problematic medical spokesperson for the right-to-die patient autonomy movement. On the other hand, he helped to bring physician-assisted death out of the shadows, where bioethicists suspect it had been hiding for a long time (PBS 1996).

Post-Kevorkian, there is an important discussion about the role of a qualified healthcare provider when assistance in dying is sought by autonomous patients who determine that they are ready to die. In these cases, the healthcare provider has the burden of determining whether the patient's request is beneficence-based or whether it violates the Principle of Non-Maleficence.

In healthcare ethics, there is a sharp distinction between moral actors who merely remove the barriers to a natural death (withdrawal of life support and treatment), known as "passive euthanasia", and those who actively facilitate death, known as "active euthanasia". The former scenario is perceived to be ethically and morally permissible for physicians of all cultures because non-interference with natural death supports the Principles of Autonomy, Respect for Persons, Beneficence, Non-Maleficence and even the Principle of Justice, which deals with resource allocation issues. The latter scenario, however, presents moral dilemmas for many practitioners, as it is perceived as violating the Principle of Non-Maleficence, long-held notions of professional duties to treat, or personal/religious values and views related to "killing" another human being. There is also a space in-between in which relief of suffering may hasten death invoking the "Principle of Double-Effect". In this scenario, it is widely accepted that a healthcare provider may prescribe/administer palliative, but not curative, treatments for comfort, to relieve suffering, which may also hasten death; in recent years, the concept of terminal or "palliative sedation" has become an accepted standard of care in cases of severe physical suffering when patients are near the end of life (AAHPM 2014). Here, because the intention to relieve suffering is of primacy, with an unintended consequence of hastening death, the action falls within the Principle of Beneficence. What medications or agents constitute "relief" is left up to medical judgment and the patient-practitioner relationship and goals of care. Additionally, there is moral diversity and clinical equipoise (Freedman 1987) over whether patient requests for assistance in dying are beneficence-based or whether they would be in violation of the Principle of Non-Maleficence. That distinction, too, may depend on medical judgment, and there may be differing opinions about what constitutes "medical futility" (Brody and Halevy 1995; Schneiderman 2011). For example, "quantitative futility" means that there is a poor chance an intervention will benefit the patient (Jecker 2014), while "qualitative futility" means there is either a poor chance that quality of life from an intervention will improve, or it could worsen (Jecker 2014).

In cases where there is medical expert consensus that a request to die would absolutely violate the Principle of Non-Maleficence and is a medically unsound request, the healthcare provider still has an obligation to help the patient find comfort. A common example of this situation is seen with burn patients. In many cases, burn patients request to die because their pain is too great, but they may have decades left of life with absolutely no life-threatening illness. In such cases, clinical ethicists will advise that it is ethically permissible to provide pain relief that may have higher risks involved.

In general, in the vast majority of cases where the capacitated patient's request to die would violate the Principle of Non-Maleficence, there is a correlation between patients feeling abandoned with insufficient social or other supports, palliative care or comfort care (Nelson and Ramirez 2017; McCue 2017). Ultimately, the healthcare provider does not have an obligation to accede to a patient's demand that s/he violate the Principle of Non-Maleficence, but the Principle of Beneficence does obligate the same healthcare provider to exhaust therapeutic and psychosocial options that may help the patient to live more comfortably so that premature death is not more appealing than continuing to live.

Conscientious Objection

When patient autonomy is challenged surrounding procedures guaranteed by the state in the interests of granting patients control over their bodies, this may be in tension with healthcare providers' religious beliefs, or perceptions of their ethical duties to the patient and society at large. Such conflicts may produce ethical imperialism, where healthcare providers are imposing values the patients do not share. This is discussed more in Chap. 5.

Assisted Death Versus Withdrawal/Withholding of Life Support

The Principle of Respect for Autonomy obligates healthcare providers to honor autonomous patient preferences to *refuse any treatment*, but do have beneficence-based obligations to help any patient to comfortably withdraw from medical treatment or life sustaining therapies, including nutrition and hydration, a right that was established in the Cruzan case (see earlier). Such obligations exist in the absence of, or regardless of, physician assistance in dying legislation. In other words, patients who wish to stop life-sustaining therapies, nutrition and hydration but whose natural deaths are not "reasonably foreseeable" *always have that option* in states where there are no death with dignity laws. Palliative care or comfort care can facilitate a natural death without interference or medical/nutritional rescue. In reality, from the lens of a healthcare provider or clinical ethicist, anyone with a "serious illness or disability" (Bill C-14) for whom withdrawal/withholding of life sustaining treatments is not a medical option, can always refuse treatment or medical rescue. Since so few patients requesting to die would meet this definition, healthcare providers in these situations would need to explore whether patients who wish to end their lives early can merely do so by withdrawal/withholding of life-sustaining therapies, aided by comfort care. Frequently, such conversations could lead to meaningful goals of care discussions with patients, who may change their minds. Notably, all patient directives regarding withdrawal/withholding of treatment, nutrition and hydration can be part of an advance directive.

Refusal of Life-Sustaining Treatment, or Nutrition and Hydration

With respect to the general patient population, "passive assistance" in dying is the de facto reality for *any* patient who relies on some form of medical treatment or medical intervention to live. Insulin-requiring diabetics can refuse to take insulin; patients on dialysis can stop dialysis; patients with pacemakers can decline to replace the batteries, or have them removed; patients in respiratory distress can refuse oxygen or intubation or be extubated from ventilators. In the simplest form of refusal of treatment, patients with or *without* swallowing difficulties, or appetite, can just stop eating and drinking through VSED (voluntary stopping of eating and drinking) or can refuse feeding tubes or any form of nutritional rescue.

When autonomous patients make the request to stop living and begin to refuse treatment or nutrition/hydration, the healthcare provider must honor the patient's wishes and goal, but can introduce comfort care and palliative care to ease any symptoms and create a code status order to "Allow Natural Death" through non-intervention. When healthcare providers and their patients understand these options are ethically and legally available and permissible in the absence of any specific death with dignity/physician-assistance in dying law, patients can always set their own expiration dates by refusing medical treatment, nutrition and hydration while their healthcare providers can facilitate a better death experience for them. In this way, the healthcare provider respects patient autonomy and uses beneficence to prescribe comfort care measures to help the patient meet his/her goals of an easier death.

The Role of Advance Directives

Advance Directives are a "patient's rights" tool designed to maximize patient autonomy; it is a document that speaks for the patient when s/he has lost decision-making capacity or is permanently unconscious. The Advance Directive varies depending on regional laws, but typically it is activated when a patient has a life-threatening or terminal condition. The Advance Directive should clarify three basic decisions: (a) who is the patient's surrogate decision maker; (b) what levels of medical intervention or life support does the patient want in various circumstances; (c) code status decisions.

There is a large literature on the problems with Advance Directives (see, for example, Levi and Green 2010), as they frequently do not account for the nuances of intensive care, or the fact that many patients may change their minds or make different decisions when they are "there"—vulnerable and in life-threatening situations. The trend in recent years is to educate clinicians about having Advance Care Planning discussions while modified advance directive documents have emerged, such as "The Five Wishes" document, which is a values history (Doukas and McCullough 1991) or Physician/Medical Orders for Life Sustaining Treatment, known as POLST/MOLST (see: www.polst.org). What may seem reasonable to the patient in his/her former healthy body may no longer be his/her position when s/he is in an intensive care unit (ICU). For these reasons, the most recent decisions a patient makes before loss of capacity—even if only verbally communicated to a healthcare provider—may supersede what is written in an Advance Directive/Living Will so long as it is properly recorded or documented. In many situations, Advance Directives may not be authentic, or are coerced documents. In the United States, for example, nursing homes may require that residents have an Advance Directive that makes clear they wish to be "DNR" because such facilities may not have emergency care; yet when such patients are transferred to a tertiary care center and asked about their DNR, some may be unaware of the document they signed or change their code status to "full code". Ideally, Advance Directives are designed to guide practitioners when patients are no longer able to voice or communicate medical decisions, but clinical ethicists typically advise that the selection of a surrogate decision maker is perhaps the most important decision, as these documents may not take all medical options into account. That said, they do reflect broad preferences and values. For example, patients may stipulate "DNI" in such a document—meaning they do not wish any intubation. Yet, some interventions are temporary or resolve urgent problems such as air hunger. Many patients may mean they do not want to be *permanently* intubated, but they could tolerate short-term interventions. Advance Directives are typically designed to guide end of life care and goals of care when a patient is diagnosed with a terminal condition, meaning that they have a condition for which there is no available treatment or available treatment will not offer any medical benefit or reasonable quality of life. When used as a guide to end of life care, practitioners can anticipate critical junctures to discuss either withdrawal of life support, transfer to comfort care, or the withholding of treatment. In several jurisdictions, physician-assisted death cannot be stipulated in an Advance Directive.

Justice Issues and Societal Implications

Within the healthcare ethics context, it's important to discuss how physician-assisted death legislation may affect vulnerable populations classically referred to as populations at risk of health disparities or health inequalities due to socio-economic status or group-specific attributes such as, geography, gender, age or gender orientation (CDC 2011). Such populations are less able to negotiate healthcare due to a number of socio-economic barriers, and they are also less able to advocate for their own needs. According to the Urban Institute (2014): "Vulnerable populations are groups that are not well integrated into the health care system because of ethnic, cultural, economic, geographic, or health characteristics. This isolation puts members of these groups at risk for not obtaining [appropriate] medical care, yet more at risk for mental health complications and thus constitutes a potential threat to their health." Unique disparities are described by both Frohlich et al. (2006) and Patrick et al. (2018), which comprise minority populations as well as anyone experiencing poverty, isolation, discrimination, and social disruption. Pediatric, senior, pregnant and psychiatric populations are globally considered to be vulnerable, while new immigrants—particularly those seeking asylum—are particularly vulnerable.

One Canadian report considered the specific vulnerable population of persons with mental disorders and notes (CCA 2018):

> Mental disorders can affect people in all socioeconomic and demographic categories, but the presence of many mental disorders is strongly correlated with certain social, economic, and environmental inequalities, such as poverty, unemployment, homelessness, and violence....[D]ata show that women, youth, and [minority populations] have higher rates of mental health problems....Additionally, there are variable rates of mental health problems for different immigrant, refugee and racialized groups. Many people with a mental disorder do not receive the necessary treatment for their condition.... Globally, mental healthcare services are poorly funded compared with other health sectors, and it is more common for patients to feel unsupported or to be unable to access such care on a timely and frequent basis. In addition, people may be reluctant to seek mental healthcare due to stigma, or they may be unable to access mental healthcare for a variety of reasons, including geographical unavailability, long wait times, lack of financial means (or inadequate insurance) for medications and/or outpatient treatment, and lack of social support.

Herein lies the argument for the "slippery slope" of too-permissive physician-assisted death laws.

The Slippery Slope

In the context of physician-assisted death, in which a healthcare provider either prescribes lethal medication or injects it, there is a moral imperative to protect vulnerable populations from being perceived as socially futile, burdensome, or too expensive. Requiring the patient who makes the request to die be fully capacitated with informed/valid consent is one important societal "check". Restricting the law to patients who meet more traditional hospice/end of life criteria is an equally important "check" (with no requirement of prognosticating the timing of death.).

Without these important caveats, unforeseen socio-political changes could "weaponize" physician-assisted death in certain groups, based on the recorded history of bioethics. It's important to prevent weaponizing these types of laws against vulnerable populations. The field of bioethics arose in response to medical abuses, and so a future where history repeats itself is more, not less plausible, and must be considered in the context of changing socio-political conditions. The facts surrounding future demographic groups potentially affected by such laws are these: Baby Boomers are predicted to age alone at unprecedented rates (McCue 2017) without defined family members or caregivers. Several countries have now confronted "loneliness" and social isolation as a significant public health problem (Mohdin 2018; Fifield 2018). Coercion, toxic family systems, lack of effective social supports, and predicted financial and climate-related displacements may augment chronic health challenges, while patient-perceived "social deaths" (see Chap. 2) can trigger a number of mental health problems that make death more appealing than living.

Conclusions

Physician-assisted death is a mainstream topic in any healthcare ethics class or syllabus, and for these reasons, *You Don't Know Jack* and Kevorkian's role should be taught. Kevorkian may be best presented as an iconoclast. But in conclusion, it's instructive to examine some transcripts during his time in the sun.

In a 1990 transcript from a *PBS Newshour* interview with Kevorkian and a health law and medical ethics expert, Kevorkian stated the following about physician-assisted death (PBS 2011):

> It's up to a physician, with his medical expertise, combined with logic and common sense and a totally free mind, to evaluate whether this person's desire and wish is medically justifiable. And, if it is, then I think it's a physician's duty to offer that option to the patient. And I wanted to demonstrate also that now there's a great need for it in this country.

On that same PBS program in 1990, lawyer and medical ethicist Susan Wolf responded with this (PBS 2011):

> I think it's blatantly unethical. The medical profession has repeatedly rejected suggestions that physician-assisted suicide be legitimated. Doctors are supposed to be healers, not killers. Beyond that, I think it's most probably illegal.

Today, healthcare providers and clinical ethicists follow Kevorkian's 1990 prescrip-tion, while there is consensus in the medical ethics community, state legislatures, the health law community and the patient community that Wolf's comments above are antiquated.

In 1996, bioethicist Arthur Caplan was interviewed during one of Kevorkian's trials (for assisting Sherry Miller to die). He noted this (PBS 1996):

> I think before Kevorkian began to do what he does and draw attention to himself, the issue of assisted suicide, euthanasia, was present in our society. It was more on the back burner. People were aware of events in Holland. They knew that the Hemlock Society was pressing to legalize some forms of assisted suicide and books like "Final Exit" were in the works and people were thinking about should we give information to people who want to die. The HIV/AIDS epidemic had led people to think about assisted suicide more seriously. The issue was being considered, it just wasn't at the same volume, pace, velocity as it is today….

> It looks to most people that Jack kind of exploded onto the scene, but his work and his writings have been around and he's certainly popped up on my radar screen as someone who had some interesting views. To the rest of the world he was a nova, he was an explosion that came out of nowhere. A feisty doc that basically said I'm gonna take on the medical establishment. I'm gonna take on the legal establishment and I'm gonna do something unheard of. I'm going to assist people in dying publicly above board and say that's what I'm gonna do and then I'm going to dare somebody to come and prosecute me for it. …He was by no means a part of the death and dying movement as it had been known by the late 80's. And I think people were kind of puzzled saying, who is this guy and where did he come from and what led him to sort of stand forward and do this?

> …I think Kevorkian deserves credit for two things. He pushed the issue of assisted suicide from the back burner to the front burner. There is no doubt in my mind that this nation eventually would have engaged in a serious heated public dialogue about the right to die that engaged all of us. He just made it happen faster. He said I'm not going to let you look away from this, you must pay attention here.

> The other thing that Kevorkian realized was that society has tolerant attitudes about mercy killing and assisted suicide. I think that most doctors didn't understand that there has never been a successful prosecution of any doctor, anywhere, for an assisted suicide.…

Three years after Caplan made these comments, Kevorkian made medical history and was indeed convicted. As for Kevorkian's ideas about "terminal research" and organ harvesting on death row prisoners which he proposed in his 1991 book (Kevorkian 1991), an academic reviewer stated (Tardiff 1995):

> The reader will find no coherent theoretical basis for Dr. Kevorkian's proposals. Dr. Kevorkian is an ethical relativist who holds that no action is right or wrong in itself, but right or wrong depending on the circumstances. Yet he does not consistently maintain this.

Robert Levine, research ethicist and coauthor of the Belmont Report stated this in 1991: "What Kevorkian is talking about is research nobody wants to do anyway" (Priest 1993).

How do we judge Kevorkian today? It's best to say there is moral equipoise regarding the Kevorkian Chapter, but the more distant we get from 1999 the closer we get to Kevorkian's vision. As of 2019, a Canadian panel of health professionals representing critical care, transplant experts, nurses and palliative care experts recommended organ transplants from posthumous donors requesting active euthanasia from Canada's MAID law (Medical Assistance in Dying) (https://nationalpost.com/health/canadian-mds-to-restart-hearts-of-the-recently-dead-as-new-source-of-donor-hearts).

Theatrical Poster

You Don't Know Jack (2010)

> Director: Barry Levinson
> Producer: Lydia Dean Pilcher, Steve Lee Jones, and Scott Ferguson
> Screenplay: Adam Mazer
> Based on: The screenplay was based on Between the Dying and the Dead by Neal Nicol and Harry Wylie
> Starring: Al Pacino, Brenda Vaccaro, Susan Sarandon, Danny Huston, and John Goodman
> Music: Marcelo Zarvos
> Cinematography: Eigil Bryld
> Editor: Aaron Yanes
> Production Company: Bee Holder Productions
> Distributor: HBO
> Release Date: April 24, 2010
> Run time: 134 min.

References

Ad-HocCommittee of Harvard Medical School. (1968). A definition of irreversible coma: report of the Ad-Hoc Committee of Harvard Medical School to examine the definition of brain death. *Journal of the American Medical Association, 205,* 337–340.

Aghajanian, L. (2010). HBO's You Don't Know Jack. *Ianyan Magazine,* April 25, 2010. http://www.ianyanmag.com/review-hbos-you-dont-know-jack-kevorkian/.

Albom, M. (1997). *Tuesdays with Morrie.* New York: Doubleday.

Altman, L. (1992). U.S. doctors affirm Kennedy autopsy report. *New York Times,* May 20, 1992. https://www.nytimes.com/1992/05/20/us/doctors-affirm-kennedy-autopsy-report.html.

American Academy of Hospice and Palliative Medicine (AAHPM). (2014). *Statement on palliative sedation.* http://aahpm.org/positions/palliative-sedation.

Anonymous. (1988). It's over, Debbie. *JAMA, 259*(2), 272. https://jamanetwork.com/journals/jama/article-abstract/2623992.

Associated Press. (2010). Al Pacino takes on Jack Kevorkian's story in HBO film. *Michigan Live,* April 22, 2010. Retrieved from https://www.mlive.com/tv/2010/04/al_pacino_takes_on_kevorkians.html.

Beauchamp, T. L., & Childress, J. L. (2008). *The principles of biomedical ethics* (6th ed.). New York: Oxford University Press.

Beecher, H. K. (1966). Ethics and clinical research. *New England Journal of Medicine, 274*(24), 1354–1360.

Bhanoo, S. (2011). Dr. Death, Jack Kevorkian dies. *Washington Post*, June 3, 2011. https://www.washingtonpost.com/local/obituaries/dr-death-jack-kevorkian-dies-at-age-83/2010/12/03/AGhktuHH_story.html.

Biography.com. (2019). *Biography of Jack Kevorkian.* https://www.biography.com/scientist/jack-kevorkian.

Brody, B., & Halevy, A. (1995). Is futility a futile concept? *Journal of Medicine and Philosophy, 20,* 123–144.

Buck v. Bell. (1927). 274 U.S. 200.

Byok, I. (2014). Foreword. In E. Kubler-Ross, *On death and dying* (2014 edition). New York: Scribner.

Canyon News. (2010). *HBO's "You Don't Know Jack" a must see,* April 18, 2010. Retrieved from http://www.canyon-news.com/hbos-you-dont-know-jack-a-must-see/23916.

CBC News. (2018). July 11, 2018. https://www.cbc.ca/news/canada/british-columbia/latimer-pardon-murder-justice-1.4743353.

CBS. (1996). Interview with Jack Kevorkian. *60 Minutes,* May 16, 1996.

CBS. (1998). Interview with Dr. Jack Kevorkian. *60 Minutes,* November 22, 1998.

Centers for Disease Control and Prevention. (2011). CDC health disparities and inequalities report—United States, 2011. *MMWR, 60*(Suppl), 3.

Chesler, P. (1972). *Women and madness.* New York: Doubleday.

Childress, S. (2012). The evolution of America's right to die movement. *PBS,* November 13, 2012. https://www.pbs.org/wgbh/frontline/article/the-evolution-of-americas-right-to-die-movement/.

Cohn, R. (1988). *The autobiography of Roy Cohn.* New York: Lyle Stuart.

Connors, A. F., et.al. (1995). A controlled trial to improve care for seriously Ill hospitalized patients. The study to understand prognoses and preferences for outcomes and risks of treatments (SUPPORT). *The Journal of the American Medical Association, 274*(20), 1591–1598. https://jamanetwork.com/journals/jama/article-abstract/391724.

Cruzan v. Director. (1990). Missouri Department of Health, 497, U.S. 261.

Death with Dignity. (2019). *Assisted dying chronology.* https://www.deathwithdignity.org/assisted-dying-chronology/2019.

Donnelly, M. (2015). Healthy retired nurse ends her life. *The Telegraph,* August 2, 2015. https://www.telegraph.co.uk/news/health/11778859/Healthy-retired-nurse-ends-her-life-because-old-age-is-awful.html.

Doukas, D. J., & McCullough L. B. (1991). The values history: The evaluation of the patient's values and advance directives. *Journal of Family Practice, 32*(2), 145–153.

Farnsworth, C. H. (1994). Woman who lost a right to die case in Canada commits suicide. *New York Times,* February 15, 1994. https://www.nytimes.com/1994/02/15/world/woman-who-lost-a-right-to-die-case-in-canada-commits-suicide.html.

Federal Register. (1982). Proclamation 4966. September 13, 1982. *Reagan Library Archives.*

Fifield, A. (2018). So many Japanese people die alone. *Washington Post,* January 24, 2018.

Freedman, B. (1987). Equipoise and the ethics of clinical research. *The New England Journal of Medicine, 317*(3), 141–145. https://doi.org/10.1056/NEJM198707163170304.

Frieden, B. (1963). *The feminine mystique.* New York: W.W. Norton.

Frohlich, K. L., et al. (2006). Health disparities in Canada today: some evidence and a theoretical framework. *Health Policy, 79*(2–3), 132–143. https://www.ncbi.nlm.nih.gov/pubmed/16519957.

Girgis, L., et al. (2018). Physician experiences and barriers to addressing the social determinants of health in the Eastern Mediterranean Region: A qualitative research study. *BMC Health Services Research, 18,* 614. https://www.ncbi.nlm.nih.gov/pmc/articles/PMC6081851/.

Gostin, L., & Curran, W. (1987). AIDS screening, confidentiality, and the duty to warn. *American Journal of Public Health, 77*(3), 361–365.

Grisso, T., & Appelbaum, P. A. (1998). *The assessment of decision-making capacity: A guide for physicians and other health professionals.* Oxford: Oxford University Press.

Hopkins, C. (2011). Dr. Jack Kevorkian remembered for his advocacy. *Daily Times*, June 3, 2011. https://www.delcotimes.com/news/dr-jack-kevorkian-remembered-for-his-advocacy-the-media-circus/article_d6f836f0-e0c2-5f61-bb41-9b72df5c7f40.html.

Hosenball, M. (1993). The real Jack Kevorkian. *Newsweek*, December 5, 1993. https://www.new sweek.com/real-jack-kevorkian-190678.

House of Lords. (1969). Voluntary Euthanasia Bill [H.L.]. *HL. Deb.* 25 March 1969, Vol. 300. cc1143-254. http://hansard.millbanksystems.com/lords/1969/mar/25/voluntary-euthanasia-bill-hl.

Humphry, D. (1991). *Final exit*. New York: Dell Publishing.

In re Quinlan 70 N.J. 10,355 A2d 647, cert. denied sub nom. Garger v New Jersey, 429 U.S. 922 (1976).

Jecker, N. (2014). *Ethics in medicine*. University of Washington School of Medicine. https://depts.washington.edu/bioethx/topics/futil.html.

Johnson, D. (1999). Kevorkian sentenced to 10 to 25 years in prison. *New York Times*, April 14. https://www.nytimes.com/1999/04/14/us/kevorkian-sentenced-to-10-to-25-years-in-prison.html.

Kerr, M. E. (2000). *One family's story: A primer on Bowen Theory*. The Bowen Center for the Study of the Family. Available at: http://www.thebowencenter.org.

Kesey, K. (1962). *One flew over the cuckoo's nest*. New York: Viking Press.

Kevorkian, J. (1956). The Fundus Oculi and the determination of death. *The American Journal of Pathology, 32*(6), 1253–1269.

Kevorkian, J. (1959, May–July). Capital punishment or capital gain. *The Journal of Criminal Law, Criminology, and Police Science, 50*(1), 50–51.

Kevorkian, J. (1960). *Medical research and the death penalty*. New York: Vantage Books.

Kevorkian, J. (1985a). A brief history of experimentation on condemned and executed humans. *Journal of the National Medical Association, 77*(3), 215–226.

Kevorkian, J. (1985b). Medicine, ethics and execution by lethal injection. *Medicine and Law, 4*(6), 515–533.

Kevorkian, J. (1987). Capital punishment and organ retrieval. *CMAJ, 136*(12), 1240.

Kevorkian, J. (1991). *Prescription: Medicide. The goodness of planned death*. New York: Prometheus Books.

Kevorkian, J. (1992). A fail safe model for justifiable medically assisted suicide. *American Journal of Forensic Psychiatry, 13*(1).

Kevorkian, J., & Bylsma, G. W. (1961). Transfusion of postmortem human blood. *American Journal of Clinical Pathology, 35*(5), 413–419.

Kim, S. (2008). "Dr. Death" plans to run for Congress. *Reuters*, March 24, 2008.

Kirkus Reviews. (1991). Review: Prescription medicine. *Kirkus Reviews*, July 15, 1991. https://www.kirkusreviews.com/book-reviews/jack-kevorkian/prescription-medicide/.

Kubler-Ross, E. (1969). *On death and dying*. New York: The Macmillan Company.

Kubler-Ross. (1966). The dying patient as teacher: An experiment and an experience. In *The Chicago theological seminary register* (vol. LVII, no. 3, pp 1–14).

Lattin, D. (1997). Expert on death faces her own death. Kubler-Ross now questions her own life's work. *San Francisco Chronicle*, May 31, 1997. https://www.sfgate.com/news/article/Expert-On-Death-Faces-Her-Own-Death-Kubler-Ross-2837216.php

Lauren, G., et al. (2019). Social determinants of health training in U.S. primary care residency programs: A scoping review. *Academic Medicine, 94*(1), 135–143. https://doi.org/10.1097/acm.0000000000002491.

Lee v. State of Oregon, 891F. Supp. 1439 (D. Or. 1995).

Levi, B. H., & Green, M. J. (2010). Too soon to give up? Re-examining the value of advance directives. *American Journal of Bioethics, 10*(4), 3–22. https://www.ncbi.nlm.nih.gov/pmc/art icles/PMC3766745/.

McCormick, R. (1974). Proxy consent in the experimental situation. *Perspectives in Biology and Medicine, 18*, 2–20.

McCue, R. E. (2017). Baby boomers and rational suicide. In R. E. McCue & M. Balasubramanian (Eds.), *Rational suicide in the elderly*. Cham, Switzerland: Springer International.

McCue, R.E., & Balasubramanian M. (Eds.). (2017). *Rational suicide in the elderly*. Cham, Switzerland: Springer International.

McNamara, M. (2010). Dr. Death brings new life. *Los Angeles Times*. June 16, 2010. https://www.latimes.com/archives/la-xpm-2010-jun-16-la-en-0616-vaccaro-20100616-story.html.

Mehta, A., & Cohen, S. R. (2009). Palliative care: A need for a family systems approach. *Palliative and Supportive Care, 7*, 235–243. https://www.ncbi.nlm.nih.gov/pubmed/19538807.

Mohdin, A. (2018). Britain now has a minister for loneliness. *Quartz,* January 18, 2018. https://qz.com/1182715/why-the-uk-appointed-a-minister-for-loneliness-epidemic-leads-to-early-deaths/.

Moore, C., et al. (1962). Present status of cadaver blood as transfusion medium: A complete bibliography on studies of postmortem blood. *The Archives of Surgery, 85*(3), 364–370. https://jamanetwork.com/journals/jamasurgery/article-abstract/560305.

Müller, S. (2009). Body integrity identity disorder (BIID)—Is the amputation of healthy limbs ethically justified? *The American Journal of Bioethics, 9*(1), 36–43. https://doi.org/10.1080/15265160802588194.

Nancy B. v. Hotel-Dieu de Quebec. Dom Law Rep. 1992 Jan 6;86:385-95. Canada. Quebec. Superior Court.

National Public Radio (NPR). (2013). Interview with Keith Lowe, author of "Savage Continent". *Fresh Air*, July 24, 2013. https://www.npr.org/2013/07/24/204538728/after-wwii-europe-was-a-savage-continent-of-devastation

New York Daily News. (2010). Dr. Death reviews his life. *New York Daily News*, April 17, 2010. https://www.nydailynews.com/entertainment/346gossip/dr-death-reviews-life-kevorkian-al-pacino-jack-article-1.167595

Nelson, L. J., & Ramirez, E. (2017). Can suicide in the elderly be rational? In R. E. McCue & M. Balasubramanian (Eds.), *Rational suicide in the elderly*. Cham, Switzerland: Springer International.

Nicol, N., & Wylie, H. (2006). *Between the dying and the dead: Dr. Jack Kevorkian's life and the battle to legalize Euthanasia*. Madison: University of Wisconsin Press.

Obituary. (1994). Kevorkian's Sister, 68, Dies. *New York Times*, September 12, 1994. https://www.nytimes.com/1994/09/12/obituaries/kevorkian-s-sister-68-dies.html.

Oransky, I. (2004). Obituary of Elisabeth Kubler-Ross. *The Lancet* (Vol. 64), September 25, 2004. https://www.thelancet.com/pdfs/journals/lancet/PIIS0140-6736(04)17087-6.pdf

Patrick, K., Flegel, K., & Stanbrook, M. B. (2018). Vulnerable populations: an area will continue to champion. *Canadian Medical Association Journal, 190*(11), E307–E307.

PBS. (1996). *Interview with Arthur Caplan*. https://www.pbs.org/wgbh/pages/frontline/kevorkian/medicine/caplan2.html.

PBS. (2011). Jack Kevorkian dies. *PBS Newshour*, June 3, 2011. https://www.pbs.org/newshour/show/jack-kevorkian-doctor-who-brought-assisted-suicide-to-national-spotlight-dies.

PBS. (2014). *Dr. Kevorkian timeline*. https://www.pbs.org/wgbh/pages/frontline/kevorkian/chronology.html.

Peligian, E., & Davey, M. (2007). Kevorkian released from prison. *New York Times*, June 1, 2007. https://www.nytimes.com/2007/06/01/us/01cnd-Kevorkian.html.

Pinto, A. D., & Bloch, G. (2017). Framework for building primary care capacity to address the social determinants of health. *Can Fam Physician, 63*(11), e476–e482. https://www.ncbi.nlm.nih.gov/pmc/articles/PMC5685463/.

Priest, D. (1993). Hillary Clinton parries criticism of health plan in hill testimony. *Washington Post*, September 30. https://www.washingtonpost.com/archive/politics/1993/09/30/hillary-clinton-parries-criticism-of-health-plan-in-hill-testimony/ca8d8483-44b5-4b0e-b394-a7b0fd31d278/?utm_term=.a74c20987e4f.

Quill, T. (1991). Death and dignity: A case for individualized decision making. *New England Journal of Medicine, 324*, 691–694. https://www.ncbi.nlm.nih.gov/pubmed/1994255.

Rodriguez, T. (2015). Descendants of Holocaust Survivors have altered stress hormones. *Scientific American*, March 1, 2015. https://www.scientificamerican.com/article/descendants-of-holocaust-survivors-have-altered-stress-hormones/.

Rodriguez v. British Columbia (Attorney General), [1993] 3 S.C.R. 519.

Rosen, J. (2004). The final stage. *New York Times*, December 26, 2004. https://www.nytimes.com/2004/12/26/magazine/the-final-stage.html

Rosenthal, M. S. (2018). *Clinical ethics on film*. Switzerland: Springer International.

Roth, L. H., et al. (1977). Tests of competency to consent to treatment. *The American Journal of Psychiatry, 134*(3), 279–284.

Schneiderman, L. J. (2011). Defining medical futility and improving medical care. *Journal of Bioethics Inquiry, 8*(2), 123–131.

Shilts, R. (1987). *And the band played on: Politics, people, and the AIDS epidemic*. New York: St. Martin's Press.

Stanley, A. (2010). A doctor with a prescription for headlines. *New York Times*, April 22, 2020. Retrieved from https://www.nytimes.com/2010/04/23/arts/television/23jack.html.

Stout, D. (1997). Janet Good, 73; Advocated the right to die. *New York Times*, August 27, 1997. https://www.nytimes.com/1997/08/27/us/janet-good-73-advocated-the-right-to-die.html.

Tardiff, A. (1995). Review of prescription: Medicide. The goodness of planned death, by Jack Kevorkian. *The Thomist: A Speculative Quarterly Review, 59*(1), 167–171.

The Council of Canadian Academies (CCA). (2018). *State of knowledge reports on: Medical assistance in dying for mature minors; advance requests; and where a mental disorder is the sole underlying medical condition*. https://www.scienceadvice.ca/wp-content/uploads/2018/12/MAID-Summary-of-Reports.pdf.

Tohid, H. (2016). Robin Williams' suicide: A case study. *Trends Psychiatry Psychother, 38*(3), 178–182. http://www.scielo.br/scielo.php?script=sci_arttext&pid=S2237-608920160003 00178&lng=en&nrm=iso&tlng=en.

United Nations Department of Economic and Social Affairs. (2019). https://www.un.org/development/desa/disabilities/history-of-united-nations-and-persons-with-disabilities-a-human-rights-approach-the-1970s.html. Accessed January 22 2019.

Urban Institute Health Policy Center. (2014). *Vulnerable populations*. https://web.archive.org/web/20140106005409/http://www.urban.org/health_policy/vulnerable_populations/.

Voluntary Euthanasia Society. (1971). *Doctors and Euthanasia*. [Note: these reports are referenced here: http://journals.sagepub.com/doi/pdf/10.1177/002581727304100103].

Wilkerson, I. (1988). Essay on Mercy Killing reflects conflict on ethics for physicians and journalists. *New York Times*, February 23, 1988. https://www.nytimes.com/1988/02/23/us/essay-mercy-killing-reflects-conflict-ethics-for-physicians-journalists.htmls.

Part II
Justice and Healthcare Access

Chapter 4
The Unethical American Healthcare System: *Sicko* (2007)

When teaching about healthcare systems, and particularly, the American healthcare system, the documentary film, *Sicko*, released in 2007 during the second Bush era, is an excellent start. As a healthcare ethics film, *Sicko* falls squarely into the "Justice and Healthcare Access" box. This factual, yet still-controversial film by Michael Moore, lays bare the injustices of American healthcare and financial rationing prior to the passage of the *Affordable Care Act* (ACA) in 2010. *Sicko* is the perfect follow-up to any of the previous films about "Medical Harms", and explains how the scourge of health disparities and discrimination widened due to immoral frameworks for rationing and priority-setting in a corporate for-profit healthcare model bioethicists considered the worst system when compared to other democratic countries.

Sicko is a documentary film under a genre known as "documentary advocacy" in which the documentarian puts himself into the film and has a distinct point of view. Moore's signature style was presented in his 2002 film, *Bowling for Columbine*, on gun violence in the United States, and his 2004 film, *Fahrenheit 9/11*, which reviewed the post-9/11 political landscape and the false premise of the Iraq war. Moore's auteur style influenced a range of documentarians, including Morgan Spurlock, whose Moore-esque *Supersize Me* (2004), was an excoriation of the fast food industry's contribution to obesity—particularly in low income groups.

When *Sicko* was released in 2007, 46 million Americans had no health insurance. The film was explosive and damaging, and led to a campaign by the health insurance industry to stop its release, and create a negative publicity machine against the film. *Sicko* presents the American healthcare system as it was prior to the passage of the ACA, (a.k.a. "Obamacare"), in which insurance companies had no legal restrictions on discriminatory practices. Such practices comprised: (a) banning "high risk" individuals from getting health insurance through unaffordable premiums; (b) refusing to cover "pre-existing conditions" or paying for evidence-based treatments; (c) dropping customers from coverage when they got sick through "caps"; or (d) issuing "junk policies". *Sicko* is an important film that helps students understand current threats to the dismantling of the ACA, which will result in a return to the healthcare system that *Sicko* depicts, where patients are denied coverage, care and abandoned. As the

© Springer Nature Switzerland AG 2020
M. S. Rosenthal, *Healthcare Ethics on Film*,
https://doi.org/10.1007/978-3-030-48818-5_4

first decade of the ACA has passed, the law has become so woven into the fabric of American society, many current healthcare students do not have a living memory of the pre-ACA era, and will find *Sicko*'s content illuminating, alarming, and even hard to believe. In fact, the pre-ACA era in American healthcare is even at-risk of being forgotten, similar to recalling a pre-Social Security, pre-Medicare or pre-Medicaid United States. In essence, *Sicko* should be treated as much as a History of Medicine film as it is a healthcare ethics film. Threats to repeal the ACA persist, and in 2017, the law would have been successfully repealed by Congress but for a single Senate vote by the late John McCain—once an opponent of the law and the President who passed the law.

Prior to the ACA, the American healthcare system is shown to be an unjust, immoral system that rationed based on income, and pre-existing conditions. *Sicko* follows the effects of underinsurance, no insurance and unethical policies on average working Americans who are too young to be on Medicare, and do not qualify for Medicaid, which includes some who were even first responders on 9/11 denied healthcare for respiratory illnesses caused by 9/11 dust. *Sicko* was released just as the 2008 Presidential election was gearing up, and helped to put healthcare reform back on the map as a major 2008 election issue during the contentious Democratic primary in which Hillary Clinton and Barack Obama debated more than 20 times. Obama and Clinton provided the country with different visions of healthcare reform. Ultimately, it is Barack Obama's championing of a market-based system—initially a Republican plan—that wins out over Hillary Clinton's vision of universal healthcare, which President Bill Clinton's administration failed to pass in 1993. Moore's progressive and powerful critique of the terrible healthcare system that existed in 2007 is credited for making healthcare reform the signature Democratic campaign promise in the 2008 election. This translated into the most significant healthcare reform law in a generation, passed by the first African American President, whose name was tied to the law, which became branded as "Obamacare". This chapter will review the origins of *Sicko*, discuss its socio-political location, history of medicine context and the healthcare ethics issues raised in the film.

From Roger to *Sicko*: The Film Career of Michael Moore

Michael Moore made his reputation as a documentary filmmaker who broke the documentary film "rules" by narrating his own films, and putting himself in the film.

Unlike the classic cinema verite style of documentary, particularly in American documentaries such *Grey Gardens* (1975), Moore debuted with *Roger and Me* (1989) about the CEO of General Motors (GM) and the demise of Flint, Michigan's manufacturing jobs.

Moore also took on the gun lobby with *Bowling for Columbine* (2002) and the Iraq War with *Faharenheit 9/11* (2004). *Sicko*, released in 2007, during the Bush Administration, when healthcare reform was not on the "menu" of Congress. It was done in the same style as his other documentary films, which have now been copied by other documentarians. The film inspired Americans to become more curious about why the U.S. healthcare system was so terrible. Moore's films were heavily influenced by the socio-political events of his time, discussed in detail in the section on the Socio-Political location of *Sicko*.

Early Roots and Career

Michael Moore is best understood as a political activist before he became a multi-media journalist and filmmaker. A devout Catholic, Moore was born April 23, 1954 in Flint, Michigan to working class roots. His father worked on the assembly line for General Motors. Upon graduating from high school in 1972, Moore ran for a school board position so he could oust the principal, and Moore became the youngest school board member ever elected in the United States. Moore briefly attended the University of Michigan, but he did not graduate, and instead, became much more active in student journalism, and eventually left to pursue journalism full time. From 1976–1985, he was Editor of a weekly radical newspaper called *Flint Voice/Michigan Voice,* and hosted the radio show, *Radio Free Flint.* He began to gain national prominence by making frequent appearances on National Public Radio (NPR). In 1986, Moore spent four months as editor at *Mother Jones*, and left over a disagreement about the treatment of an article (Encyclopedia of World Biography 2019).

As a Flint native and resident, he was affected and outraged over GM layoffs, which led his self-financing of his first documentary, *Roger & Me*, (1989), which he made for $250,000. The film established a new genre of provocateur documentary film making, in which the point of view of the documentarian is part of the film. It was the highest grossing documentary ever made. It also led to techniques that would later be employed by a wide array of journalists covering absurdity in the news. *Roger & Me* received much praise, and reframed the power of documentary film, becoming Moore's signature formula. Moore married film producer, Kathleen Glynn, in 1991 (divorcing in 2014), and then spent most of the 1990s experimenting with television: *TV Nation* and *The Awful Truth*—both early hybrids of satirical takes on the news, such as the *Daily Show*, which exposed injustices, including insurance companies not covering patients. Moore also published his first book, Downsize This! Random Threats from an Unarmed American (Moore 1996) about corporate greed. Then came pivotal events that led to his foray into extremely influential work: the 1999 Columbine massacre, which led to *Bowling for Columbine* (see further), and 9/11, which led to *Farhenheit 9/11* (see further) (Encyclopedia of World Biography 2019; The Famous People 2019).

Synopsis

Sicko harshly examines the U.S. healthcare system, and its socio-political history. It covers the failed Clinton *Health Security Act*, and Hillary Clinton's role in drafting that legislation. Moore interviews several patients who were underinsured; health insurance insiders and politicians, and also compares the U.S. system to several of other democratic countries, such as Canada, France, and Britain, who are close allies of the United States Situated also as a "post-9/11 film", Moore also looks at the denial of health coverage to 9/11 first responders in a pre-*James Zadroga Act* U.S., in which there was no healthcare compensation for those suffering from the effects of 9/11 dust in the aftermath of the collapse of the World Trade Center. Moore's political advocacy antics include trying to go to Guantanamo Bay for "government-sponsored free healthcare" so that the 9/11 victims can at least get the same healthcare access as the 9/11 perpetrators held at Guantanamo Bay, Cuba, which is U.S. territory. Ultimately, Moore takes his 9/11 first responder group to Cuba, to get them properly assessed and taken care of, demonstrating Cuba's excellent universal healthcare system, long-studied by healthcare ethics students as a superior model.

Sicko debuts just in time for the 2008 election season, in which Hillary Clinton is the presumed front-runner as the Democratic nominee, in a "Democratic Year" where anti-Bush sentiment is high, and his poll numbers are below 25%. For these reasons, *Sicko* is also a political science film, conceived within the intense partisan politics that dominated the Clinton era and the Bush 43 era. The film's impact, however, inadvertently led to the passage of Obamacare during the Barack Obama Presidency, and its terminal battle for survival during the Trump era. *Sicko* has broad appeal, and is as much at home in a medical school class or health law class as it is in any American social science or humanities class.

Sickness and Health in Post-9/11: The Socio-Political Location of *Sicko*

The film, *Sicko* is a distinct post-9/11 film in which the health and wellbeing of Americans is presented within that specific time frame and political landscape, which deeply affected Moore in his twenty-first century work. Moore weaves 9/11 into the film several times, but the most dramatic reference to this time frame is his choice to focus part of the film on 9/11 first responders suffering from World Trade Center dust-related illnesses, who cannot get adequate healthcare. He juxtaposes that moral absurdity against the "free government-sponsored healthcare" prisoners like Khalid Sheik Mohammed (principal architect of 9/11) received at Guantanamo Bay. Although controversial, the contrast symbolized the depths of the U.S. healthcare system's immorality. But not all current viewers of *Sicko* may appreciate the differences between pre-9/11 and post-9/11 United States. Yet this appreciation is necessary in order to fully comprehend *Sicko*.

For Americans with a living memory of the events of September 11, 2001, the country in the previous decade had been enjoying one of its most prosperous international peacetime eras in the wake of the fall of the Berlin Wall (November 9, 1989), victory in the Gulf War (August–February, 1991), and hopefulness during the Clinton Administration (1993–2001). To understand the social climate of the country post-9/11, when *Sicko* is conceived, it's important to review the socio-political climate of the country pre-9/11, which shapes and informs Moore's worldview, his conviction that the country had lost its way, and his twenty-first century mandate to fix it. This section reviews the Clinton era; the 2000 election, which denies the Presidency to the progressive winner of the election, Al Gore; 9/11 and the intense post-9/11 landscape where the film is produced and launched.

The Clinton Era

When Bill Clinton was sworn in as President on January 20, 1993, the country had turned inward and became focused on domestic issues and domestic policy throughout the 1990s. A number of critical events took place during the new President's first year in office—some were planned, and some were unplanned. As Commander-In-Chief, Clinton's opening act was to review the Pentagon's policy on gays in the military, which led to "Don't Ask, Don't Tell" (see Chap. 2)—a law that would come in handy after 9/11. On the legislative front, Clinton wanted to pass sweeping healthcare reform by introducing the Health Security Act, which was not universal healthcare but universal coverage; but invoking the 1935 *Social Security Act* led critics and the public into thinking it as a single-payer universal healthcare program, when it was not (see under History of Medicine).

By the end of his first full month in office, Clinton already had problems with the women in his political life. When he sought to appoint a female Attorney General, his first nominee Zoe Baird had embarrassing immigration problems that were also a comment on the dilemma of women with careers. Zoe Baird had to withdraw her nomination on January 22, 1993 because she had hired undocumented immigrant women as nannies. Clinton's next nominee, Kimba Wood, had the same problem, and dropped out as a nominee February 6, 1993. Clinton's third nominee, Janet Reno was single with no children, and was nominated February 11, 1993 (confirmed March 11, 1993). Reno was tasked with confronting two major simultaneous events that took place about week before she was confirmed. On February 26, 1993 the first bombing of the World Trade Center occurred. On February 28, 1993, a bizarre cult operating in Waco Texas by David Koresh, whose followers called themselves the "Branch Davidians" (see further) turned into a tense 51-day standoff and siege in Waco where multiple cult members were shot. The late February events were covered simultaneously, but the World Trade Center story wound up taking a back seat to the situation in Waco. The terrorists responsible for the 1993 World Trade Center bombing were found and prosecuted by the Southern District of New York in the same time frame as the O.J. Simpson trial. Unbeknownst to most Americans at the time, the 1993 bombers were discovered to be connected to a growing terrorist organization known as al-Qaeda. In the wake of the 1993 bombing, the Clinton administration set up a distinct anti-terrorism unit within the FBI, which reported its activities to Reno. The experiences of the FBI and the CIA during this time frame were the subject of Lawrence Wright's book, The Looming Tower (2006), which was published while *Sicko* was in production. Wright's book is also the subject of the 2018 Hulu miniseries, *The Looming Tower* (Ultimately, turf wars between the FBI and CIA interfered with preventing 9/11, a plot in full swing by 1999.).

With respect to the legislative landscape, armed with a Democratic House and Senate, the new Democratic President could not get enough support for his Health Security Act to pass, (see further under History of Medicine). But the real problem was that despite their majorities in both houses in Congress, Congressional Democrats had too many conflicts of interest to do something bold, and were not equipped to handle the mounting opposition campaign against the centrist health-care bill. Millions of dollars were unleashed by the Conservative right and their supporters to kill the health reform bill, and misinform voters about the bill's content. It worked; eventually, Democrats caved to pressure and did not unanimously support the bill. Nonetheless, Clinton signed three major health-related bills into law during his presidency: The *Family and Medical Leave Act* (1993), granting employees 12 weeks of unpaid medical leave without penalty; the *Health Insurance Portability and Accountability Act*, or HIPAA (1996), and the *Children Health Insurance Program* (1997)— each critical programs for people with HIV/AIDS (see Chap. 2).

Democrats lost their majority in Congress in the 1994 mid-term elections, which led to a Republican crusade to politically decimate Bill Clinton. But by January 1995, when the new Congress was sworn in, public attention was diverted to the simmering racial tensions epitomized by the O.J. Simpson trial (1994–6). Then on April 19, 1995, Bill Clinton became a consoler-in-chief when he found himself dealing with the largest domestic terrorist attack on U.S. soil to date with the Oklahoma City Bombing (see further). In the mid-term elections of 1994, Republicans won control of the House of Representatives for the first time in 40 years, and shut the government down throughout the fall and winter of 1995–1996: November 14 through November 19, 1995; December 16, 1995–January 6, 1996.

The shutdowns only helped Clinton's popularity, and in 1996, he was riding high as he began his re-election campaign, with his motto: "Bridge to the twenty-first Century". Despite the failure of the Health Security Act, the majority of Americans were feeling good overall as the economy boomed, and the first generation of Americans to be connected to the Internet were exploring the information superhighway thanks to Clinton's Vice President, Al Gore, who introduced the internet law while he was a senator. The *High Performance Computing Act* of 1991, which led to the information "superhighway" during the Clinton Administration, as well as Microsoft Windows '95 which created seamless internet access. In fact, Janet Reno would bring a major anti-trust lawsuit against Microsoft Corporation for cornering the "internet search" market by bundling Internet Explorer with Microsoft Office. (Google was not founded until 1998.) With the dot com boom beginning, 1996 was the first year that any presidential campaign had a dedicated website. That year, the Dole/Kemp campaign (Republican Senator, Bob Dole and Republican House Representative, Jack Kemp) unveiled a signature "market based" healthcare reform plan—the counterplan to the Health Security Act—on its 1996 website, which few took notice of, but which would become ingredients in the *Affordable Care Act* 14 years later (see under History of Medicine). Bob Dole's lackluster appeal and uninspiring message other than "it's my turn" led to an overwhelming victory for Bill Clinton, who sailed to his second term with 379 electoral college votes. Clinton's second term was eventually a legislative success; he passed a bi-partisan budget that led to a surplus. However, Clinton's second term would become haunted by the 1995–6 shutdown, when the use of volunteer White House interns became a workaround for furloughed employees. One of the most competent interns working at that time was Monica Lewinsky, who began to confide in the wrong woman over an alleged affair she had started with Bill Clinton during the shutdown, when there was a skeleton staff in the West Wing.

Whitewater, Sex, Lies and Videotape

For the duration of most of the Clinton presidency, the "Whitewater Investigation" was ongoing. It began as an investigation to clear Bill and Hillary of charges trumped up by their political opponents alleging they were somehow involved in a shady real estate deal surrounding a property known as "Whitewater". Eventually, the investigation expanded into a range of business dealings during their Arkansas days, including Hillary's work at the Rose Law Firm, in Arkansas. In 1994, the Clinton Administration appointed its own Special Prosecutor, Robert Fiske, to investigate the Whitewater matter as well as the suicide of deputy White House counsel, Vince Foster. Fiske issued a 1994 report clearing the Clintons of any connections to shady business in Arkansas or the death of Vince Foster. Fiske was then appointed, under the *Independent Counsel Reauthorization Act* (1994) as Independent Counsel, but was replaced by Kenneth Starr in August 1994 because Fiske had been appointed by Janet Reno, which was considered to be a conflict of interest by 1994 standards. When Starr took over, he greatly expanded the Whitewater investigation, which eventually included 1994 allegations against Bill Clinton of sexual assault by Paula Jones. In 1994, Jones alleged that when Clinton was governor of Arkansas, he ordered a state trooper to bring her to his hotel room, where he proceeded to make unwanted sexual advances toward her, in addition to exposing himself. Ironically, 23 years later, similar accusations would bring down film mogul, Harvey Weinstein, whose production company made *Sicko*, and who was a close friend of Bill and Hillary Clinton.

In January 1996, Kenneth Starr subpoenaed Hillary Clinton in a criminal probe because of suspicions she had withheld documents; this was the first time a sitting First Lady would be subpoenaed, and led to a decades-long pattern of Hillary being the subject of an investigation.

On January 17, 1998, one of the first web-based magazines, The Drudge Report, broke the Monica Lewinsky story, which had been leaked to reporters by Lewinsky's "frenemy", Linda Tripp, who secretly taped her conversations with Monica in 1997. *Newsweek* reporter, Michael Isikoff, had the story for months, but had been sitting on it. *The Washington Post* ran their story on the affair four days later on January 21, 1998. When asked for a comment on January 26, 1998, Bill Clinton said the following (Washington Post 1998):

> Now, I have to go back to work on my State of the Union speech. And I worked on it until pretty late last night. But I want to say one thing to the American people. I want you to listen to me. I'm going to say this again: I did not have sexual relations with that woman, Miss

Lewinsky. I never told anybody to lie, not a single time; never. These allegations are false. And I need to go back to work for the American people. Thank you.

The Lewinsky affair became the basis for 11 obstruction of justice charges against Bill Clinton, who denied the affair under oath in video-taped testimony with his famous "it depends on what the meaning of 'is' is" dodge; this served as the basis for his impeachment. Lewinsky revealed that she lied about the affair, too, when interviewed by the Paula Jones attorneys under oath. By April 1, 1998, The General Accounting Office announced that Starr had spent nearly $30 million on his investigation as of September 1997 (Washington Post 1998). The Starr Report was released to the public, notably, on September 11, 1998. That morning, as Americans were drinking their coffees, they began to consume the juicy, sexually sensational report—co-authored by Brett Kavanaugh—by downloading it slowly on their dial-up internet from several websites that posted it in its entirety. No one could predict that exactly three years later, by 10:28 AM, many of those same Americans would be choking on their morning coffees as they watched the World Trade Center completely collapse from the worst foreign attack on America soil since Pearl Harbor, while some of the Starr-gazers that day would be dead three years later. Ironically, by the Starr Report's third anniversary, all any American wanted was to go "back to the '90s" when the worst political news was about a cigar, a semen-stained blue dress, and an intern with poor taste in girlfriends.

By the Starr Report's twentieth anniversary, Brett Kavanaugh, who took great pains to construct an impeachment case resting solely on the Lewinsky affair, was himself accused of sexual assault while drunk by a California professor of neuroscience, Christine Blasey Ford. Kavanaugh's testimony denying the allegations suggested it was just a "conspiracy" seeking revenge for what he did to Bill Clinton in the Starr Report. According to a 2019 "lookback" on the impact of the Starr Report, the *New York Magazine* stated (Kilgore 2019)

> An estimated 20 million people read the Starr Report online (or from printed downloads) within two days. Government servers crashed, and media sites scrambled to offer alternative feeding tubes for the insatiable demand. This was when less than half the U.S. population was online, and there was no social media to goose things along.

University of Southern California law professor, Orin Kerr, mused the following in 2019 (Correll 2019):

> Imagine if the Starr Report had been provided only to President Clinton's Attorney General, Janet Reno, who then read it privately and published a 4-page letter based on her private reading stating her conclusion that President Clinton committed no crimes.

The Lewinsky scandal indeed led to the Clinton impeachment hearings in December 1998, which he survived. In fact, Clinton's popularity rose during impeachment, a consideration in the impeachment of Donald Trump in 2019–2020. After impeachment, Clinton's approval was close to 70% in the polls, and the majority of Americans felt that impeaching him for "lying about a blow job" was ridiculous, considering what a great job he was doing. However, the scandal may have been enough to damage Vice President Al Gore in a razor-thin 2000 election that was ultimately decided in a 5–4 Supreme Court decision by the conservative majority of justices (see further).

From a legislative agenda standpoint. while the Clinton era was marred by the failure of the Health Security Act, the political climate that really killed the bill lay in growing "anti-government" sentiments by an expanding fringe of angry white males. It was during the Clinton era that the epidemic of disaffected white males who became "self-radicalized" would begin to commit shocking acts of domestic terrorism enabled by the National Rifle Association (NRA) lobby and lax gun laws. With virtually no restrictions on gun access, such laws made it easy for anyone to buy weapons. Those who were mentally ill or psychologically unstable; ideological extremists, psychopaths, and sociopaths could easily amass and store reservoirs of military-grade weapons and bomb materials intended for mass destruction of innocent victims. These acts informed Moore's first film of the twenty-first century: *Bowling for Columbine* (2002), in which Moore began to articulate his thesis that the United States was not a very nice place to live; it lacked the minimum basic social safety nets such as healthcare and gun control, which other democracies had, which made people feel safe and secure, and produced a less violent society. Instead, according to Moore's thesis, the society that had evolved in the United States by the end of the 1990s, was a greed-based, ethically-challenged country that placed guns at greater importance than children, and which punished the poor and rewarded the rich. Moore begins to work out some of his *Sicko* content in *Bowling for Columbine* when he has an epiphany: when countries offer universal healthcare and other basic needs, it creates a better society where citizens have a greater sense of wellbeing, including mental health. By contrast, in a society that makes clear the wellbeing of its citizens is not as important as profits, the citizenry starts to act out, and disaffected males, in particular, become even more disaffected, delusional/grandiose, and may get more aggressive. Moore further concludes that the country's sense of safety and wellbeing was already very fragile by September 10, 2001, but 9/11 further augmented and magnified the societal problems and sense of fear that were already there. He noted that in the aftermath of 9/11, gun sales went through the roof. Thus, to understand Moore's *Sicko*, it's important to understand the events that led to *Bowling for Columbine*.

"Sickos" in the 1990s: Waco, Oklahoma and Columbine

The confluence of unfettered access to guns to psychologically unstable or dangerous people led to a particular problem of "domestic terrorism" in the United States during the Clinton era. In 1993, the Waco FBI siege against the cult known as the "Branch Davidians" led by David Koresh, who had over 140 "wives" and untold children, became a confrontation because the Davidians had been stockpiling illegal weapons, in the delusional belief that it was a chosen sect immune to the laws of the United

States. Reports of child abuse led to the investigation in the first place. Ultimately, the Davidians were "smoked out" and several cult members who were women and children were killed in the standoff, which finally ended on April 19, 1993.

Two years later, on the same date—April 19, 1995—Timothy James McVeigh (1968–2001), a radicalized American white male who was an anti-government extremist, blew up an Oklahoma City government building using "weapons of mass destruction". The building also housed a daycare within it. The bombing killed 168 people and injured over 680. It remained the deadliest act of terrorism within the United States pre-9/11, and is still the deadliest act of domestic terrorism in American history. McVeigh was a Gulf War veteran who apparently sought "revenge" on the government for the Waco siege, which killed 86 people. The Oklahoma Bombing took place on the second anniversary of the Waco siege. McVeigh apparently wanted to inspire a "revolt" against the federal government. McVeigh was arrested after the bombing and found guilty on 11 charges, and sentenced to death; he was executed by lethal injection on June 11, 2001—exactly three months shy of 9/11. Ironically, McVeigh was on death row at the same time Jack Kevorkian (see Chap. 3) was in prison awaiting his appeal (scheduled for the morning of September 11, 2001) for his 1999 conviction for actively euthanizing a patient using lethal injection, having gone on the record that lethal injection should be used for death row prisoners. In fact, the night before McVeigh's execution, *60 Minutes* replayed its tape of Jack Kevorkian euthanizing Thomas Youk on camera (see Chap. 3) via lethal injection (James 2001).

Waco and Oklahoma had involved adult perpetrators. But the most chilling act of violence was the Columbine massacre, the first school shooting of its kind, in which the perpetrators were students themselves. Dylan Klebold, 17 years old and Eric Harris, 18 years old, were disaffected white male high school students who didn't fit in and made chilling video diaries. They obtained automatic weapons, and on April 20, 1999,—four years after Oklahoma—they entered Columbine High School in Littleton, Colorado and shot up anyone they could find before killing themselves. Clearly, it was a planned teen suicide in which they decided they would go out with a bang; they also set off propane bombs in the school. Investigators revealed that the teen shooters were inspired by the Oklahoma City bombing and were initially planning to bomb the school; they had been planning the attack for at least a year, and had placed a number of bombs in strategic locations, which did not detonate. It was "planned as a grand, if badly implemented, terrorist bombing" (Toppo 2009; Brown 1999). They killed 12 students, one teacher and wounded 20. Both had previously been involved in a car break-in and were undergoing mental health treatment. The teens had also thought about their escape after their plan. Harris left many journals in which he made clear that he hated everyone and was filled with rage. Eerily, he actually wrote this (CNN 2001): [sic], bold emphasis mine:

> If by some wierd as s–t luck [we] survive and escape we will move to some island somewhere or maybe mexico, new zelend or some exotic place where Americans cant get us. if there isnt such a place, then we will hijack a hell of a lot of bombs **and crash a plane into NYC with us …as we go down**.

As for Dylan Klebold, born September 11, 1981, 9/11 occurred on what would have been his twentieth birthday. He had been named for poet, Dylan Thomas, who famously wrote: "And death shall have no dominion."

Guns v. Gore

Al Gore, the Vice President of the United States, was running as the Democratic Presidential Nominee in the 2000 election. Although the Lewinsky scandal had soured many voters from the Democratic ticket, Gore was really running a post-Columbine campaign. On the first anniversary of Columbine, Gore said this (CBS News 2000; Dao 2000).

> We have to address…the physical fact that there are too many guns…I think one of the lessons of Columbine is that we have to stand up to the N.R.A. and the gun industry and get guns out of the hands of people who shouldn't have them.

At the time, political analysts saw gun control as a major 2000 election issue, dwarfing other domestic issues, such as healthcare. Gore was campaigning on gun reform laws, such as requiring photo ID for handgun buyers, as well as a gun registry to help monitor guns and gun owners. George W. Bush, on the other hand, ran on enforcing existing laws, and was opposed to a gun registry. In May 2000, the *New York Times* stated (Bruni 2000):

> The lines that divide the presidential candidates on the issue of gun control came into sharper focus today, as Vice President Al Gore vowed to veto any measure prohibiting cities from suing gun manufacturers and Gov. George W. Bush suggested he might support it.

Gore's positioning as a gun control candidate was actually the reason he lost his own state of Tennessee in 2000; Gore would have won the Presidency had he won his own state (Perez-Pena 2000). On November 9, 2000, the *New York Times* said this: "While Tennessee has moved to the right in national politics, Mr. Gore has moved to the left since his days as a congressman, particularly on issues like abortion and gun control…" (Perez-Pena 2000). Gore had a long record of being pro-second amendment. Journalist Karen Tumulty wrote February 7, 2000 (Tumulty 2000):

> In an interview last weekend, the Vice President said his early views of the [gun] issue reflected the perspective of a Congressman from a rural part of the South where "guns did not really present a threat to public safety but rather were predominantly a source of recreation." As a young representative of a conservative Tennessee district, Gore opposed putting serial numbers on guns so they could be traced, and voted to cut the Bureau of Alcohol, Tobacco and Firearms budget….What is likely to be more troublesome now are the votes he took in 1985 when the Senate – taking its first major stand on gun control in almost two decades–significantly weakened the gun law it had put into place after the assassinations of Martin Luther King Jr. and Robert F. Kennedy.

Oddly, Congress during the Reagan era had argued for weakening gun laws despite an assassination attempt on Reagan by a delusion-suffering John Hinkley Jr., who permanently disabled Reagan's Press Secretary, James Brady.

By the 2000 election, if you were a marginalized American during the Clinton years, health and prosperity was relative for predictable vulnerable populations. But in the wake of Columbine, everyone was now a vulnerable population, and parents were worried that school shootings would become routine—which, of course, they did. Malcolm Gladwell put it this way in 2015 (Gladwell 2015):

> In April of 1999, Eric Harris and Dylan Klebold launched their infamous attack on Columbine High, in Littleton, Colorado, and from there the slaughter has continued, through the thirty-two killed and seventeen wounded by Seung-Hui Cho at Virginia Tech, in 2007; the twenty-six killed by Adam Lanza at Sandy Hook Elementary School, in 2012; and the nine killed by Christopher Harper-Mercer earlier this month at Umpqua Community College, in Oregon. Since Sandy Hook, there have been more than a hundred and forty school shootings in the United States.

Current students do not have a memory of the 2000 election, which was a dead heat. Aggravating the tight race, Ralph Nader, who had run as an Independent third candidate in 2000, took just enough of the vote from Gore to make Florida the decisive state, but systemic issues with Florida's dysfunctional and irregular voting and ballots, led to a long Florida recount process, which ended in one of the most controversial Supreme Court decisions to date: *Bush v. Gore*, in which late Justice Antonin Scalia wrote the majority opinion. (When Scalia died in February 2016, Merrick Garland had been nominated by President Obama to replace him, but was denied a Senate confirmation hearing by Mitch McConnell.) Scandalously, Scalia was a friend of the Bush family (hence ethically conflicted) yet decided the 2000 Presidential Election in favor of Bush by declaring the recount be stopped, despite the fact that Gore had won the popular vote. This left Gore supporters—particularly Michael Moore—feeling that Bush was an illegitimate president. In the Summer of 2001, at around the same time that al-Qaeda terrorists were learning how to fly at U.S. flight schools (9/11 Commission 2004), political scientist, Gerald Pomper, wrote this (Pomper 2001):

> The presidential election of 2000 stands at best as a paradox, at worst as a scandal, of American democracy. Democrat Albert Gore won the most votes, a half million more than his Republican opponent George W. Bush, but lost the presidency in the electoral college by a count of 271-267. Even this count was suspect, dependent on the tally in Florida, where many minority voters were denied the vote, ballots were confusing, and recounts were mishandled and manipulated. The choice of their leader came not from the citizens of the nation, but from lawyers battling for five weeks. The final decision was made not by 105 million voters, but by a 5-4 majority of the unelected U.S. Supreme Court, issuing a tainted and partisan verdict. That decision ended the presidential contest, and George W. Bush now heads the conservative restoration to power, buttressed by thin party control of both houses of Congress. The election of 2000, however, will not fade. It encapsulates the political forces shaping the United States at the end of the twentieth century. Its controversial results will affect the nation for many years of the new era.

But no one could predict how consequential the decisions made by the Bush Administration prior to, during, and after 9/11, would alter the democracy of the United States and the geo-political infrastructure for decades to come. Michael Moore was instrumental in telling that story; Moore also surmised that the Bush "war machine" was dependent on depriving American citizens of basic societal needs, including higher education and healthcare.

From 9/11 to Sicko: *2001–2007*

What would become one of the most haunting stories of 9/11 was entangled in the political subterfuge that engulfed the 2000 election. In *Bush v. Gore*, George W. Bush (Bush 43) was well represented by experienced attorney, Theodore Olson, born on September 11, 1940. Olson, who had served as White House Counsel during the Reagan administration, had argued dozens of cases before the Supreme Court in the past; he was appointed Solicitor General of the United States in the Bush 43 Administration in February 2001.

Ted Olson's wife, Barbara Olson, had been a former federal prosecutor and attorney in the Department of Justice in the Bush 41 Administration, and then worked as an investigator for Congress during the Clinton era, tasked with looking into "Whitewater" related issues. She had also become a well-known conservative "Clinton-bashing" pundit on both Fox News and CNN during the Clinton era. In 1999, she had released the first *New York Times* bestselling anti-Hillary book, Hell to Pay: The Unfolding Story of Hillary Rodham Clinton (Olson 1999), as many had speculated about Hillary's "Lady Macbeth" hidden agenda to rise in politics using her philandering husband to her own ends. The book was released before Hillary formally announced her run for the Senate while she was still serving as First Lady in 2000, but Olson predicted in the book it was a certainty, as a forewarning of Hillary's unbridled quest for power to remake America into a socialist "Marxist" state. Olson also saw Hillary's numerous international trips as First Lady (seen as unprecedented and excessive), as her seeding the ground as an eventual Secretary of State (CSPAN 1999). A review of the book noted this (Amazon Reviews 1999):

> As the Clinton presidency draws to a close…Olson predicts the Senate won't be enough [for Hillary], just the next step toward becoming the first woman president: "Hillary Clinton seeks nothing less than an office that will give her a platform from which to exercise real power and real world leadership." While Olson admits that "Bill Clinton has always excited the greatest passion not among his supporters, but among his detractors," the same could certainly be said of his wife–whose supporters will probably consider Hell to Pay a rehash of a too-familiar story, but whose detractors will no doubt savor every page.

After the book was released, Hillary indeed announced her bid for a New York Senate seat in the Summer of 2000, and accepted victory in the 2000 Congressional election "six black pants suits later" (CNN 2000). As Bill Clinton began to plan his legacy, Hillary indeed would be joining the Senate in January 2001. By then, Barbara Olson had become a very popular conservative political analyst and media guest. Her last appearance was on CSPAN's *Washington Journal* on September 9, 2001 (CSPAN 2001), in which she was promoting her forthcoming book on the Clinton Administration's final days (Olson 2001). In that appearance, she predicted Hillary would run for President in 2008 (not 2004), discussed Janet Reno's run for

Florida Governor to defeat Jeb Bush, and discussed Bush 43's Education Bill and his promotional tour for it beginning on "Tuesday [September 11th]". Her appearance also included debates surrounding the legitimacy of *Bush v. Gore* and the Clinton impeachment. During the "call in" segment, one caller tells Olson she is filled with "hate for the Clintons", has the "devil in her" and "won't survive too long" (CSPAN 2001). Two days later Barbara Olson was one of the passengers on American Airlines Flight 77, which crashed into the Pentagon.

Olson had been invited to tape the September 11, 2001 episode of *Politically Incorrect*, hosted by Bill Maher (1993–2002), a pre-cursor to his HBO weekly politics show, *Real Time with Bill Maher*. Barbara Olson was supposed to fly to Los Angeles on September 10, 2001 for the appearance, but wanted to make sure she could wake up to wish Ted Olson "Happy Birthday" and changed her flight to September 11th instead. As the events of that morning unfolded, Ted Olson was relieved when Barbara called him from her plane (using an airfone), but she was not calling to say she was okay. Barbara called Ted to let him know that she was on American 77, which had been hijacked, and provided some of the first eyewitness details about the hijacking before it flew into the Pentagon a few minutes later, at 9:37 AM. The Olsons' joint effort to make sure that George W. Bush would become the 43rd President of the United States was met with "hell to pay". What the 9/11 Commission would uncover two years before *Sicko* was released is that the Bush Administration ignored clear warnings from the counter-terrorism unit that almost foiled the 9/11 plot (Wright 2006; 9/11 Commission Report 2004). Robert Mueller, who was the new FBI director, had only started on September 4, 2001, and was not yet up to speed. (Mueller replaced Louis Freeh and was succeeded by James Comey in 2013.)

By 10:28 AM on September 11, 2001—the third anniversary of the Starr Report—the North Tower of the World Trade Center (Tower 1) had collapsed, and George W. Bush became a wartime President, now presiding over the worst attack on U.S. soil since Pearl Harbor. This fact made Michael Moore sick, which he would later make crystal clear.

Between 2001–2, the first year post-9/11, a hyper-patriotism swept the country and no one was talking about *Bush v. Gore* anymore. In fact, Bush enjoyed close to an 80% approval rating, while stores were unable to keep up with the consumer demand for American flags. After Bush's bullhorn speech on Friday, September 13, 2001, standing on the "pile" (remnants of the World Trade Center) with an array of first responders who were not wearing protective masks, polls showed that the country was in favor of going to war to retaliate by close to 90%. The Bush Administration was essentially greenlighted by the public to make an array of decisions within a month's time: the creation of the Department of Homeland Security; the Transportation Security Administration (TSA); the passage of *The Patriot Act* (2001, which granted broad powers and eroded personal liberties; the invasion of Afghanistan (ongoing), and establishing the Bush Doctrine of pre-emptive attack on countries that "harbor terrorists". Slowly, a new war hawk era had begun in which criticism of the government and the Bush Administration was met with harsh retaliation. Moore references 9/11 in *Bowling for Columbine* to make the point that the country's sense of panic had augmented, leading to more gun sales. But he did not criticize the Bush

Administration's handling of 9/11 until the invasion of Iraq in March 2003. At the 75th Academy Awards ceremony, *Bowling for Columbine* won for best documentary, and in his speech, Moore denounced the Bush Administration on stage, while his next documentary, *Fahrenheit 9/11* was well underway. Moore made the following acceptance speech (Oscars.org 2003):

> I have invited my fellow documentary nominees on the stage with us … They're here in solidarity with me because we like nonfiction. We like nonfiction and we live in fictitious times. We live in the time where we have fictitious election results that elect a fictitious president. We live in a time where we have a man sending us to war for fictitious reasons…we are against this war, Mr. Bush. Shame on you, Mr. Bush, shame on you. And any time you've got the Pope and the Dixie Chicks against you, your time is up. Thank you very much.

Moore was booed when he spoke up, but the booing would not last long. When *Fahrenheit 9/11* was screened May 17, 2004, the following year at the Cannes Film Festival, the film received a 20 minute standing ovation. And then the damning *9/11 Commission Report* was released two months later.

The 9/11 Commission Report

In the aftermath of 9/11, the families affected began to ask tough questions, and lobby the reluctant Bush Administration for an independent commission to look into why their loved ones were murdered by terrorists. The effort was spearheaded by four New Jersey widows, known as the "Jersey Girls" whose husbands had been killed in the World Trade Center (Breitweiser 2006). In November 2002, the bi-partisan 9/11 Commission was formed, and its report was released July 22, 2004. The report essentially points to incompetency at all layers, involving immigration. airline security, airline industry protocols, military and air defense systems, evacuation failures at the World Trade Center and communication problems at the highest level of government intelligence agencies—chiefly, the FBI and CIA. These failures made the United States not just vulnerable to the attack, but too incompetent to stop it. *Fahrenheit 9/11* premiered May 17, 2004,—before the 9/11 Commission Report—wound up being the perfect companion work, and excoriated the Bush Administration. In it, Moore tells a convincing and disturbing narrative of the stolen 2000 election and coup by the conservative Supreme Court; the Bush Administration's ties to the Saudis, Big Oil, and ignoring warnings of the attack; using the attack as an excuse to invade Iraq for the fictional ruse that Saddam Hussein had ties to al-Qaeda and weapons of mass destruction (WMD). Moore began the film in 2003, and makes military families a big part of the story. Moore stated: "My form of documentary is an op-ed piece. It presents my opinion that's based on fact. I am trying to present a view of the last three-and-a-half years that I don't feel has been presented to the American public" (Hernandez 2004).

The famous footage of Bush looking stunned as he's reading to schoolchildren on 9/11 was footage that Moore retrieved, as noted by Roger Ebert in his review (Ebert 2004):

Although Moore's narration ranges from outrage to sarcasm, the most devastating passage in the film speaks for itself. That's when Bush, who was reading My Pet Goat to a classroom of Florida children, is notified of the second attack on the World Trade Center, and yet lingers with the kids for almost seven minutes before finally leaving the room. His inexplicable paralysis wasn't underlined in news reports at the time, and only Moore thought to contact the teacher in that schoolroom – who, as it turned out, had made her own video of the visit. The expression on Bush's face as he sits there is odd indeed….[Moore] remains one of the most valuable figures on the political landscape, a populist rabble-rouser, humorous and effective; the outrage and incredulity in his film are an exhilarating response to Bush's determined repetition of the same stubborn sound bites.

Between Iraq and a Hard Place

Fahrenheit 9/11 is also released around the time that the Abu Ghraib story hit, in which young American soldiers put in charge of Iraqi detainees at the Abu Ghraib prison, replicate the behavior demonstrated in the famous "Stanford Prison Study" from the 1970s, in which ordinary college students turned into abusive "guards" within a few days. Similarly, young American guards, with confusing orders, begin to abuse and torture the detainees. Photos are leaked, leading to the resignation of Donald Rumsfeld, Secretary of Defense. By this point, *Fahrenheit 9/11* started its own genre of anti-Iraq war films, with the premise that the Bush Administration is not only putting America's volunteer soldiers in harm's way without moral justification, but that the war was misguided and mismanaged with trillions of dollars of taxpayer money being diverted for war profiteering and illegitimate purposes. Familiar themes from *Bowling for Columbine* are also revisited: that Americans are being lied to, manipulated, and not taken care of.

In 2004, the Democratic Convention nominated Senator John Kerry as the Presidential nominee, and at the convention, Democratic Senate nominee, Barack Obama gave a speech on the convention floor. The speech was so eloquent, electrifying and inspiring, few forgot his name, and the new Senator, Barack Obama became the new shiny object when he began his Senate career in January 2005. But not enough Americans were ready to jump ship in the 2004 election, and Bush was decisively re-elected with a huge wave of Evangelical Christians, who supported the War on Terror as anti-Muslim sentiment foments, and the Fox News channel solidifies as a "War on Terror/Bush Doctrine" propaganda channel (Sherman 2017). John Kerry, running on his Vietnam War record, had made a misstep by not responding to a string of false attacks by other Vietnam veterans he did not serve with, who claimed Kerry was not a war hero and had lied about his record as a swift boat commander. The incident led to the term "Swiftboating"—a political smear campaign that blindsides the opponent. Kerry's campaign handled it badly, and the 2004 election was Kerry's to lose, leading to four more years of war and quagmire.

A year after *Fahrenheit 9/11* was released, Hurricane Katrina occurred August 31, 2005, devastating New Orleans. The aftermath led to one of the most horrendous displays of economic and health disparities in the United States when the city was abandoned by the Bush Administration, leaving mostly African American victims of the hurricane without food, shelter, water or basic resources in the stifling August heat of New Orleans. Over the Labor Day weekend holiday, pictures of Katrina haunted Americans who initially thought they were looking at scenes from a third world country. What they were viewing on CNN, noted for its extensive coverage at the time, was real poverty in a population below "see level". Those with means, mobility or good health, evacuated New Orleans. But those who were left behind were old, sick, or just too poor to evacuate. Echoes of Tuskegee (see Chap. 1), or abandoned AIDS patients (see Chap. 2) were familiar scenes playing out. Stranded hospitals practiced extreme rationing at the bedside (Fink 2013), and some patients received medical aid in dying (see Chap. 3). Katrina became as much of a socio-political disaster as it was a natural disaster. But the health disparities story demonstrated that people with poor healthcare access are far less likely to survive climate-change induced events. Indeed, the heavily diabetic population in New Orleans who could not afford to stockpile their insulin stayed behind, and many died from no access to their insulin after the storm. Hurricane Katrina, in fact, was only eclipsed in racially-motivated rescue disparities by the response to Puerto Ricans stranded in the wake of Hurricane Maria in 2017, which led to a similar death toll that occurred on 9/11 (Newkirk 2018).

By 2006, five years post-9/11, the dust from the World Trade Center had not yet settled; in fact, hundreds of reports of illnesses in people exposed to the dust—particularly first responders working on the "pile"—began to make news. But many who suffered from 9/11 dust-related illnesses either had no access to healthcare, or were told that treatment for their illnesses was not covered by their insurance plans. The *James Zadroga Act* (see under History of Medicine), would not be signed into law until a decade after 9/11. Moore's decision to shine a spotlight on sick first responders was significant. It was a different story being told this time. Moore demonstrated "healthism" (Leonard 2008)—a new kind of discrimination against people in poor health. Moore posits that if this is how America treats its best—discriminating against the heroes of 9/11 because they got sick—then "who *are* we"? It was a powerful allegory about a healthcare system that was morally unsustainable. Moore's evolving thesis about sickness and health in post-9/11 America came to fruition in *Sicko*, which debuted May 19, 2007, after both Hillary Clinton and Barack Obama had announced their runs for President, and the Democrats had won control of both houses of Congress in the 2006 mid-term elections.

Inconvenient Truths: Bush's Second Term

As *Sicko* is in production, another major documentary about political "sickness and health" debuts. Former Vice President Al Gore's *An Inconvenient Truth* (2006), wins the Academy Award for Best Documentary at the 79th Academy Awards ceremony held on February 25, 2007. Gore also wins a Nobel Peace Prize in 2007 for his work with the Intergovernment Panel on Climate Change. The dust had finally settled on *Bush v. Gore*. By 2007, the Bush Presidency had become a slow-moving car crash, as it becomes mired in the chaos and quagmire of the Iraq War with no exit strategy; struggles with the under-resourced Afghanistan war, where the hunt for Osama bin Laden had gone cold. The all-volunteer army is stretched with military personnel asked to do multiple tours. Veterans return home to an overwhelmed Veteran's Administration (VA), with the signature wound of the Iraq War: blast-induced traumatic brain injury. Over 2000 soldiers are dead by now, and mothers such as Cindy Sheehan start to speak up about the war, and camp out at Bush's ranch in Texas. Bush is also dealing with fallout from Katrina, runaway spending that eventually leads to the 2008 financial crisis, and toxic poll numbers. In contrast, Al Gore has never been more popular: he wins an Academy Award for one of the most important films about the threats of climate change, while millions of Americans who voted for him wonder what might have been in a Gore administration. *Sicko* rides the wave of progressive documentary films, and gives its viewers something to hope and fight for: healthcare reform. On the sixth anniversary of 9/11, a Moore-esque documentary film, which is the perfect complement to Moore's body of work, debuts at the Toronto International Film Festival: *Body of War* (2007) co-produced by Phil Donahue. The film is about one of the first wounded enlisted veterans of the Iraq war, his paralyzed body and broken life. But the film also excoriates the Senators who voted for the Iraq war without sufficient debate, and does a lot of damage to one of the candidates running for President: Senator Hillary Clinton, who also voted for the Iraq War in 2003.

Sicko's *Impact: The 2008 Election*

By January 2007 the majority of Americans had 9/11 fatigue, counting the days until the darkness of the Bush Administration would end. Americans were impatient for the post-Bush era, leading to a very early start to the 2008 campaign season. Hillary Clinton, as predicted by Barbara Olson years earlier, indeed announced her run for the Presidency on January 20, 2007. Hillary was the presumed frontrunner, and Bill Clinton soared in popularity in his post-Presidency with the Clinton Global Initiative. In fact, Bill Clinton had also grown close to the Bush family, as he and Bush 41 did humanitarian work together. Everything was going according to the Hillary Clinton campaign "coronation plan" until February 11, 2007, when Barack Obama announced his run for the presidency. The *New York Times* summarized it this way (Nagourney and Zeleny 2007):

> Senator Barack Obama of Illinois, standing before the Old State Capitol where Abraham Lincoln began his political career, announced his candidacy for the White House on Saturday by presenting himself as an agent of generational change who could transform a government hobbled by cynicism, petty corruption and "a smallness of our politics." "The time for that politics is over," Mr. Obama said. "It is through. It's time to turn the page." The formal entry to the race framed a challenge that would seem daunting to even the most talented politician: whether Mr. Obama, with all his strengths and limitations, can win in a field dominated by Senator Hillary Rodham Clinton, who brings years of experience in presidential politics, a command of policy and political history, and an extraordinarily battle-tested network of fund-raisers and advisers.
>
> Mr. Obama, reprising the role of Mr. Clinton, on Saturday presented himself as a candidate of generational change running to oust entrenched symbols of Washington, an allusion to Mrs. Clinton, as he tried to turn her experience into a burden. Mr. Obama is 45; Mrs. Clinton

is 59…But more than anything, Mr. Obama's aides said, they believe the biggest advantage he has over Mrs. Clinton is his difference in position on the Iraq war.

Hillary Clinton lost the Iowa Caucus to Barack Obama, which threw her campaign into turmoil (Green 2008). Although she squeaked out a victory in New Hampshire, she struggled to lead in the polls thereafter. Meanwhile, Clinton-era baggage and the Hillary-bashing narrative that the late Barbara Olson had begun in 1999, made Hillary Clinton a polarizing and unlikeable figure even for many women who self-identified as feminists. But she also failed to properly explain her vote for the Iraq war, which had also been based on faulty intelligence. She evaded direct answers when questioned about it and tried to justify her vote without simply saying it was wrong. (It became a preview of her failure to properly explain her private email server in 2016.) Hillary Clinton's scripted prose and talking points were also uninspiring. As Democratic voters began to view her as a face from the past, they began to fall in love with the future they saw with Barack Obama, whose time had come. As the primary season dragged on, and it was clear Barack Obama was now the front-runner, Hillary Clinton refused to drop out—something she would later blame Bernie Sanders for in the 2016 campaign. But thanks to Michael Moore's *Sicko*, the signature domestic issue debated during the Democratic primary season was healthcare reform, while Moore was a frequent commentator about the primary debates. The healthcare debate was not who was for it or against it, but which model would pass. Hillary Clinton was promoting universal healthcare, and a revised attempt to dust off the Clinton era bill that failed to pass in 1993 (see under History of Medicine), while Barack Obama was promoting something different: a market-based solution touted by Republicans years earlier, which he thought would get bi-partisan support and actually pass. And Obama could prove that it worked: one Republican governor was riding high on that same healthcare reform model. In 2006, Mitt Romney, governor of Massachusetts, had passed a version of Bob Dole's 1996 healthcare model with great success, covering nearly 100% of Massachusetts residents (see under History of Medicine). The "Romneycare" model would eventually become the *Affordable Care Act*.

Hillary Clinton ended her 2008 bid for the Presidency June 7, 2008, and post-campaign analysts uncovered her campaign as one of entitlement, and "inevitability" (Green 2008). In fact, Hillary Clinton's 2008 campaign was discovered to be one of the worst-managed Democratic campaigns in history for a front-runner (Green 2008), and unfortunately, the campaign resulted in *Citizen's United v. Federal Election Commission* (2010). To combat being "Swiftboated" (see above) the Clinton Campaign sued a conservative non-profit organization called "Citizens United" for plans to air an "anti-Hillary" film during that primary season, which was in violation of the *Bipartisan Campaign Reform Act* (2002); that law prevented corporations from spending money to promote or oppose any candidate. But the lawsuit backfired, and the Conservative Supreme Court upheld that "corporations are people" and had free speech to oppose or promote any candidate they wanted. The decision led to an avalanche of undisclosed donors (a.k.a. "dark money") into American elections, which, of course, ran the risk that some of that money would likely come from foreign entities wanting to influence elections in the future. Eight years later, Hillary Clinton would run virtually the same "coronation campaign" with even worse miscalculations and consequences that Michael Moore would later document in his film, *Fahrenheit 11/9* (2018).

McCain/Palin v. Obama/Biden

Arizona Senator, John McCain, had an unimpeachable record as a war hero, and was also an unapologetic war hawk, who felt strongly that the United States needed to maintain a military presence in Iraq and Afghanistan. Foreign policy, in fact, was the signature issue of the McCain campaign, which was unable to present anything "new" other than to build off of McCain's foreign policy chops and war record. "Country First", his campaign slogan, was having a hard time competing with Obama's campaign slogans, "Change We Can Believe In" and "Yes We Can". As the Bush Administration wore on and had hitched itself to religion and Christian conservatism, the Republican party was changing. An extremist, right-winged, white supremacy populist movement had begun to infiltrate the Republican Party. When Republican nominee John McCain saw the enthusiasm for Barack Obama, who was now being endorsed even by some moderate Republicans—including Bush's own Secretary of State, Colin Powell (2001–5), the campaign decided to pick a "game changing" Vice Presidential (VP) candidate (Heilemann and Halperin 2010). Originally, McCain had wanted Senator Joe Lieberman on a bi-partisan ticket, which he thought would strike the right notes for bridging the partisan divide. Instead, the campaign recommended a right-winged woman for the VP spot, Alaska Governor, Sarah Palin, whose populist appeal and xenophobic dog whistles, were lapped up by an increasingly rabid crowd. However, Palin had not been properly vetted, and McCain's campaign would come to have moral distress over selecting her (CBS 2010) once they became acquainted with how extreme and ignorant of American history she really was. But Palin had star power: attractive, folksy, a military mother with sons in active duty, Palin was far more popular than McCain, and became a *Saturday Night Live* mainstay channeled by Tina Fey. Palin helped to promote "birtherism" and painted Obama as a dangerous Marxist/socialist. But there was an "October Surprise"—Wall Street began collapsing and the 2008 financial crisis was threatening to destroy the American economy, leading President Bush to call for emergency Congressional funding to head off a Great Depression. The financial crisis dominated the rest of the 2008 campaign season, and Obama became the most articulate candidate surrounding policies and plans to deal with it. Barack Obama won a decisive and historic victory in November 2008, as the first African American President. With a Democratic super majority in Congress, and an inherited financial crisis where millions of people were now losing their jobs and health insurance, he was determined to get healthcare reform passed in a time frame he saw as his window. *Sicko* would continue to make the Obama Administration's argument for fixing healthcare.

The History of Medicine Context

When using *Sicko* to teach about the American healthcare system, the overall history of federal healthcare policy in the United States is what you'll need to cover, which supplements the History of Medicine content presented in *Sicko*. What health law experts will point out is that healthcare law and policies in the United States are a patchwork of different laws that were designed to close gaps or address specific populations, in recognition that the for-profit health insurance industry was the model. When *Sicko* was released in 2007, the U.S. healthcare system was a hybrid of several government-sponsored healthcare programs for vulnerable populations and veterans, as well as private insurance plans offered through employers, which widely practiced "healthism". Such plans routinely underinsured patients, cut them off if they got "too sick" and did not cover "pre-existing conditions". (See more under Healthcare Ethics Issues). Essentially, they were "junk policies". *Sicko* does offer a brief History of Medicine on how the U.S. healthcare system became so terrible, but generally focused on Americans with insurance and the insurance industry's discriminatory practices. The film does not cover any the existing universal U.S. healthcare programs, such as Medicare, Medicaid, or the Veterans Administration (VA) healthcare system—the latter having been established by Abraham Lincoln during the Civil War. But after *Sicko* was released, its impact led to two more significant health laws: The *Affordable Care Act* (2010) as well as the *James Zagroda Act* (2011), which covered all Americans suffering from 9/11-related illnesses due to exposure to the dust from the collapse of the World Trade Center. This section will review all of the key pieces of legislation in the "history of American healthcare"—notes you'll need to hit when teaching *Sicko*: the significant government-sponsored healthcare programs in the United States, as well as the universal healthcare laws that got away: failed attempts at universal healthcare. Since *Sicko* already focuses on how the for-profit insurance industry operated pre-ACA, I will not review that again here, but instead, discuss how the ACA made "Healthism" illegal. Finally, you may wish to review band-aid health legislation, such as the *Emergency Medical Treatment and Labor Act* (EMTALA), which mandates emergency care (see further) or the *Ryan White CARE Act* for HIV/AIDS patients, covered in Chap. 2.

Abraham Lincoln and the Veterans Administration (VA)

At one point, Americans led the way in universal healthcare by being the first to provide government-sponsored healthcare benefits for its veterans and their families, which dates back to the country's founding pre-Revolution (VA 2019), compensating colonists injured in battles with America's indigenous tribes (aka American Indians). The formal establishment of the Veterans Administration as we know it, traces its founding to a speech made by Abraham Lincoln on March 4, 1865, which officially became its motto in 1959 (VA 2019):

> With malice toward none, with charity for all, with firmness in the right as God gives us to see the right, let us strive on to finish the work we are in, to bind up the nation's wounds, to care for him who shall have borne the battle and for his widow, and his orphan, to do all which may achieve and cherish a just and lasting peace among ourselves and with all nations.

The VA makes clear even now:

> With the words, "To care for him who shall have borne the battle and for his widow, and his orphan," President Lincoln affirmed the government's obligation to care for those injured during the war and to provide for the families of those who perished on the battlefield... Today, a pair of metal plaques bearing those words flank the entrance to the Washington, D.C. headquarters of the Department of Veterans Affairs (VA 2019). VA is the federal agency responsible for serving the needs of veterans by providing health care, disability compensation and rehabilitation, education assistance, home loans, burial in a national cemetery, and other benefits and services.

The Veterans Health Administration went through many versions, but began to function as it is now after World War 1, and then grew from operating about "54 hospitals in 1930 to 1600 health care facilities today, including 144 VA Medical Centers and 1232 outpatient sites of care of varying complexity".

When Franklin Roosevelt was elected President in 1932—the same year the Tuskegee study began (see Chap. 1)—only 14 years had passed since the end of World War 1 (then called "The Great War"). This is relative in time span to how much time had passed since 9/11 and 2015, when President Obama was in his second to last year in office. But strikingly, in 1932, hundreds of Civil War veterans were still alive, and the last surviving veteran had lived until 1959. Thus, benefits that had been established under Lincoln were still being paid out to Civil War veterans under President Eisenhower.

When viewing *Sicko*, many students typically assume that all 9/11 "heroes" Moore presents were taken care of through the VA, which is not so. No civilian first responder to the attacks on 9/11 was categorized as serving in the military, nor were first responders on 9/11 who had professional unions. Firefighters, police, emergency medical technicians, etc., were not eligible for health benefits from the VA for any 9/11-related exposures (see further). Moreover, even families of civilian passengers on United 93, who fought the hijackers and tried to take back the plane, and were acknowledged by the Bush Administration as "the last line of defense" against an attack on the U.S. Capital, did not qualify for any VA benefits as "citizen soldiers".

The VA health benefits, however, remain government-sponsored healthcare based on a justice framework of "merit", which is one defined benefit of enlisting in the U.S. military. That said, after 9/11, the VA system indeed became overwhelmed, and it took years to accommodate women's healthcare into the VA.

FDR and LBJ

The only Presidents who managed to pass significant healthcare-related laws in the twentieth century were Franklin Roosevelt and Lyndon Johnson. In the same way that current students may not remember the United States pre-ACA, no one alive today remembers the United States before the *Social Security Act* (1935) which was passed at a time when 50% of the country was unemployed and older Americans were literally starving to death and abandoned, or died prematurely from lack of medical care. As summarized by *Health Affairs* in 1995 (Skocpol 1995):

> Some officials involved in planning the 1934 Social Security legislation wanted to include a provision for health insurance, but President Roosevelt and his advisers wisely decided to set that aside. Because physicians and the AMA were ideologically opposed to governmental social provision, and were organizationally present in every congressional district, Roosevelt feared that they might sink the entire Social Security bill if health insurance were included. Instead, Social Security focused on unemployment and old-age insurance and public assistance.

Roosevelt, like Obama 75 years later (see further) decided not to push things too far. By providing a guaranteed income to every American over 65, some income could be used for their health-related needs (e.g. nutrition) and was still a monumental progressive law that became part of the American society. Later, Social Security disability coverage was passed in 1956, providing benefits to Americans who cannot work due to disability.

President Lyndon B. Johnson's vow to create the "great society" built on Social Security by passing Medicare in 1965, which offered universal healthcare coverage to all Americans over 65, and completed what Roosevelt had started 30 years earlier. Medicaid was healthcare coverage for the impoverished and extremely vulnerable populations, also passed in 1965. "Care for the Old" was Medicare, while "Aid to the Poor" was Medicaid (Huberfeld 2006). When *Sicko* debuted, Medicare covered 43 million Americans regardless of income or medical history, and also covered those under age 65 with disabilities, which was added to the bill in 1972, representing roughly 15% of the population. But the problem with Medicare was that beneficiaries used four times more health services than the average American, and it paid hospitals at a lower rate, while the median income of most Medicare recipients was roughly $17,000 per year.

Medicaid (Public Law 89-97) sends funds directly to the States to finance healthcare for vulnerable populations, such as single parents with dependent children; the aged, blind, and disabled. When *Sicko* debuted, Medicaid covered about 51 million people, comprising roughly 39 million people in low-income families, 12 million elderly and persons with disabilities, and paid for nearly 1 in 5 health care dollars and 1 in 2 long term care (nursing home) dollars. In several states, "Work for Medicaid" programs are being illegally established (Huberfeld 2018). The ACA expanded Medicaid and redefined income criteria (see further).

The Nixon-Kennedy Debate

Sicko discusses the *Health Maintenance Organization* Act (1973), what most working Americans who do not qualify for Medicare or Medicaid are dealing with when *Sicko* debuts. Health Maintenance Organizations (HMOs) were an "insurance pool" in which basic health services are offered to an enrolled population that "pre-pays" (instead of fee-for-service healthcare), and has access to a number of health-care providers who are contracted by the HMO network. Thus, it was a complete myth that Americans could "choose their doctors"; they were frequently limited in choice by the HMO. The HMO is essentially a system of healthcare rationing to curb medical spending, but rationing is based on profit not morality (See under Health-care Ethics Issues). The HMO system served to sanction discriminatory practices in health insurance such as denying coverage for "pre-existing" conditions.

What is less known about the Nixon Administration, responsible for passing the *Environmental Protection Act*, is that Nixon was willing at one point to pass a version of the ACA because when he was young, his two brothers died from treatable illnesses due to lack of medical care in the 1920s. Senator Ted Kennedy admitted years later that he had actually blocked such a plan. The *Boston Globe* summarized it like this (Stockman 2012):

> When Nixon, a staunch Republican, became president in 1969, he threw his weight behind health care reform. "Everybody on his cabinet opposed it," recalled Stuart Altman, Nixon's top health aide. "Nixon just brushed them aside and told Cap Weinberger 'You get this done.'"
>
> Nixon had other reasons, beside his dead brothers, to support reform. Medicare had just been passed, and many Americans expected universal health care to be next. Ted Kennedy, whom Nixon assumed would be his rival in the next election, made universal health care his signature issue. Kennedy proposed a single-payer, tax-based system. Nixon strongly opposed that on the grounds that it was un-American and would put all health care "under the heavy hand of the federal government."
>
> Instead, Nixon proposed a plan that required employers to buy private health insurance for their employees and [provided a public option or subsidies] to those who could not afford insurance. Nixon argued that this market-based approach would build on the strengths of the private system. Over time, Kennedy realized his own plan couldn't succeed. Opposition from the insurance companies was too great. So Kennedy dispatched his staffers to meet secretly with Nixon's people to broker a compromise. Kennedy came close to backing Nixon's plan, but turned away at the last minute, under pressure from the unions. Then Watergate hit and took Nixon down. Kennedy said later that walking away from that deal was one of the biggest mistakes of his life.
>
> "That was the best deal we were going to get," Kennedy told [Boston Globe] before he died. "Nothing since has ever come close." Until Obama.

Many analysts conceded that we could have had the ACA (or Nixoncare) under Nixon decades ago, had it not been for Ted Kennedy. According to one political scientist (Newsweek 2009):

Nixon would have mandated that all employers offer coverage to their employees, whi e creating a subsidized government insurance program for all Americans that employer coverage did not reach…It was a rare moment in [Ted Kennedy's] Senate career where he made a fundamental miscalculation about what was politically possible—a lot of libera.s did…What was not recognized by anyone at the time was that this was the end of the New Deal era. What would soon come crashing over them was the tax revolts…

The article continued:

In fact, when Kennedy cosponsored another unsuccessful reform effort called Health America in 1991 that combined an employer mandate with a new public program, Princeton University health economist Uwe Reinhardt reportedly asked one of Kennedy's aides if they intended to cite Richard Nixon in a footnote.

Kennedy may have always preferred the government insurance approach, and he introduced legislation called "Medicare for All" as recently as 2006, perhaps trying to push the envelope when a Democratic resurgence was creating a new opening for progressive legislation. But this didn't stop him from enthusiastically championing the employer-based approach put forward by Barack Obama when he became the party's leader.

Reagan: The Emergency Medical Treatment and Labor Act (EMTALA)

EMTALA is a federal law that passed in 1986 that "requires anyone coming to an emergency department to be stabilized and treated, regardless of their insurance status or ability to pay" (ACEP 2020). This law was supported by emergency physicians concerned with widespread "patient dumping" that had been going on throughout the Reagan era, and is known as an "anti-dumping" law (Zibulewsky 2001; ACEP 2020).

In particular, two articles about Chicago's Cook County Hospital patient dumping (Zibulewsky 2001) led to outrage in the medical community and the passage of EMTALA. Patient dumping was originally defined as: "the denial of or limitation in the provision of medical services to a patient for economic reasons and the referral of that patient elsewhere" (Zibulewsky 2001). Lack of insurance was cited as the reason for denying emergency care to patients and transferring them elsewhere in 87% of the cases at Cook County hospital, and found to be widespread in "most large cities with public hospitals" throughout the 1980s.

EMTALA was "designed to prevent hospitals from transferring uninsured or Medicaid patients to public hospitals without, at a minimum, providing a medical screening examination to ensure they were stable for transfer [and] requires all Medicare-participating hospitals with emergency departments to screen and treat the emergency medical conditions of patients in a non-discriminatory manner to anyone, regardless of their ability to pay, insurance status, national origin, race, creed or color" (ACEP 2020).

As the gap in health insurance grew, EMTALA became the only way for many uninsured Americans to receive care, as *Sicko* documents extensively, which then began to overwhelm Emergency Departments with non-emergent patients, and escalate healthcare costs as EMTALA care was not covered. It also led the AMA to conclude that many emergency medicine physicians were essentially providing "charity care" in the United States. By 2003, the situation was becoming unmanageable, and many health insurance plans routinely denied claims for legitimate emergency departments visits (ACEP 2020), or even required "preauthorization" which made no sense. The American College of Emergency Physicians (ACEP) concluded then that: "These managed care practices endanger the health of patients and threaten to undermine the emergency care system by failing to financially support America's health care safety net." In 1986, physicians who wanted to stop patient dumping were initially worried that vague definitions of emergency care and stabilization would make EMTALA enforcement difficult. But by 2001, Zibulewsky (2001) notes:

> EMTALA has created a storm of controversy over the ensuing 15 years, and it is now considered one of the most comprehensive laws guaranteeing nondiscriminatory access to emergency medical care and thus to the health care system. Even though its initial language covered the care of emergency medical conditions, through interpretations by the Health Care Financing Administration (HCFA) (now known as the Centers for Medicare and Medicaid Services), the body that oversees EMTALA enforcement, as well as various court decisions, the statute now potentially applies to virtually all aspects of patient care in the hospital setting.

By 1996, the "uncompensated costs to emergency physicians for services provided under EMTALA" rose to $426 million in 1996, while unpaid in-patient care costs rose to $10 billion (Zibulewsky 2001). Meanwhile, tracking with the AIDS crisis, between 1988 and 1996, employer-sponsored health care coverage decreased from 72 to 58% (Zibulewsky 2001). Ultimately, EMTALA exposed the degree of underinsured and uninsured Americans, which Moore covers extensively in *Sicko*.

Clinton's Health Security Act: "Hillarycare"

As discussed earlier, the Clinton Administration drafted the 1993 Health Security Act to model in appeal, the *Social Security Act*. What President Clinton didn't anticipate was the vitriolic reaction he got when he put his wife, First Lady Hillary Rodham Clinton, in charge of drafting the bill "behind closed doors" along with adviser, Ira Magaziner. Due to the intense partisan politics that prevailed during the Clinton era, discussed in detail earlier, this singular incident made Hillary a lightning rod for the Republican party ever since. Dubbed "Hillarycare" there were putative complaints that Hillary had drafted the bill without bi-partisan input, which was not exactly the case. The task force did engage hundreds of stakeholders and held many meetings and hearings. But it did not make any of these stakeholders official representatives of the task force, and kept the deliberations confidential (Skocpol 1995).

The 1993 bill, in fact, resembled much of the current *Affordable Care Act*, and had three components to it: universal coverage through employers or regional state "health alliances" and federal subsidies (exactly the same as the current "state exchanges" and expanded Medicaid); bans on coverage denial for pre-existing conditions; the requirement for everyone to have coverage (Amadeo 2019).

However, there was vast misunderstanding of the bill, as most thought it was a single-payer universal health care system, because it invoked Social Security (Skocpol 1995):

> When he introduced his 1993 Health Security bill, President Clinton tried to invoke the Social Security precedent once again. This time, however, the analogy was purely rhetorical; it held only for the goal of universal, secure coverage. There was no relevant analogy to Social Security with regard to how governmental mechanisms in the proposed system would actually work.

> The key mechanism was the mandatory purchasing cooperative, something the Clintonites labeled the "health care alliance."

But the opposition seized on the fact that the bill was complex. According to a 1995 post-mortem (Skocpol 1995):

> Republican Dick Armey called the Clinton plan "a bureaucratic nightmare that will ultimately result in higher taxes, reduced efficiency, restricted choice, longer lines, and a much, bigger federal government....Centrist Democrat Bill Clinton had done his best to define a market-oriented, minimally disruptive approach to national health care reform, and his plan was initially well received. Nevertheless, by midsummer 1994 [voters] had come to perceive the Clinton plan as a misconceived "big-government" effort that might threaten the quality of U.S. health care for people like themselves.

The Task Force on National Health Care Reform, chaired by Hillary Clinton with Bill Clinton's advisor friend and business consultant, Ira Magazine, started January 1993. Complaints that the Task Force was working in secret started immediately from physician groups and Republicans in Congress. In May 1993, Hillary presented the plan to 52 senators, but "Republicans felt the administration had already gone too far in developing the plan without them and the task force dissolved" (Amadeo 2019).

The bill failed when it was brought to the floor November 20, 1993 (S.1757 (103rd): Health Security Act) and declared "dead" by September 1994. Two remnants of the bill were passed under Clinton: *The Health Insurance Portability and Accountability Act* (1996), which allowed employees to keep their company-sponsored health insurance plan for 18 months after they lost their jobs, but was more famous for its privacy laws in 2003. Hillary Clinton also convinced Senators Kennedy and Orrin Hatch to pass the *Children's Health Insurance Program* (1997), or CHIP (Amadeo 2019).

Ultimately, as *Sicko* makes clear, it was vitriol for Hillary that killed the bill: the country did not want the First Lady, an unelected person, to be put in charge of drafting a major bill, even though Bill Clinton campaigned on "two for the price of one". Cries of nepotism in the setting of Whitewater investigations doomed this universal health coverage bill. The 1993 Health Security Act would essentially be retrofitted into the *Affordable Care Act*, borrowing a few seasonings from Bob Dole's 1996 plan (see further).

The 1996 Dole/Kemp Healthcare Plan

The kernels of the current ACA can be seen on the Dole/Kemp 1996 campaign website (Dole/Kemp 1996). which told Americans the following in 1996, listed here verbatim:

Bob Dole Favors Common Sense, Free Market Health-Care Reforms

Losing health insurance coverage or facing obstacles to obtaining such coverage is a real concern for many Americans. We need insurance reform – but the answer is not a federal government takeover of the health care industry. Bob Dole supports consumer choice, open enrollment and portability of medical insurance.

In 1994, Dole introduced a health care reform bill which had no price controls, no mandates and no taxes. The bill took on a common sense, free market approach to health-care reform, focusing primarily on insurance reform, while offering subsidies to help low-income Americans buy health insurance.

As President, he will:

- Seek ways to make health care more accessible and affordable for all Americans.
- Ensure that individuals who change jobs do not lose their coverage or face pre-existing condition limitations.
- Give self-employed individuals the same tax deductions that large corporations have to buy health insurance.
- Make Medical Savings Accounts a real option available to all Americans.
- Support efforts to make community and home based care more readily available.

2006: RomneyCare

In 2006, Governor Mitt Romney signed into law the Massachusetts health care insurance reform law. The law mandates every resident to obtain a minimum level of healthcare insurance requiring residents to purchase insurance, and provided tax credits to businesses or pay a tax, but it provided free healthcare to those who were at 150% of the poverty level. The law ultimately covered 95% of the state's half a million uninsured residents within three years. The law was deemed a success story, and was wildly popular. Mitt Romney promoted the law as one of his crowning achievements and made clear in numerous interviews that it should be a model for the country. Barack Obama agreed. The Massachusetts law had many features of Dole's Healthcare Reform proposals from 1996. It was a model based on car insurance—everyone would be required to get health insurance, which would bring down costs but keep it market based.

From Sicko *to the Affordable Care Act (a.k.a. "Obamacare")*

Without *Sicko's* role in educating Americans about what "good healthcare" looks like, the window of opportunity that arrived in 2009, and an ambitious Democratic President determined to get healthcare reform passed, may not have occurred. The term "Obamacare" was established before the *Affordable Care Act* was even drafted—in 2007 when journalists began to attach various healthcare reform plans to the 2008 candidates (Reeve 2011). When Barack Obama became President in 2009, over 46 million Americans were uninsured. Barack Obama begins transparent bipartisan discussions about healthcare reform, making it clear that he is interested in a market-based model similar to RomneyCare, and very open to Republican ideas. He also appoints his former rival, Senator Hillary Clinton, to serve as Secretary of State, and she is not part of the healthcare bill's crafting or debates—hence removing déjà vu and "Hillarycare" from the lexicon. Although he has a Democratic supermajority in Congress, Obama makes many attempts to include Republicans and Independents in discussion. But it was too late. On election night 2008, the Republican minority met in a panic to develop a new strategy in Congress: obstruct everything (Kirk 2017), while Minority Leader Mitch McConnell was committed to making Barack Obama a one-term President. Meanwhile, unbridled racism abounded within the Republican Party, which condoned and encouraged "birtherism"—the conspiracy theory that Barack Obama is not a U.S. citizen, despite the fact that he was born in Hawaii to his white American mother, Anne Dunham. Suddenly, the market-based model Obama championed—the very same model Romney championed and Dole had proposed—became "socialism" while powerful Republican "brands" such as Alaska Governor Sarah Palin began to stoke fears about Obama's "death panel" (Gonyea 2017).

Republicans in Congress treated Obama with disrespect, and openly used racist language, even daring to call him a "racist against white people" (The Week 2009). What was once a fringe element in the Republican Party was now the base, and Obama's healthcare reform law became hopelessly mired in racism, when it was essentially a Republican model of healthcare reform. Key features of the law were requiring businesses of a certain size to cover their employees through group insurance rates; expansion of Medicaid; allowing adult children to stay on their family's plan until age 26; requiring everyone to purchase health insurance through an individual mandate by using a state insurance exchange; and requiring all health insurance policies to be compliant with strict non-discriminatory practices banning "healthism" and providing primary care and screenings. Eventually, nobody called the *Affordable Care Act* by its proper name except academics or conscientious Democrats. For everyone else, it was "Obamacare", which also helped to couple racist sentiments with the healthcare law. Two aspects of the law in the first version of the bill were met with feverish backlash: a "Public Option," in which those who couldn't afford insurance could get a form of government-sponsored healthcare; and reimbursement to healthcare providers to provide end of life counseling, given how much healthcare was usurped at the end of life with poor outcomes. The latter became the basis for cries of "Death Panels" by Sarah Palin. The final bill dropped those two provisions.

But what was kept, was still major reform, which comprised getting everybody into the "pool" to lower costs, and forcing insurance companies to remove all discriminatory practices: they had to cover pre-existing conditions and could not drop or raise premiums once the patient got sick. All insurance plans had to be ACA-compliant, however, which meant that millions who had the *Sicko*-styled "junk policies" were not able to keep their awful plans, which made Obama's promise that "if you like your plan, you can keep it" ring hollow, but the intention of the statement made the presumption that most employer insurance plans were ACA-compliant, when the majority were not. ACA plans also mandated coverage of contraception, leading to numerous religious group challenges. State Exchanges began to go up from the first day the law was active, and all had different names, even though the Exchanges were all part of the *Affordable Care Act*. However, the term "Obamacare" inspired such a knee-jerk reaction that in Kentucky, patients were actually fooled into thinking that "Kentucky Kynect" was wonderful (its State exchange set up by Democratic Governor Steve Beshear), but "Obamacare" was terrible (Kliff 2014). When Mitch McConnell ran for re-election to the Senate in 2014, Kentucky voters loved the *Affordable Care Act* so much that he vowed to keep "Kentucky Kynect" but promised he would repeal "Obamacare"—when Kentucky Kynect was, in fact, "Obamacare".

The Individual mandate was challenged multiple times, and upheld by the Supreme Court in 2012. It comprised a minor tax for refusing to buy insurance, and enacted as part of the Commerce Clause (Huberfeld 1997). The argument was that everyone needs healthcare; not everyone has health insurance; hospitals must treat and absorb costs of those who cannot pay; costs rise in anticipation of uncollectible debt and are passed on to private health insurance plans; premiums rise $1000/year due to uncompensated healthcare services. Therefore, mandating insurance was justified by the Commerce Clause (Huberfeld 1997). In turn, it would increase access to health insurance by unifying insurance products and markets, leveling the playing field, making everyone insurable and broadening the risk pool to include all citizens. This time, the bill passed with Democratic votes only, and Speaker Nancy Pelosi was the midwife. The law was hugely misunderstood at first, but over time, many Americans who had voted against Obama, realized the law was a matter of life and death. Many Democrats were "punished" for voting for the ACA in the midterms of 2010, but the law prevailed despite many attempts by Republicans to repeal the law, or have the courts declare it unconstitutional. The law was upheld in 2012 by the Supreme Court as well, just as the 2012 campaign season was heating up, in which Barack Obama would compete with Mitt Romney, who had difficulty explaining why he was opposed to "Obamacare" when it was genetically identical to "Romneycare". Although Mitt Romney attempted to slander the ACA by referring to it as "Obamacare"—Barack Obama made clear on the campaign trail that he liked the name "...because I *do* care" (Washington Post 2012). Eventually, for millions of Americans, Obamacare turned from a pejorative term into an endearing tribute.

By the time Obama left office in 2017, the law was integral for millions of Americans, and was highly popular, leading to almost 90% of Americans covered by health insurance, and reduced health care spending by $2.3 trillion. According to bioethicist Ezekiel Emmanuel (2019):

Despite constant criticism and occasional sabotage, the Affordable Care Act has successfully expanded health insurance coverage — even though it included individuals with pre-existing conditions — and controlled runaway health care costs. We need to build on its tremendous cost-control success.

Obama's political calculation in 2009 was that the country was not ready for single-payer universal healthcare and he did not want a repeat of the failed Health Security Act. The *Affordable Care Act* was not the single-payer system Moore had wanted, which he made that clear in an Opinion piece in which he excoriated the bill for being a Republican model (Moore 2013):

Now that the individual mandate is officially here, let me begin with an admission: Obamacare is awful.

That is the dirty little secret many liberals have avoided saying out loud for fear of aiding the president's enemies, at a time when the ideal of universal health care needed all the support it could get.

But by the 2016 election, Bernie Sanders led the Democratic Primary debates back to single-payor universal healthcare as one of his signature issues, while Sanders made clear that he was a key author of *The Affordable Care Act*. Much has been written since about the dire consequences of the 2016 election, which is beyond the scope of this chapter. However, in the Trump era, it's important to discuss with students how the major attempts to "Repeal and Replace" the ACA, made the law even more popular, leading thousands of Americans—including Michael Moore—into massive protests in the streets in order to keep the law intact. Additionally, Sanders run in the 2020 election focused on "Medicare for All" as did several other democrats running in the crowded primary field that year.

Repeal and Replace

In 2017, repeal and replacement of the *Affordable Care Act* (ACA) was proposed by the Republican-led Congress and White House with a return to "*Sicko*-like" insurance with the proposed American Health Care Act (AHCA). Under that proposed law, it was estimated that 24 million Americans who currently have health insurance and access to healthcare, would have lost their coverage. As proposed, the AHCA would have removed the current requirement under the ACA of mandated coverage for "essential health benefits" comprising 10 services: outpatient care, emergency room visits, hospitalization, maternity and newborn care, mental health and addiction treatment, prescription drugs, rehabilitative services, lab services, preventive care and pediatric services. Republicans argued that such coverage leads to higher insurance premiums (Belluz 2017). In short, the AHCA would have removed the requirement to cover what most would define as "healthcare". Some columnists referred to the proposed bill as "cruel" (Willis 2017). The bill would also allow for non-coverage of pre-existing conditions again (Kodjak 2017). But there were many other problems with the bill that would have removed healthcare access, which included restructuring Medicaid, tax cuts, and eliminating the lubricant that allows the ACA to work in the first place: the individual mandate (Gebelhoff 2017). Ultimately, only 17% of Americans were in favor of the proposed replacement bill (Bruni 2017; New York Times Editorial Board 2017).

In a second attempt, the Better Care Reconciliation Act of 2017, also a return to "Sicko-Insurance", would have killed thousands of American citizens by denying them affordable access to healthcare, and dramatically altering Medicaid. Drafted by only white, Republican males, this bill targeted vulnerable populations: the poor, the old—particularly those in nursing homes; the mentally ill, the disabled, pregnant women and their unborn children, and even neonatal patients (Pear and Kaplan 2017; New York Times Editorial Board 2017). The bill would have allowed states to opt out of covering vulnerable populations and essentially defunded Medicaid. It also eliminated caps on what insurance companies could charge people with complex health needs, and would have allowed insurers to at least double what older people pay (Jacobson 2017; Gawande 2017).

The Republican Party argued this bill would have been "better" because it gave "choices": the freedom to choose from a variety of unaffordable or inaccessible plans that would punish patients for their pre-existing conditions, and punish women for requiring prenatal and maternity care services. Again, only 17% of voters approved of it, and even *Republican* senators were against it (Bruni 2017; New York Times Editorial Board 2017). The American Medical Association, in a letter to Senators Mitch McConnell and Charles Schumer, stated this about the second repeal bill: "Medicine has long operated under the precept of Primum non nocere, or "first, do no harm." The draft legislation violates that standard on many levels" (AMA 2017). The bill recalled a poignant line from *Sicko*: "May I take a moment to ask a simple question? Who *Are* We?" The law almost passed; one single Senate vote made the difference: the late Senator John McCain, who was dying from glioblastoma at the time, and had refused to vote for the ACA, saved the law and voted against the repeal. When Senator McCain died in 2018, he made specific arrangements to have both George W. Bush and Barack Obama deliver the Presidential eulogies, banning the sitting President from his funeral. Notwithstanding, when the 2018 tax reform law was passed by the Republican Congress, the individual mandate that was integral to the architecture of the ACA was repealed.

Taking Care of 9/11 Responders: The James Zadroga Act (2011)

As mentioned earlier, *Sicko* is also a 9/11 film. When Moore is filming *Sicko*, it is 2006, the fifth anniversary of 9/11, and two years after his damning documentary, *Fahrenheit 9/11*. Michael Moore states in *Sicko* that we can "judge a society by how it treats its best," bringing attention to an environmental illness that was plaguing the first responders on 9/11 or anyone with prolonged exposure to Ground Zero or the "pile". Although there were beginning to be scattered news reports about illness from the World Trade Center dust exposure, the Bush Administration handled it much the same way the Soviet government handled questions about the air quality from the Chernobyl fiasco (April 26, 1986): they lied about the air quality to quell panic.

Sicko called attention to one of the most ethically egregious gaps in healthcare: coverage of illness from 9/11-related dust for first responders and others exposed to the toxic dust that permeated New York City for almost a year after 9/11. The failure of the EPA to accurately report the risks of dust exposure, or to supply the right safety equipment, led to a rolling health disaster caused by environmental exposure to what became known as "9/11 dust", which comprised diesel exhaust, pulverized cement, glass fibers, asbestos, silica, benzene from the jet fuel, lead, decaying human remains, and burning fires until December 2001 (Rosenthal 2016; Lioy 2010).

The Pile and Clean up

Immediately after the collapse of the World Trade Center towers, heroic efforts of first responders in New York City, inspired a "critical mass" of altruistic risk from volunteers who flocked to Ground Zero for rescue and recovery. The complete number of the first responders are unknown, and may have been as high as 80,000. Eventually, 60,000 became the widely used estimate. Workers came from all over the country, including many private engineering and construction workers and contractors, in addition to formal first responder services. There were also many private citizen volunteers never formally registered as recovery workers (Rosenthal 2016). Recovery and clean-up of the site lasted over a year with crews working around the clock in 12-hour shifts. In fact, construction workers found human remains as late as 2006. The worksite was divided into four zones, each with an assigned lead contractor, team of three structural engineers, subcontractors, and rescue workers. The worksite was essentially a "giant toxic waste site with incomplete combustion".

Between September 12, 2001, and May 30, 2002, (removal of the last column of the WTC), a myriad of occupational and safety issues emerged in which workers were not wearing personal protective equipment such as masks (some viewed it as "unpatriotic"), or were not sufficiently informed about the risks. Additionally, the EPA issued many misleading statements about the air quality throughout the first few weeks after the collapse of the towers (Rosenthal 2016; Lioy 2010).

Previous frameworks for evaluating the dust were not applicable, and studies began indicating health concerns in 2002; workers exposed to the dust began to report symptoms, with many becoming quite ill with "World Trade Center cough" by 2006 (Lioy 2010). Eventually, WTC dust and its health effects became its own separate field of research for study, but there were issues with self-report and sampling. Early symptoms included severe respiratory or sinus symptoms (within 24 hours), sore throats, severe shortness of breath, cough, and spitting grey mucus laced with solid particles of grit. Without adequate health insurance coverage for their symptoms, it became unclear which authority was in charge of taking care of them. There were multiple cases of pulmonary fibrosis, or chronic obstructive pulmonary disease, and as the years wore on, many developed cancers tied to the dust exposure.

James Zadroga was 29 on 9/11. He was an NYPD first responder who was a non-smoker and died in 2006. He couldn't work beyond October 2002, and died from respiratory and cardiac failure. His autopsy revealed "innumerable foreign body granulomas" on his lungs (Rosenthal 2016). His was the first case in which the autopsy tied his death to the WTC dust exposure. Another common illness related to the dust exposure was sarcoidosis, which spiked in the exposed population of first responders. Beyond the first responders, thousands of residents were exposed as the EPA did not even test indoor air of buildings in the WTC area for five months (Rosenthal 2016). According to Depalma (2011): "Contaminated schools, poisoned offices, and apartments where babies crawled along carpets that could be laced with asbestos remained a problem … Any movement—even someone plopping onto a sofa—would propel the thin fibers back into the air."

There were a myriad of resident clean-up issues, while the EPA ignored a panel of technical experts regarding minimizing risks. Undocumented workers who got sick were never covered for healthcare at all, while many schools in the area opened prematurely, which caused pediatric health issues in children. Residential clean ups were disorganized and many homes were never properly dealt with. Ultimately, this process dragged on through 2007, when samples were still being taken, and *Sicko* was released, which brought the problem of dust-related illness to the attention of the American public.

Before the *James Zadroga Act* was passed, worker compensation claims, and the September 11th Victim Compensation Fund, covered a number of personal injuries and deaths caused by 9/11, but victims had to make claims by December 2003. There were also federal grants to healthcare institutions for the World Trade Center Medical Monitoring and Treatment Program, which provided limited coverage for treatment. In 2006, the *James Zadroga 9/11 Health and Compensation Act* was proposed but didn't pass in the Bush era. It finally passed in 2010, and became active on the 10th anniversary of 9/11, four years after *Sicko* was released. The act expanded health coverage and compensation to first responders and individuals who developed 9/11-related health problems, setting aside $2.775 billion to compensate claimants for lost wages and other damages related to the illnesses. But it required reauthorization, and in 2019, former *Daily Show* host and comedian, Jon Stewart, a strong proponent for the Act, gave an impassioned speech to Congress to reauthorize it in 2019 (Iati 2019). Ultimately, in addition to all of the respiratory problems, 58 types of cancer were added to the "list of WTC-related Health Conditions" published in the Federal Register.

In one of the final scenes of the film, Moore makes the point that the 9/11 terrorists, including the plot's mastermind, Khalid Sheik Mohammed, were getting better healthcare than the 9/11 first responders. The scene plays out like this:

MOORE: There is actually one place on American soil that has free universal health care. Which way to Guantanamo Bay?

UNIDENTIFIED MALE: Detainees representing a threat to our national security are given access to top notch medical facilities.

MOORE: Permission to enter. I have three 9/11 rescue workers. They just want some medical attention, the same kind that the evildoers are getting.

In a 2007 interview with Larry King Live, Moore states (CNN 2007):

But the irony of the fact that the Al Qaeda detainees in GITMO, who we accused of plotting 9/11, receive full free medical, dental, eye care, nutrition counseling. You can get the list from the Congressional records of how many teeth cleanings they've done and how many colonoscopies – the whole list of this. And I'm thinking geez, they're getting better care than a lot of Americans. And, in fact, I knew these 9/11 rescue workers who weren't getting any care at all. They have now respiratory ailments as a result of working down at ground zero…And it just seemed highly ironic to me that the people who tried to save lives on 9/11 weren't getting help. The people who helped to plot 9/11 were getting all this free help. So I thought why don't we take them down there and see if we can get the same kind of help.

Healthcare Ethics Issues

Sicko is about violations of the Principle of Justice with respect to distributive justice, in which the burdens and benefits of societal goods, such as healthcare, ought to be evenly distributed within the population. That means that discussions for a healthcare trainee audience should focus on the ethics of healthcare access, rationing, and health disparities. Foundational philosophical concepts surrounding the Principle of Justice (e.g. materials principles of Justice, such as each according to need, merit, etc.) could be introduced with *Sicko*, but will not be discussed here, with the exception of the Rawlsian theory of Justice (see further), which Moore himself raises in a scene surrounding "patient dumping": we judge a society by how we treat its most vulnerable.

Healthcare Rationing

Distributive justice generally requires rationing, but the rationing framework for commonly available primary care and medical treatment in the wealthiest country in the world (the United States) would need to be based on the Principle of Benef-icence—maximizing benefit while minimizing harm, which may place limits on Autonomy. This generally excludes the withholding of emergency care, primary care, pediatric care, or any other medically necessary treatment for a new or pre-existing condition. In the United States, if the insurance model were based on the Principle of Beneficence, this would still ethically justify withholding medically unnecessary or non-beneficial treatments. For example, in critical care frameworks, aggressive therapy in a patient for whom death is imminent, who has no capacity or quality of life and who would not benefit from a surrogate-requested treatment, could be denied on ethical grounds. Rationing based on cost may also be ethically justified in some cases; for example, in patients with an end-stage cancer, who may live another three months with a tyrosine kinase inhibitor with an average retail price of $15,000 per month, would be denied based on a cost/benefit analysis in both Canada and Britain.

Instead, *Sicko* demonstrates that healthcare rationing in the American system is based solely on income, wealth, and discrimination against people in poor health or with pre-existing conditions, which can be defined as "Wealthcare" and "Healthism". Such discrimination clearly leads to premature death in the United States. Moore aptly put it this way (CNN 2007):

> We have a tragedy taking place every year now. Eighteen thousand people a year – these are the – these are the actual official statistics – 18,000 people a year die in this country for no other reason other than the fact that they don't have a health insurance card. That's six 9/11 s every single year in America. Forty-seven million without health insurance.

Wealthcare

Sicko led to a new definition by healthcare scholars of the U.S. health insurance model, known as "Wealthcare", which Moore brilliantly puts on full display in the film, with the poignant story of middleclass professionals, Larry and Donna, whose premiums were raised so high when they got sick, they became homeless, which any viewer is forced to conclude is morally indefensible. Initially, the health insurance industry tried to deny this was its model, but the moral distress of industry insiders channeled the famous moral acts of last resort—whistleblowing—which Jeffery Wigand did in the 1990s, when he blew the whistle on Big Tobacco. After *Sicko*, the parallel figure to Wigand—health insurance insider Wendell Potter—validated *Sicko's* accuracy (Potter 2010). Ultimately, practicing extortion on the insured was "wealthcare". Larry and Donna's ordinary tale went something like this: Now that you're sick, you can stay on your plan, but have to pay us more and more, or else we'll drop you, and then you'll never find insurance again. Clearly, "wealthcare", while an excellent business model, was completely immoral and a form of social Darwinism. Wealthcare is also practiced through drug costs (see further).

Healthism

Healthism was another new term introduced by scholars after *Sicko* (Skocpol 1995). Healthism is discrimination against someone in poor health, or against an insured person who gets sick. Here, people with pre-existing conditions are either denied insurance altogether, or coverage of an insured person's illness is denied if it can be connected—in any way—to a "pre-existing condition". Healthism is an important definition because it describes discrimination based solely on a health condition, rather than social position or race, or stigma of a particular disease, such as AIDS discrimination (see Chap. 2).

Racism and Health Disparities

Health disparities are multifactorial, aggravated by healthism and wealthcare; rationing based on income is also another way of discriminating against minority groups. Poverty, lack of access to primary care or early screening, and various genetic predispositions lead to health disparities, which then lead to healthism due to "pre-existing conditions", which then lead to wealthcare due to higher premiums, if insurance is even offered.

Sicko clearly demonstrates rationing based on overt racism and health disparities. In a particularly wrenching health narrative surrounding a mixed-race couple, a grieving white widow describes how her African American husband was abandoned by the healthcare system when he had end stage kidney disease, a disease that is far more prevalent in African Americans due to higher rates of type 2 diabetes and thrifty genes. In this story, a newer evidenced-based drug prescribed was denied, and labelled "experimental", and his widow stated that she was told her husband's treatment was also denied based on his race, echoing themes of Tuskegee (see Chap. 1). One is also forced to wonder about the role of racism in two other narratives. In one case, the denial of a cochlear implant for a 3-year-old—the progeny of a mixed-race couple—was probably based on race. In another tale, when an African American little girl is spiking a dangerous fever, her insured African American mother is turned away from the closest hospital and sent to one in a lower-class neighborhood that is in "her network" clearly in violation of EMTALA (see above). The little girl seizes and dies because she did not receive timely emergency care.

Price Gauging and Drug Costs

Sicko demonstrated in 2007 that U.S. drug prices were much higher than any other wealthy democratic country. Countries such as Canada, Britain or France negotiate drug pricing to make them more accessible or reasonably priced. In the U.S., drug prices are not controlled, which can be seen in post-*Sicko* examples such as the insulin rationing example from 2019 (Fralick and Kesselheim 2019). This reflects the influence of the pharmaceutical lobby. When Moore takes 9/11 first responders to Cuba, one of them cries (Reggie) when she discovers that a $125.00 inhaler can be obtained for five cents, for example. Drug pricing, however, is also entangled with the health insurance lobby. The *British Medical Journal* had this to say in 2007, when the film premiered (BMJ 2007) :

> *Sicko* criticises the US health insurance lobby, which, it says, paid huge sums to the campaign funds of leading politicians—nearly $900 000 to President Bush, for example—to support a bill requiring elderly Americans in the Medicare insurance plan to sign up to one of a confusing number of plans offering drug discounts. The bill, passed in the middle of the night nearly four years ago, prohibited Medicare from negotiating drug prices with manufacturers, leading to higher prices for Medicare users and the Medicare administration.

Ultimately, drug price gauging affects Americans on Medicaid, Medicare, VA benefits or private health insurance, and is another form of wealthcare. The *Affordable Care Act* significantly improved some of this by mandating that insurance companies cover prescription drugs as one of 10 Essential Health Benefits (HealthCare.gov 2020), which also includes contraception for women (see Chap. 5), leading to religious objections.

"Who Are *We"? The Rawlsian Theory of Justice*

Towards the end of *Sicko*, viewers watch egregious examples of immoral healthcare in the United States in which vulnerable patients are dumped, violating the tenets of safe discharge, Moore interjects: "May I take a moment here to ask a question? Who *ARE* we? They say you judge a society by how it treats its most vulnerable…" At that moment, Moore is summarizing philosopher John Rawls' "veil of ignorance" argument in his political Theory of Justice (1971), which holds that the test for a just and fair society is whether the architect of such a society would be willing to live on the bottom societal rung. In other words, one must build the society wearing a veil of ignorance regarding whether one is the winner or loser of a genetic or social lottery. According to Rawls, inequalities of a social good such as healthcare is only permitted if to the person who is the "worst off" could still benefit, rather than just be penalized for a social position that is beyond that person's control, such as being born into poverty or having a "pre-existing condition".

Veil of Ignorance: How to Build a Good Society

Rawls makes use of a thought experiment other philosophers previously invoked, called the "veil of ignorance". It goes like this: You are building a just society but don't know what part you're playing in that society, so you need to design the society in a way that permits you to enjoy minimum benefits, human rights and dignity should you wind up being the person who is the worst off socially or in the poorest health. According to Rawls (1971): "no one knows his place in society, his class position or social status; nor does he know his fortune in the distribution of natural assets and abilities, his intelligence and strength, and the like". Designing the society with the "veil of ignorance" means you can't let conflicts of interest or commitment or bias to interfere with your design. In other words, what society, in 1932, would have been designed by the architect, if s/he risked playing the part of an African American sharecropper in Macon County Alabama (see Chap. 1)? Or waking up as a 14-year-old rape victim who is pregnant in 1972, before abortion was legal (see Chap. 5)? Or waking up as gay male with AIDS in 1982? Or waking up with ALS in 1998, and not allowed to ask for a physician's assistance to end your suffering and die sooner (see Chap. 3)? Or waking up in New York City on September 11, 2001 as a first responder who then inhales large quantities of toxic dust when the

World Trade Center collapses, and is denied healthcare? Or waking up with any chronic pre-existing health condition in 2007, when *Sicko* was released? Or, in 2020, being hospitalized with COVID-19 (see Afterword) without health coverage. Moore channels Rawls by showing us not only the unjust society that Americans live in, but he shows us the architects who designed it, too, in which conflicts of interest and bribes led to the immoral healthcare system we found ourselves in by 2007.

In *Sicko*, Moore also challenges the costs of higher education, which would enable those who were the least advantaged to maximize their opportunities. Rawls essentially asks us to use a double-blinded method to build a just society; no one knows who they will be when the project is built. In a Rawlsian framework, only when principles of justice are chosen behind a veil of ignorance—or double-blindedness—will it lead to a design where even the worst off have maximum opportunities to at least improve because then even the most selfish would require equal opportunity as the minimum starting point (Rawls 1971).

Rawls' work broke through because it was applied philosophy to overt political problems. It was published just around the time the Tuskegee study was making headlines, and particularly resonated with emerging bioethics frameworks, eventually borrowing from Rawls in articulating the bioethics Principle of Justice. The *New York Times* book review wrote this just two weeks after the Jean Heller story about the Tuskegee study (see Chap. 1) broke (Cohen 1972):

> In the opinion of John Rawls, professor of philosophy at Harvard… it is no accident that the Founding Fathers looked into [social contract theory] for their main philosophical doctrines….But the revival of political philosophy [is] also a response to the moral obtuseness and the debased political rhetoric of our time….[It] is therefore not surprising that Rawls's penetrating account of the principles to which our public life is committed should appear at a time when these principles are persistently being obscured and betrayed. Rawls offers a bold and rigorous account of them and he is prepared to argue to a skeptical age that in betraying them it betrays reason itself.

The *New York Times* makes clear that Rawls argues against utilitarianism because "[s]uch a philosophy could justify slavery or, more to the present point, a suppression of the very political rights that are basic to our notions of constitutional government." With respect to *Sicko*, it is Rawls second principle of Justice…(Cohen 1972):

> …which applies to the distribution of wealth and to arrangements of power and authority. In this domain Rawls is not an equalitarian, for he allows that inequalities of wealth, power and authority may be just. He argues, however, that these inequalities are just only when they can reasonably be expected to work out to the advantage of those who are worst off. If, however, permitting such inequalities contributes to improving the health or raising the material standards of those who are least advantaged, the inequalities are justified. But they are justified only to that extent—never as rewards for "merit," never as the just deserts of those who are born with greater natural advantages or into more favorable social circumstances….The natural and social "lottery" is arbitrary from a moral point of view…There is another way to deal with them. As we have seen, they can be put to work for the benefit of all and, in particular, for the benefit of those who are worst off….

Even if these speculations are sound, however, they cannot justify the inequalities we now accept or the impairment of our liberties that we now endure. For our politics are inglorious and our high culture does not enjoy extravagant support. Whatever else may be true it is surely true that we must develop a sterner and more fastidious sense of justice.

Conclusions

Sicko is one of the most important and influential films made in the post-9/11 era which fundamentally led to enduring healthcare reform resulting in the *Affordable Care Act*, as well as lifelong coverage for 9/11-related illnesses. However, in a current time frame in which historical amnesia is setting in more quickly, reviewing the socio-political origins of *Sicko* in a pre-ACA, intensely partisan era will help current students understand and appreciate what it would mean if the law were to be weakened or repealed. Teaching *Sicko* also helps to inform students about why the ACA is still not "universal healthcare". And in the wake of the Coronovirus Pandemic of 2020 (see Afterword), the vulnerabilities of the disparities wrought by the American healthcare system have made Sicko only more relevant. *Sicko* documented what America looked like before the ACA, so that viewers can say "never again". In 2007, *Sicko* provided the counter narrative to the lie that the U.S. healthcare system was the best in the world. Moore explained the film this way (CNN 2007):

> My feeling is, is that for two hours, I'm going to come along and say here's maybe another way to look at it. Here's maybe a story that isn't told. And so that's – and that's what I do with most of my movies….Sometimes I go too far in advance and people aren't ready to hear it or listen to it. But my contribution to this country is to make these films in the hopes that we can get things right, make them better and aspire to everything that I think that we're capable of doing here….
>
> Illness isn't Democrat. It isn't Republican. And I've made a film where I'm hoping to reach out and you saw that in the film, reaching out to people who disagree with me politically but saying to them we can have some common ground on this issue. We should come together on this issue and let's fix this problem.

Theatrical Poster

Sicko (2007)

Directed by: Michael Moore
Produced by: Michael Moore and Meegan O'Hara
Written by: Michael Moore
Starring: Michael Moore
Narrated by: Michael Moore
Music: Erin O'Hara
Production Company: Dog Eat Dog Films
Distributed by: Lionsgate and The Weinstein Company
Release Date: June 22, 2007

References

Amadeo, K. (2019). *Hillarycare, the Health Security Act of 1993*, March, 2020. https://www.thebal ance.com/hillarycare-comparison-to-obamacare-4101814.

Amazon Reviews. (1999). *Hell to pay.* https://www.amazon.com/Hell-Pay-Unfolding-Hillary-Cli nton/dp/0895262746.

American College of Emergency Physicians (ACEP). (2020). *EMTALA fact sheet.* https://www. acep.org/life-as-a-physician/ethics–legal/emtala/emtala-fact-sheet/.

American Medical Association (AMA). (2017). *Letter to senate majority leader*, June 26, 2017. https://searchlf.ama-assn.org/undefined/documentDownload?uri=%2Funstructured%2Fb inary%2Fletter%2FLETTERS%2FBCRA-Letter.pdf.

Belluz, J. (2017). The Republican attack over essential health benefits explained. *Vox*, May 4, 2017. http://www.vox.com/2017/3/23/15031322/the-fight-over-essential-health-benefits-explained.

Breitweiser, K. (2006). *Wake-up call: the political education of a 9/11 widow.* New York: Warner Books.

British Medical Journal (BMJ). (2007). US health professionals demonstrate in support of Sicko. *BMJ*, June 30, 2007. *334*(7608), 1338–1339. https://dx.doi.org/10.1136/bmj.39258.421111.DB

Brown, J. (1999, April 23). Doom, Quake and mass murder. *Salon.*

Bruni, F. (2000). Gore and Bush clash further on firearms. *New York Times*, May 6, 2000. https://www.nytimes.com/2000/05/06/us/the-2000-campaign-the-gun-issue-gore-and-bush-clash-further-on-firearms.html.

Bruni, F. (2017). Trump's Trainwreck. *New York Times*, March 24, 2017. https://www.nytimes.com/ 2017/03/24/opinion/sunday/trump-and-ryan-lose-big.html.

CBS (2010). Inteview with Steve Schmidt. *60 Minutes*, January 10, 2010.

CBS News. (2000). *Gore and Bush recall Columbine.* April 20, 2000. https://www.cbsnews.com/ news/gore-bush-recall-columbine/.

Cohen, M. (1972). The Social contract explained and defended. *New York Times*, July 16, 1972. https://www.nytimes.com/1972/07/16/archives/a-theory-of-justice-by-john-rawls-607-pp-cambridge-mass-the-belknap.html?_r=0.

CNN. (2000, November 4). *Clinton senate acceptance speech.*

CNN. (2001). Colombine killer envisioned crashing plane in NYC. *CNN*, December 6, 2001. https://web.archive.org/web/20111006163130/http://archives.cnn.com/2001/US/12/05/ columbine.diary/.

CNN. (2007). *Larry King Live.* June 30, 2007. http://transcripts.cnn.com/TRANSCRIPTS/0706/ 30/lkl.02.html.

Correll, D. (2019). Monica Lewinsky: If F-king only. *The Washington Examiner*, March 17, 2019. https://www.washingtonexaminer.com/news/monica-lewinsky-if-f-king-only-a-four-page-summary-of-the-starr-report-was-released.

CSPAN. (1999). *Barbara Olson*, November 15. https://www.c-span.org/video/?153644-1/hell-pay.

CSPAN. (2001). *Olson on Washington journal*, September 9, 2001. https://www.c-span.org/video/? 165914-2/news-review.

Dao, J. (2000). Gore tables gun issue as he courts Midwest. *New York Times*, September 20, 2000. https://www.nytimes.com/2000/09/20/us/the-2000-campaign-the-strategy-gore-tables-gun-issue-as-he-courts-midwest.html.

Depalma, A. (2011). *City of dust: Illness, arrogance and 9/11.* New York: Pearson Education.

Dole/Kemp 1996 campaign website. (1996). http://www.dolekemp96.org/agenda/issues/health.htm

Ebert, E. (2004). Review: Fahrenheit 9/11. *Roger Ebert.com*, June 24, 2004. https://www.rogere bert.com/reviews/fahrenheit-911-2004.

Emmanuel, Z. (2019). Name the much-criticized federal program that has saved the U.S. $2.3 trillion. Hint: it starts with Affordable. *Stat News*, March 22, 2019. https://www.statnews.com/ 2019/03/22/affordable-care-act-controls-costs/.

Encyclopedia of World Biography. (2019). https://www.notablebiographies.com/news/Li-Ou/ Moore-Michael.html#ixzz6FkD1j2Sy

Fink, S. (2013). *Five days at memorial*. New York: Crown Publishing.

Fralick, M., & Kesselheim, A. S. (2019). The U.S. insulin crisis—Rationing a lifesaving medication discovered in the 1920s. *New England Journal of Medicine, 381*, 1793–1795. https://www.nejm.org/doi/full/10.1056/NEJMp1909402?query=NC.

Gawande, A. (2017). How the Senate's healthcare bill threatens the nation's health. *New Yorker*, June 26, 2017. http://www.newyorker.com/news/news-desk/how-senates-health-care-bill-threatens-nations-health.

Gebelhoff, R. (2017). *Your guide to the most contentious parts of the GOP health plan*. March 24, 2017. https://www.washingtonpost.com/blogs/post-partisan/wp/2017/03/24/your-guide-to-the-most-contentious-parts-of-the-gop-health-care-plan/?tid=a_inl&utm_term=.0d36a1dbe10f.

Gladwell, M. (2015). Thresholds of violence. *The New Yorker*, October 12, 2015. https://www.newyorker.com/magazine/2015/10/19/thresholds-of-violence.

Gonyea, D. (2017). From the start, Obama struggled with fallout from a kind of fake news. *NPR*, January 10, 2017. https://www.npr.org/2017/01/10/509164679/from-the-start-obama-struggled-with-fallout-from-a-kind-of-fake-news.

Green, J. (2008). The Frontrunner's fall. *The Atlantic*, September, 2008. https://www.theatlantic.com/magazine/archive/2008/09/the-front-runner-s-fall/306944/.

Heilemann, J., & Halperin, M. (2010). *Game change*. New York: HarperCollins.

Hernandez, E. (2004). Is "Fahrenheit 9/11" a documentary film? *Indiewire*, July 2, 2004. https://www.indiewire.com/2004/07/is-fahrenheit-911-a-documentary-film-or-what-is-a-documentary-film-78811/.

Huberfeld, N. (1997). The commerce clause post-Lopez: It's not dead yet. *Seton Hall Law Review*, January 1, 1997.

Huberfeld, N. (2006). *Medicare and medicaid*. Presentation to the University of Kentucky Program for Bioethics.

Huberfeld, N. (2018). Can work be required in the medicaid program? *New England Journal of Medicine, 378*, 788–791. https://www.nejm.org/doi/full/10.1056/NEJMp1800549.

Iati, M. (2019). You should be ashamed of yourselves. Watch Jon Stewart tear into congress over 9/11 victims fund. *Washington Post*, June 12, 2019. https://www.washingtonpost.com/politics/2019/06/11/jon-stewart-blasts-congress-first-responders-fund/.

Jacobson, L. (2017). What's in the Senate healthcare bill? *Politifact*, June 25, 2017. http://www.politifact.com/truth-o-meter/article/2017/jun/25/whats-senate-health-care-bill-here-are-five-key-pr/.

James, C. (2001). The McVeigh execution. *New York Times*, June 12, 2001. https://www.nytimes.com/2001/06/12/us/mcveigh-execution-critic-s-notebook-coverage-oklahoma-bomber-s-final-hours-are.html.

Kilgore, E. (2019). Remembering the Starr report as we await the Mueller Report. *New York Magazine*, April 17, 2019. http://nymag.com/intelligencer/2019/04/remembering-the-starr-report-as-we-await-the-mueller-report.html.

Kirk, M., et al. (2017). Producer. The divided States of America. *Frontline*, January 17, 2017. https://www.pbs.org/wgbh/frontline/film/divided-states-of-america/.

Kliff, S. (2014). Kentuckians only hate Obamacare if you call it 'Obamacare'. *Vox*, May 12, 2014. https://www.vox.com/2014/5/12/5709866/kentuckians-only-hate-obamacare-if-you-call-it-obamacare.

Kodjak, A. (2017). GOP health bill changes could kill protections for people with pre-existing conditions. *NPR*, March 23, 2017. https://www.npr.org/sections/health-shots/2017/03/23/521220359/gop-health-bill-changes-could-kill-protections-for-people-with-preexisting-condi.

Lioy, P. (2010). *Dust: The inside story of its role on September 11th*. New York: Roman and Littlefield.

Leonard, E. W. (2008). Teaching Sicko. *Journal of Law, Medicine and Ethics*, 139–146.

Moore, M. (1996). *Downsize this! random threats from an unarmed American*. New York: Harper Perennial.

Moore, M. (2013). The Obamacare we deserve. *New York Times*, December 31, 2013. https://www.nytimes.com/2014/01/01/opinion/moore-the-obamacare-we-deserve.html.

Nagourney, A., & Zeleny, J. (2007). Obama formerly enters presidential race. *New York Times*, February 11, 2007. https://www.nytimes.com/2007/02/11/us/politics/11obama.html.

New York Times Editorial Board. (2017). The TrumpRyanCare debacle. *New York Times*, March 24, 2017. https://www.nytimes.com/2017/03/24/opinion/the-trumpryancare-debacle.html.

Newkirk, V. (2018). A year after Hurricane Maria, Puerto Rico finally knows how many people died. *The Atlantic*, August 28, 2018. https://www.theatlantic.com/politics/archive/2018/08/pue rto-rico-death-toll-hurricane-maria/568822/.

Newsweek. (2009). Echoes of Kennedy's battle with Nixon in healthcare debate. *Newsweek*, August 26, 2009. https://www.newsweek.com/echoes-kennedys-battle-nixon-health-care-deb ate-211550.

Olson, B. (1999). *Hell to pay: The unfolding story of Hillary Rodham Clinton*. Washington D.C.: Regenery Press.

Olson, B. (2001). *Final days: The last, desperate abuses of power by the Clinton White House*. Washington D.C.: Regenery Press.

Oscars.org. (2003). Academy Award speeches. *Awards*, 2003. https://www.oscars.org/oscars/cer emonies/2003

Pear, R., & Kaplan, T. (2017). Senate Healthcare bill includes deep cuts to Medicaid. *New York Times*, June 22, 2017. https://www.nytimes.com/2017/06/22/us/politics/senate-health-care-bill. html?hp.

Perez-Pina, R. (2000). Loss in home state leaves gore dependent on Florida. *New York Times*, November 9, 2000. https://www.nytimes.com/2000/11/09/us/the-2000-elections-tennessee-loss-in-home-state-leaves-gore-depending-on-florida.html.

Pomper, G. M. (2001). The 2000 presidential election: Why gore lost. *Political Science Quarterly, 116*(2), 201. https://www.uvm.edu/~dguber/POLS125/articles/pomper.htm.

Potter, W. (2010). My apologies to Michael Moore and the health insurance industry. *PR Watch*, November 22, 2010. https://www.prwatch.org/news/2010/11/9642/wendell-potter-my-apologies-michael-moore-and-health-insurance-industry.

Rawls, J. (1971). *A theory of justice Cambridge*. Mass.: The Belknap Press of Harvard University Press.

Reeve, E. (2011). Who coined 'Obamacare'? *The Atlantic*, October 26, 2011. https://www.theatl antic.com/politics/archive/2011/10/who-coined-obamacare/335745/.

Rosenthal, M. S. (2016). *Ethical issues with the James Zagroda 9/11 health and compensation act*. University of Kentucky Grand Rounds, September 13, 2016.

Sherman, G. (2017). *The loudest voice in the room*. New York: Random House.

Skocpol, T. (1995). The rise and resounding demise of the Clinton Plan. *Health Affairs, 14*(1). https://www.healthaffairs.org/doi/full/10.1377/hlthaff.14.1.66.

Stockman, S. (2012). Recalling the Nixon-Kennedy health plan. *The Boston Globe*, June 23, 2012. https://www.bostonglobe.com/opinion/2012/06/22/stockman/bvg57mguQxOVpZMmB1 Mg2N/story.html.

The Famous People. (2019). *Michael Moore biography*. https://www.thefamouspeople.com/pro files/michael-moore-8667.php.

The New York Times Editorial Board. (2017). The Senate's unaffordable care act. *New York Times*, June 23, 2017. https://www.nytimes.com/2017/06/23/opinion/senate-obamacare-repeal.html? action=click&pgtype=Homepage&clickSource=story-heading&module=opinion-c-col-left-reg ion&region=opinion-c-col-left-region&WT.nav=opinion-c-col-left-region.

The Week. (2009). *Glenn Beck calls Obama a racist*, July 29, 2009. https://theweek.com/articles/ 503258/glenn-beck-calls-obama-racist.

The Washington Post. (1995). *Whitewater timeline: 1995–98*. https://www.washingtonpost.com/ wp-srv/politics/special/whitewater/timeline2.htm.

The Washington Post. (1998). *What Clinton said*, January 26, 1998. https://www.washingtonpost. com/wp-srv/politics/special/clinton/stories/whatclintonsaid.htm#Edu.

The Washington Post. (2012). Obama: I like the name Obamacare. *Washington Post*, September 2, 2012. https://www.washingtonpost.com/video/politics/obama-i-like-the-name-obamacare/2012/09/02/869180aa-f543-11e1-86a5-1f5431d87dfd_video.html?utm_term=.003a9ec80f4c.

Toppo, G. (2009). *USA today "10" years later, the real story behind Columbine*, April 14, 2009.

Tumulty, K. (2000). Gore's gun problem. *CNN*, February 7, 2000. http://www.cnn.com/ALLPOLITICS/time/2000/02/07/gun.html.

Veterans Affairs. (2019). *About VA*. https://www.va.gov/about_va/vahistory.asp.

Wright, L. (2006). *The looming tower*. New York: Alfred A. Knopf.

Willis, J. (2017). Trumpcare exposed the cruel apathy of the Republican Party. *GQ*, March 25, 2017. http://www.gq.com/story/trumpcare-exposed-gop-cruelty.

Zibulewsky, J. (2001). The emergency medical treatment and active labor Act (EMTALA): What it is and what it means for physicians. *Baylor University Medical Center Proceedings, 14*(4), 339–346. https://www.ncbi.nlm.nih.gov/pmc/articles/PMC1305897/.

Chapter 5
"Nobody Puts Baby in a Corner": Reproductive Justice in *Dirty Dancing* (1987)

Dirty Dancing is an iconic 1987 film that is the top "chick flick" for most Baby Boomers and their progeny. When you think about this film, abortion may not come to mind at first, but "Baby" sure does. By the end, "Nobody puts Baby in a corner!" falls off our lips as easily as the last line from *Casablanca* (1942): "Louis, this looks like the beginning of a beautiful friendship." When the final cut of *Dirty Dancing* was screened, 39% of those viewing it got so caught up in the film's music and energy, they didn't realize that abortion was, in fact, the main plot (Dawn 2017). If you're tackling the topic of reproductive justice and abortion access, this film is a wonderful entry point for a necessary discussion about pre-Roe America, post-Roe America, and the potential repeal of *Roe v. Wade* (1973) in the twenty-first century.

In 1987, the first generation of women had grown up with the legal right to safe abortion; *Roe v. Wade* was now 14 years old, and decided as part of guaranteed rights granted under the Fourteenth amendment surrounding personal liberty, and the Ninth amendment, surrounding reservation of rights of the people (similar to physician-assisted suicide—see Chap. 3). *Roe v. Wade* became the law of the land on January 22, 1973—towards the start of Nixon's second term. The courts had finally recognized under the law that women had the constitutional right to seek safe and legal abortion for unwanted pregnancies with few restrictions on the timing of the abortion, claiming broad definitions of "welfare of the woman". But *Dirty Dancing* takes place a decade before—in 1963—just as Civil Rights is heating up. The film is based on the real experiences of screenwriter Eleanor Bergstein, the daughter of a Jewish doctor in Brooklyn, who went to the Catskills each summer with her family. Indeed, the summer of 1963 was, for Baby Boomers, the "before times"—when John F. Kennedy was still President for another few months, Martin Luther King would give his "I have a dream" speech, and several thousand women would die from sepsis caused by unsafe abortions. To a 1987 audience, amidst the exploding AIDS crisis (see Chap. 2), and the publication of And the Band Played On (Shilts 1987), *Dirty Dancing* was also about the consequences of unsafe sex, classism and health disparities. If Baby has the right Daddy, the problem can go away. But if not, as the handsome medical student-waiter, Robbie, notes: "some people count,

© Springer Nature Switzerland AG 2020
M. S. Rosenthal, *Healthcare Ethics on Film*,
https://doi.org/10.1007/978-3-030-48818-5_5

some people don't." Ultimately, this "Sex, Sepsis and Dance" film is so entertaining, it will allow you to really cover the "time of our lives" in reproductive justice: TRAP laws—Targeted Restrictions for Abortion Providers, and states that have now completely banned and criminalized abortion. This chapter will review the origins of *Dirty Dancing*, its social location, the History of Medicine context of abortion access, as well as the healthcare ethics issues relevant in the film.

Where Does Baby Come from? Creating *Dirty Dancing*

Dirty Dancing is an original screenplay written by Eleanor Bergstein, based on her own experiences as a teenager spending summers in the Catskills with her sister and parents; Bergstein's father was a doctor. She completed the script in 1985, and it was rejected by several studios until she found a smaller independent studio to make it (see further). Eleanor was nicknamed "Baby" herself until she was 22 (she was the baby of the family), and the resort was based on Grossinger's Catskill Resort in Liberty, New York in the Catskill mountains. The resort was Kosher and catered to Jewish guests. Ironically, the resort closed in 1986, just when filming *Dirty Dancing* was beginning. The film was shot on location at a different resort in North Carolina (see further).

   Dirty Dancing is also a work of Jewish Americana running parallel to works by Phillip Roth, whose 1959 book, Goodbye Columbus, was made into a film in 1969 (starring Richard Benjamin and Ali Macgraw) or Mordecai Richler, whose 1959 book, The Apprenticeship of Duddy Kravitz was made into a film in 1974 (starring Richard Dreyfus and Randy Quaid). Similarly, Bergstein's experiences track along the same time frame. There is a specific slice of Jewish American life in post-war United States, after the dark cloud that followed the executions of Julius and Ethel Rosenberg. Between 1955-November 21, 1963 was a golden period for diaspora Jews living in their "Promised Land" of the United States, but also with the knowledge that the Holocaust was behind them and Israel existed "just in case". Many American Jewish Baby Boomers were being raised in secure homes, with fathers who were professionals (doctors, lawyers, accountants, engineers, or successful business people). Although their mothers were frustrated just like Lucille Ball's "Lucy" to "have a part in the show", they were good Jewish homemakers, who created uniquely "Jewish American" dishes in which creative use of common 1950s canned goods became delicious instant marinades for chicken or brisket, and strange desserts with Jello and non-dairy instant whipped topping ("Pareve") were all the rage. It was this time frame that gave birth to the term "J.A.P."—for Jewish American Princess, an acronym coined by Jews themselves in post-war America, and which describes Lisa Houseman, in *Dirty Dancing*, but generally referred to a sheltered Daddy's girl.

   In 2017, the Amazon Prime series, *The Marvelous Mrs. Maisel*, which now has a cult following, is also treading on Bergstein's turf, set in New York in the late 1950s, with several episodes taking place at the Catskills during its heyday. In fact, it's clear that many of *The Marvelous Mrs. Maisel* Catskills scenes were inspired by *Dirty Dancing*.

One Jewish journalist noted in 2011 (Butnick 2011):

> A week ago, I told Eleanor Bergstein...that when I first saw the film years ago, I hadn't realized how heavily influenced it was by Jewish culture. She beamed, as she had the entire evening, and assured me it was a seriously Jewish movie. So Jewish, in fact, that none of the characters ever need to explicitly mention their Jewishness – they're spending the summer at Kellerman's resort in the Catskills, after all, and, Bergstein pointed out proudly, milk and meat are never served in the same scene. It's a Jewish film, she explained, "if you know what you're looking at."

For example, in a critical scene where Baby first sees Penny crouched in the hotel's kitchen crying, Baby is peering at her with a visible box of Matzo Meal in the background (needed to make Matzo balls); later bagels, lox, cream cheese, capers and onions are on breakfast plates.

When *Dirty Dancing* was released in 1987, it took many viewers back to what seemed a simpler time. Bergstein would spend her summers at Grossinger's dancing while her parents played golf; she had become quite the Mambo dancer, and also did "dirty dancing"—a term for more sexualized rhythmic dancing, and competed in numerous dance competitions. She also worked as a dance instructor at Arthur Murray's dance studios when she was in college before she got married in 1966. Much of the music in the film came from Bergstein's personal music collection, but the appeal of the film's music is also in the original songs that convey the spirit of the film but were, in fact, written for the film (see further). In one interview, Bergstein recalled (Nikkhah 2009): "When my parents hit the golf course, I hit the dance floor...I was a teenage mambo queen, winning all kinds of dirty dancing competitions." Bergstein's "signature step was to put her leg around the neck of the boy she was dancing with" (Nikkhah 2009).

Bergstein is not a Baby Boomer and was born in 1938; she was at Grossinger's as a teenager in the 1950s. In 1963, she was 25 and no longer going to the resort with her parents, but she was working at some of the Catskills resorts as a dance instructor through her college years, and thus, 1963 was likely one of the last times she spent any time in the Catskills. But placing the story in the summer of 1963 was deliberate and symbolic; that year is emblazoned in the memory for anyone who was over 10 at that time because of what occurred on November 22, 1963: the assassination of John F. Kennedy. This is when the 1960s became the "Sixties"—now a euphemism for turbulence, chaos and radical socio-political change (see further under Social Location).

Dirty Dancing became an enduring film, however, because it was a real story about real people, and came from Bergstein's personal life and unique female perspective of the early 1960s. When Bergstein struggled to make a living as a professional dancer, she transitioned to a writer. One of her screenplays was finally sold—*It's My Turn* (1980) about a female math professor (played by Jill Clayburgh) and a retired baseball player (played by Michael Douglas). *It's My Turn* features a dance sequence that was cut out, which Bergstein was angry about, but it led her to write a script exclusively based on dancing and her own life, which turned into *Dirty Dancing*. By this time, Bergstein was in her forties, and by the mid-1980s, the Catskills resorts were all on "life support" and were no longer a haven for secular Jews, and began to be frequented by orthodox Jews looking for somewhere Kosher. There was no Mambo dancing at the Catskills in the 1980s; it was now only in the memories of past guests.

Bergstein and her sister were FDR babies: born during the height of the New Deal and Franklin D. Roosevelt's popularity. Eleanor Bergstein was named after First Lady, Eleanor Roosevelt, and her sister, Frances, was named after FDR's female Secretary of Labor, Frances Perkins, who was a unique choice, as she was also a well-known sociologist. In fact, it was Frances Bergstein who was the studious sister in the family, while Eleanor was the "artsy" one who was seemingly directionless. Bergstein's Frances Baby Houseman is a composite for both herself and her sister. The real Frances Bergstein became a math professor, just like Jill Clayburgh's character in *It's My Turn*.

Bergstein's Five Archetypes

Bergstein's plot, which revolves around a back-alley abortion, ties together five enduring archetypal characters that have become frequent subjects of modern popular culture scholarship, but is what also makes *Dirty Dancing* the abortion film of choice, if you will. Each character's actions and motivations are propelled by the abortion, which also reveals their true natures. Baby, Johnny, Penny, Robbie and Dr. Houseman are all entangled in a well-choreographed morality tale about reproductive justice in a pre-Roe America.

The story: Penny gets "knocked up" by Robbie, who refuses to take responsibility, and Johnny, her co-worker and dance partner, steps into support and comfort her. Penny wants to terminate the pregnancy but doesn't have the money. Baby learns of the problem and asks her physician-father for the money, but withholds the reason she needs it. He asks: "It's not [for something] illegal is, it?" She replies: "No, Daddy." Baby has already ascertained that the money solves a *moral* problem (see under Healthcare Ethics Issues). But Penny can "only get an appointment for Thursday"— when she and Johnny have to perform a Mambo dance number at a different hotel. Baby volunteers to fill in and learn the steps, and she successfully fills in. By the time she and Johnny return to the resort, Penny is septic and we learn there was a "dirty knife and folding table". Baby races to her cabin to get her father, who saves Penny's life, laboring under false assumptions that Johnny is the culprit, and furious that he inadvertently paid for an illegal abortion, hence morally complicit, because he trusted his good-hearted daughter. Despite this clear plot line, this important story is hiding in plain sight within amazingly entertaining dance numbers in a film typically synopsized as a "coming of age" romance between Baby and Johnny, who fall in love while he teaches her the dance number. They are also from different social classes, creating a Romeo-Juliet/West Side Story tension that is particularly "Reaganite" (see further).

But I argue it is the Baby-Johnny romance that is really the "subplot"; Penny's abortion is the story. Indeed, the abortion is so well-hidden, neither the *Los Angeles Times* review in 1987 (see further), or the *New York Times* review of its re-release in 1997 (see further) even mention it—even though it's really the last line of the movie, reminding us that yes, the film is all about "Baby"!

In contrast, other abortion films released after *Dirty Dancing* in the 1990s are just not as entertaining, or too mired in "the decision". There is the grueling *If These Walls Could Talk* (1996), which presents separate short stories about abortion in different time frames, but it suffers from a contrived plot construction in which the entire story revolves around women making "the decision". Many bioethicists point to *Cider House Rules* (1999) as the quintessential abortion film (from the abortionist's perspective), but it does not get the viewer emotionally involved with the characters affected. This is perhaps the enigma of Bergstein's script—which has led to as great a cult following as the film *The Sound of Music* (1965). It is what happens when the story is from a woman's perspective. Similarly, *The Sound of Music* was based on Maria von Trappe's actual improbable life. *Dirty Dancing*, as does *The Sound of Music*, makes the cut for "everyone's favorite movie" leading to a deep fan blogosphere of never-ending analysis. Everyone is talking about "Baby" which means everyone is thinking about reproductive justice, even if they don't know the history. With its built-in fans and cult following, *Dirty Dancing* is the perfect film to launch a discussion about pre-Roe America and reproductive justice. Each character also resonates as a familiar archetype viewers can relate to.

Frances "Baby" Houseman

Baby Houseman, played by Jennifer Grey (born 1960, and 27 when she played Baby), embodied most young Jewish American women's experiences. In this demographic, you were either a Frances (similar to a Ruth Bader Ginsberg type)—bound for academic greatness. Or you were a Lisa—focusing on your looks and attractiveness to marry a "Jewish Prince", which typically meant a doctor, such as the father character. If you weren't Frances, you knew one, or she was your sister. If you weren't Lisa, you knew her or she was your sister, too. In *Dirty Dancing*, the two sisters are juxtaposed: we meet Baby on the car ride up to the Catskills, engrossed in reading a book while her sister is primping in the mirror. Baby is bound for the all-women's liberal arts college, Mount Holyoke. We meet the character when she is older, introducing the film as a memory: The older Baby recalls: "That was the summer of 1963, when everybody called me "Baby", and it didn't occur to me to mind. That was before President Kennedy was shot, before the Beatles came – when I couldn't wait to join the Peace Corps."

When the film was re-released in 1997, the *New York Times* summarized our introduction to Baby this way (Kolson 1997):

Into the Borscht Belt Eden comes 17-year-old Baby, bookish, slightly awkward and earnest (played by a luminous Ms. Grey), her more beautiful princess of a sister, Lisa, and their doctor father and housewife mother….

One of the reasons Jennifer Grey's embodiment of Baby became so complete was that her appearance made her believable as Baby Houseman. Kolson (1997) notes:

Ms. Grey was best known as the malevolent sister in "Ferris Bueller's Day Off." Afterward, she was known everywhere as Baby, and that became both a blessing and a curse for her. "I became recognizable, known and loved by so many people," she said. "I didn't look like a movie star. I had a Jewish nose. People loved seeing that…' But I was never that character," she continued. "I was not a virgin; I hate to admit it. I also could dance. I was kind of a daddy's girl, though. The curse was that people think I was that character, and I'm not."

Carmon (2010) notes, Baby is "the daughter of the first generation of American Jews to reap widespread upper-middle class prosperity, if not elite cultural acceptance, she is swathed in a pre-Kennedy assassination liberalism…."

Baby becomes the advocate for the underclass in saving Penny. Abortion content is not new to Grey, as her father, Joel Grey, played the emcee in *Cabaret* (1972)—a role he perfected in its stage version. Set in Weimar Republic Germany during the rise of Nazism, *Cabaret* featured a range of taboo subjects including abortion, but also the concept of a Jewish professional middleclass that was finding itself under siege in the early 1930s. In fact, Joel Grey recalled in an interview with CNN (McLaughlin 2013):

The film's most controversial line was a lyric to the song "If You Could See Her." As Grey slow-danced with [a performer] dressed in a gorilla suit, the last line of the song went: "If you could see her through my eyes, she wouldn't look Jewish at all." The lyric and gorilla costume were meant to show how anti-Semitism was beginning to run rampant in Berlin, but Grey said it caused problems in the New York stage production.

"There were a number of Jewish groups who missed the point of that song," he said. "They just thought that we were saying Jews were ugly."

The line was replaced with "She isn't a meeskite at all," but the response was tepid. (Meeskite was an old Yiddish term to describe an unattractive woman.) Grey, who is Jewish, said the lyrics didn't offend him and he was hesitant about changing them.

In light of her background, Jennifer Grey's awareness of what she called her "Jewish nose" was also what made her a star, and audiences fall in love with her. But long known anti-semitic tropes surrounding Jewish hooknoses led thousands of American Jewish women into cosmetic rhinoplasty—one reason, for example, which made Barbra Streisand stand out when she was carving out a career in the early 1960s— around the same time frame *Dirty Dancing* is set in.

In the early 1990s, Jennifer Grey indeed succumbed to rhinoplasty; she had two done—the second to correct surgical complications from the first (Bryant 2012).

The biggest complication of her rhinoplasty, however, was what the procedure did to her career: she became anonymous and unrecognizable because it changed her face so much that it prevented her from working. Grey noted: "I went in the operating room a celebrity – and came out anonymous. It was like being in a witness protection program or being invisible." (Wenn 2006).

Grey had also been involved in a serious car accident in Ireland, prior to the release of *Dirty Dancing* with her then-boyfriend, Matthew Broderick, who she met on the set of *Ferris Bueller's Day Off* (1986). The crash killed the passengers in the oncoming car due to Broderick driving on the wrong side of the road. Her emotional trauma from that experience, combined with her anonymity, led her to withdraw from acting. Grey's career essentially ends by the early 1990s (Hamilton 1999; Laurie 2010).

Ultimately, both Baby and Jennifer Grey are part of the Jewish "swan" archetype— the Jewish girl who feels like the "ugly duckling" realizing what makes her beautiful has nothing to do with her nose.

Johnny

The character of Johnny Castle, played by the late Patrick Swayze (1952–2009), was a composite character of dance instructors Bergstein knew and interviewed, and the specific experiences of dancer, dance instructor, and choreographer, Michael Terrace, who had frequented the Catskills during that era, and was known for bringing the Mambo craze to the United States (Jones 2013). The Mambo was a new dance phenomenon in the United States at a time frame when Americans loved Cuba, and one particular Cuban "loved Lucy". In the 1950s, Cuban entertainment came into everyone's living room in the form of Desi Arnaz who played Ricky Ricardo on *I Love Lucy* (1951–7), which also broke new ground in 1952 by showing a real pregnant woman on television—Lucille Ball, who was pregnant at the time, and her pregnancy was written into the script. Deemed very controversial at the time, audiences loved Lucy pregnant because many of them were, too—in a time frame where contraception was unavailable for most (see under History of Medicine). Bergstein's portrayal of the hired-hand dancers was based on numerous qualitative interviews, including Jackie Horner (see further) "who contributed voluminous scrapbooks and details. Horner got her start with New York City's famous June Taylor Dancers before landing a full-time gig teaching at Grossinger's in 1954 (she stayed until it closed, in 1986)" (Jones 2013).

Bergstein's literary symbols were not subtle, and she named her characters aptly. Johnny Castle is, of course, the real "Prince" in the *Dirty Dancing* storyline.

Patrick Swayze was survived by his mother, Patsy Swayze (1927–2013), who was a dancer, dance instructor and choreographer. Swayze was a dancer himself, and he also co-composed and sang "She's Like the Wind", featured in the film. Swayze had made several films prior to *Dirty Dancing*, and even starred opposite Grey in the 1984 film *Red Dawn* (about a Soviet takeover of the United States). But it was *Dirty Dancing* that launched Swayze to stardom, and his career peaked two years later in *Ghost* (1990) opposite Demi Moore. Swayze also married a dancer and choreographer, and later in 2007 had a very public battle with pancreatic cancer, and died on September 14, 2009. Swayze was survived by his widow, Lisa Niemi, who met Swayze when she was 14 through his mother, who was the director of the Houston Ballet Dance Company where Niemi studied (Tauber 2008). She was married to Swayze for 34 years.

Penny

The character of Penny Johnson, played by dancer and actress, Cynthia Rhodes, was based on the experiences of Jackie Horner. According to Sue Tabashnik (2011):

> Jackie spent the summer of 1985 telling Eleanor Bergstein her story–including going through photos of the era, clothing, and hair styles and telling her various anecdotes from her experiences as a dance pro at Grossinger's from 1954-1986. Check out the *Dirty Dancing* screen credits and you will see Special Thanks to Jackie Horner. Here is one anecdote from Jackie: "Shelly Winters, bless her heart, was a dear friend and a Grossinger guest so often. One day she said, 'Jackie, you are going to get hurt practicing your lifts on these hard wood stages. Why don't you come down to the lake with me on Sunday and practice your dance lifts in the lake?'" Jackie was involved in stealing vodka-spiked watermelons for staff parties. Jackie taught a couple who was stealing and her partner was blamed for it. Sound familiar? Note though that Penny had an abortion and Jackie did not have an abortion–although per Jackie, there were other staff at the hotel who did.

Ultimately, Penny is the vehicle through which her personal story tells a much larger political story of women's struggle for reproductive autonomy and reproductive justice (see under Healthcare Ethics Issues). Bergstein's literary symbolism was not subtle here, either: "Penny" is so-named because of how she is perceived by the resort's guests: cheap and disposable—emblematic of a classism that began to infect diaspora Jews, who themselves, as a victimized minority group, should know better. To a Jewish viewer, Penny is also the "shikse" (Derogatory Yiddish for "non-Jewish woman") that Jewish men loved to have sex with but never married.

Bergstein was adamant about the Penny plot in the film, which some had wanted her to remove. *Jezebel* noted in 2010 (Carmon 2010):

> …Penny's botched abortion [is still an] incredibly rare and key plot point that Eleanor says she put in back in the mid-1980s because she was afraid Roe v. Wade would be overturned. (Helpfully, her refusal to take it out later lost the movie a pimple cream sponsor that would have forced its image onto every movie poster.) A few weeks ago, she read about a pro-choice march where a man asked a protester what exactly a coat-hanger abortion was and she snapped, "Haven't you seen *Dirty Dancing*?"

In 1985, when Bergstein was writing the script, few women were actually concerned that abortion access would revert back to the days of back-alley abortions, but Bergstein did not think reproductive justice battles had been secured; it was also a time frame in which the Moral Majority under the Reagan Administration was gaining political capital, and Bergstein's Penny investment wound up hitting the jackpot with future generations of *Dirty Dancing* fans. But when Bergstein finished her script , a shocking new dystopian book, by Margaret Atwood, has just been published, called

The Handmaid's Tale (Atwood 1986), which was made into an excellent film in 1990 (see further), and which predates the Hulu miniseries of Atwood's book at the dawning of the Trump era. The Handmaid's Tale was speculative fiction that the Christian Right would eventually wrest control of the United States' "liberal" progressive social justice march forward and take it back to the 1950s, or further into a patriarchal societal hell for women. Atwood was writing about the Reagan era, not the Trump era, but Bergstein and Atwood saw a few red flags in the Reagan era that made them uneasy about reproductive rights (See further under Social Location.) In 2017, the following was noted about Penny (Aroesty 2017):

> But 30 years later in 2017, we still empathize with Penny, the victim of a botched abortion. She isn't just some irresponsible floozy (but even if she was, as woke millennials, we wouldn't slut-shame her). Penny found herself in a situation no one is immune to: pregnant by a man she thought she loved, who turned out to be a real dick. She was scared, with no good options, the best being almost unaffordable and illegal. And she suffers as a result of this best option, because the government refused to regulate a serious medical procedure. Today, some government officials are working towards a reality that would more closely resemble Penny's in 1963. So we still relate to her. She could be any of us.

Robbie Gould: The Medical Student/Waiter

Robbie Gould, played by the late Michael Cantor (1959–1991), was an American journalist and actor, who died from an accidental heroin overdose while on assignment—ironically researching an article he was writing about addiction on New York City's lower east side for *The Village Voice* (IMDB 2019; Aronowitz 2002). Cantor was a Harvard graduate, and son of famous agent, Arthur Cantor. Cantor played a Yale graduate on his way to becoming Jewish Royalty: a doctor. He is also the villain and impregnates Penny and abandons her. As a medical student, and future doctor, he represents a coming generation of medical professionals who do not have the moral character to be in a healing profession, and he stands in contrast to Baby's father, Jake Houseman, who saves Penny's life, but inadvertently rewards Robbie with a large tip intended for his medical education (see further under Healthcare Ethics Issues). Bergstein's choice to write Robbie as a medical student was deliberate—the sociopath in disguise, while Johnny is the real "mensch" (Yiddish for someone of fine character). What doctor-to-be gets a girl pregnant, refuses to take responsibility, or even pay for the abortion? According to Bergstein, more typical than not. Eventually, Robbie winds up bedding the film's "cougar"—the character of Vivian Pressman, a Fortysomething sexually frustrated woman who is looking for sexual adventure anywhere she can find it; of course, a deeper look into Vivian's life would likely reveal a capable woman now in midlife who was never allowed to reach her potential due to sexism.

Dr. Houseman: Baby's Father

Dr. Houseman, played by Jerry Orbach (1935–2004), is the Jewish doctor archetype that many Baby Boomers remember, but also symbolizes the patriarchy that dominated American medicine, which comprised of the quota system, in particular (see Chap. 9) for Jewish students who wanted to enter medicine. Dr. Houseman—not too subtle a metaphor for "head of the house"—also saves the day because he has a duty to treat, in spite of the circumstances. As an archetype in the American Jewish diaspora, the Jewish doctor also became a symbol of excellence in medicine, and in fact, non-Jewish patients often brag that they have a "Jewish" doctor or subspecialist, which has come to denote "a good doctor". However, empathy and virtue are not necessarily traits that Dr. Houseman embodies when it comes to his handling of the Penny situation. We presume that while he treats Penny with kindness and virtue as a doctor, he reserves his punitive judgement for Johnny whom he presumes was the "cad" who impregnated Penny and refused to marry her. Within medical paternalism, doctors routinely were judgmental with their patients and family members, and inserted their opinions and biases about a range of social matters. Worse, although Johnny treats Dr. Houseman with respect, it's not reciprocated as Dr. Houseman dismisses him, and makes character assumptions that are wholly false, and mis-assigning character traits of virtue to Robbie (see further) instead.

Orbach's portrayal of Dr. Houseman is realistic, and is a character Orbach instinctively knew as the son of Jewish immigrants, who studied at the Actors Studio with Lee Strasberg when he was 20. Orbach became well known for his 11-year starring role in *Law and Order* throughout the 1990s. Orbach was diagnosed with prostate cancer in 1994, two years after he joined the cast of *Law and Order*, and died from it in 2004 while being treated at Memorial Sloan-Kettering Cancer Center in New York at age 69.

Production

Dirty Dancing is an independent film and was made outside of the typical Hollywood model. Bergstein could not find any traditional production company for her script. Initially she took it to producer Linda Gottlieb at MGM, but then Gottlieb left MGM, and no other production company would green light the script. Gottlieb then took the script to a video distribution company, Vestron, which was gearing up to begin its own film development and production wing. It was a situation akin to Netflix original films—when the distributor becomes the producer. Vestron told Gottlieb they'd do the film, "but only if she could guarantee bringing it in for $5 million, about half of what she said it would have cost to film with union crews in New York. Gottlieb, who had 16 years' experience developing and producing educational films, finally hired non-union crews and got the movie done–for $5.2 million–in right-to-work states Virginia and North Carolina" (Matthews 1987). Director, Emile Ardolino, had read the script when he was on jury duty, and was eager to come aboard (Matthews 1987).

Gottlieb was familiar with Ardolino's television and dance work, but he had never tackled a feature film. According to the *Los Angeles Times* (Matthews 1987):

> Ardolino and Gottlieb had to overcome Ardolino's image as a dance director. Although he had directed several dramatic programs for television, the bulk of his credits were associated with dance–28 programs for PBS' "Dance in America," specials featuring Mikhail Baryshnikov and Rudolf Nureyev, and the remarkable "He Makes Me Feel Like Dancing'," which won a 1984 Academy Award as best feature documentary [about the ballet dancer and teacher Jacques d'Amboise].

> When you hire a director, what you get writ large is the director's own sense of taste." [Gottlieb] said. "What we got with Emile went way beyond his love of dance. He has a kindness, a gentleness, and all of that comes through in the picture…Dirty dancing is partner dancing," Gottlieb said. "All the elements are like the foreplay of sex. Learning to dance is the central metaphor of the film."

The appeal and impact of the film also had to do with its rating; initially it got an R rating—not for the abortion content, which oddly, none of the reviewers really took note of, but for nudity. Said Gottlieb: "The really erotic sequences are the dance sequences…We felt it was important to have a PG rating, so we kept going" (Matthews 1987). Indeed, after editing the nudity, the film was rated PG-13, which was important given its history lesson regarding abortion access.

Ardolino also avoided hiring dancer doubles for the roles, and "insisted from the beginning that the actors do all their own dancing". For these reasons, hiring dancers was critical. Ardolino noted in 1987 (Matthews 1987):

> What distinguishes this movie from more recent dance movies is that it's about partner dancing…In 'Flashdance,' only women danced on that stage and it was for themselves. Even in 'Footloose,' the kids didn't dance with each other. In 'Saturday Night Fever,' the basic thrust was a guy being satisfied when he danced alone…. In "Flashdance" and "Footloose," the stars were doubled by professional dancers…Using doubles imposes a particular shooting style where you film body parts," Ardolino said. "I wanted to be able to go from full shots to faces.

But there was also an AIDS story in the production tale of *Dirty Dancing*; Ardolino, who was openly gay, died of AIDS in 1993, before the protease inhibitor era (see Chap. 2). He would never live to see the explosion of interest in *Dirty Dancing* over the years, but his artistic decisions gave the film a certain alchemy that could not be recreated. He would go on to make *Sister Act* (1992), which was also a musical hit (Kolson 1997).

Kenny Ortega had studied with Gene Kelly, and was hired as the choreographer; he had worked with Jennifer Grey before. He had choreographed a number of well-known 1980s films including *St. Elmo's Fire* (1985); *Pretty in Pink* (1986) and *Ferris Beuller's Day Off* (1986), co-starring Grey (see earlier).

Reception

Shot in 43 days in 1986 (wrapping up October 27, 1986), *Dirty Dancing* is considered the highest grossing independent film to date. Made for $5.2 million, when it opened in August 1987, it made $16.5 million in its first 17 days in what was considered the "busiest movie summer in history" (Matthews 1987). Nobody could quite understand why it performed so well at the time. The film was billed as a romance or "coming of age" film with no hint of the abortion plot mentioned anywhere in critics' reviews. A 1987 review guessed at why it was popular (Matthews 1987):

> "Dirty Dancing" seems to be working for a variety of reasons...The film's setting, and its period score, has caught a wave of resurgent popularity of early '60s, pre-synthesizer music. It's not just nostalgia; teen-agers like it as much as their parents do...Perhaps the movie's unabashed romanticism, and its pre-sexual revolution ethic, are simply tapping into a shift in the focus of teen-agers' relationships. The boast of virginity as a measure of character may never be heard again, but the concept of courtship as a progressive experience seems to be making a comeback.

The film's tight budget also created problems for permissions on old songs; to solve that problem, the film included five new songs written exclusively for the film, which led to Grammy winning songs, and a whole other life. By 1997, it was noted (Kolson 1997):

> Several songs were written for the movie, including the Grammy-winning "(I've Had) the Time of My Life," sung by Bill Medley and Jennifer Warnes, and "She's Like the Wind," written and performed by Mr. Swayze. In November 1987, the soundtrack surged to No. 1 on Billboard's pop-music chart, ahead of Bruce Springsteen's "Tunnel of Love" and Michael Jackson's "Bad." The first album has sold nearly 35 million copies, and "Dirty Dancing II," an album of more songs from the soundtrack, has sold 9 million; neither has been out of release.

After the film was released in theaters, it became a best-selling video rental, and then a best-selling DVD. The film kept making millions. In 1997, when the film was re-released for its tenth anniversary, no one could quite figure out why it had continued to resonate so deeply. The *New York Times* mused then (Kolson 1997):

> The film's sexual electricity was certainly important to its success, but today most of the principals don't believe that it was sex that made "Dirty Dancing" such a phenomenon. Mr. Swayze has a theory that he posits in this way: "The folkie Jewish girl gets the guy because of what's in her heart."

Five Presidents later, in 2009, after *Dirty Dancing* was made into a play (see further), Bergstein said (Nikkhah 2009):

> We thought it was one of those movies that would go into the [movie] theatres for a few days and then straight to video, so we made it on a shoestring...We expected mockery, failure and disaster.

At that point, "*Dirty Dancing* made more than $400 million worldwide, and became the first film to sell more than a million copies on video. When it was re-released in

1997, videos of the film sold at a rate of 40,000 a month, and following the release of a digitally re-mastered version in 2007, it has sold more than 12 million copies on DVD" (Nikkhah 2009).

Dirty Dancing was re-conceived as musical theater in 2004 (Nikkhah 2009) and called "Dirty Dancing: The Classic Story on Stage". Bergstein recounted (Lowry 2015):

> For years, I refused offers to make it into a musical…I loved musical theater. I went with my parents to the theater all the time. But I watched how people were watching the movie over and over again, owning it on VHS and LaserDisc and DVD. Then TV stations started playing it on a loop, and people would tell me they watched it every time. So it was natural for live theater.

When I first began to promote *Dirty Dancing* as a reproductive justice film, nobody would take me seriously. But things began to change when the TRAP laws emerged, and various U.S. states severely restricted abortion access (see under History of Medicine). By 2011, special screenings of the film were being set up by pro-choice advocacy groups (Butnick 2011) drawing "packed houses". Ultimately, *Dirty Dancing* has become a popular culture classic that is not categorized as a traditional musical (Tzioumakis and Lincoln 2013), but forged its own genre now known as the "*Dirty Dancing* formula". Somehow, in Reagan era America, its abortion plot was ignored, and the film was mainly touted as a Romeo-Juliet coming of age film in which lovers from different classes come together. In 1987, while the critics had apparently put the "baby in a corner", the female audience for whom it resonated, did not. Known, too, as a classic Reaganite film (see further), which pits working class against the wealthy/professional class, access to abortion was always about class—before Reagan and after.

Synopsis

The synopsis of *Dirty Dancing* has changed over the years. Typically, it was described as a coming of age summer romance story involving 17-year-old Frances "Baby" Houseman (played by Jennifer Grey), who falls in love with a dance instructor (played by Patrick Swayze) from the other side of the tracks, at a Catskills resort in the 1960s. And then, when people went to see the film, it was not quite that, and so much more.

At times, it's been described as a "musical", but despite all the music in *Dirty Dancing*, it actually does not meet the criteria for a musical because it does not involve diegetic singing or other numbers (musical numbers that help tell the story), although many argue the last number "I've Had the Time of My Life" in which the lyrics seem to convey the feelings of Baby and Johnny (Tzioumakis and Lincoln 2013).

Here's a more accurate and recent synopsis of the film beautifully written by *Tablet* magazine (Butnick 2011):

For those of you unfamiliar (shame, shame), the film centers around Frances Houseman, a 17-year-old (Jewish) New Yorker whom everyone calls Baby, who is spending the summer with her family in the Catskills at Kellerman's resort. Heading to Mount Holyoke in the fall to study economics of underdeveloped countries, the idealistic Baby – portrayed perfectly by a pre-plastic surgery Jennifer Grey – intervenes to help get Penny, a dance instructor at the resort, the money she needs for an abortion. Sheltered enough to not realize exactly how dangerously makeshift that procedure might be, but insistent on helping and convinced she can, Baby also fills in for Penny and dances with Johnny, the male dance instructor played by the delightfully swoon-worthy Patrick Swayze, for the pair's annual gig at the nearby Sheldrake resort.

Returning to find Penny in alarmingly bad shape after the primitive, unsanitary abortion, Baby calls upon her father, a doctor, for help. Jerry Orbach saves the day, as usual, though he is horrified at what his daughter has become a party to. Though Dr. Houseman expressly forbids Baby from seeing Johnny, mistakenly believing he is responsible for what is euphemistically referred to throughout the film only as getting Penny in trouble, she sneaks out to see him for, as [some writers] convincingly argues, "the greatest love scene of all time." Class tensions and scheduled activities resume, while Baby is forced to deal with her changing relationship both to Johnny and her father, challenging each man with her resolute determination (and, dare I say, complete stubbornness).

Tablet's synopsis above tells us a lot about the film, but *Dirty Dancing* has shapeshifted into a bottomless pit of popular culture discourse. Despite its professed social location of the summer of 1963, it is also a quintessential 1980s film within a quintessential 1960s story. In short, *Dirty Dancing* is a 1980s riddle wrapped in a 1960s enigma. And vice versa.

The Social Location of *Dirty Dancing*

Dirty Dancing locates itself in its opening lines by the narrator—a much older "Baby": "That was before President Kennedy was shot, before the Beatles came – when I couldn't wait to join the Peace Corps." When Bergstein was writing the script in the 1980s, setting this story in the summer of 1963 would be akin to setting a story now in the summer of 2001—just before 9/11—again, before everything changed for the worse. The audience knows the film is a "moment in time" that will radically shift, but the characters don't. Bergstein stated in one interview (Lowry 2015):

> …It still felt like a time of innocence in America…It was several months before John F. Kennedy's assassination, less than a year before the Beatles would take over America, and just before Martin Luther King Jr.'s "I Have a Dream" speech…If it had been set the following year, this story wouldn't take place. Young men would be in Vietnam, and the Beatles would happen and they wouldn't be listening to this music.

But the social location of *Dirty Dancing* is also 1986–7, when the film is produced and released. This section will discuss *Dirty Dancing* first in the context of a 1980s film, and then in the context of its story set in 1963.

Dirty Dancing *as a Reagan Era Film: 1985–1987*

Many aspects of the Reagan era are discussed at length in Chap. 2 in the context of the unfolding HIV/AIDS crisis, including apocalyptic fears of nuclear war. Between 1985–7, AIDS was becoming a public health nightmare in the wake of actor Rock Hudson's death, an explosion in heterosexually-transmitted cases and transfusion-AIDS. Thus, films about sexuality in the Reagan era uniquely reflected cultural anxieties over unprotected sex. By 1987, nobody had put "baby" in a corner. *Dirty Dancing* opened the same year as *Fatal Attraction* (1987), starring Michael Douglas and Glenn Close, which was about a one-night stand that put a man's entire family at risk when the psychotic "homewrecker" (played by Glenn Close) claims to be pregnant; the film was not just a metaphor for unsafe sex, but every man's worst nightmare. In *Three Men and a Baby* (1987), a man unwittingly fathered a baby, and must care for it when the mother is unable to do so; he and his two bachelor friends who all live together (yet are not gay), fall madly in love with "Baby Mary" as their paternal instincts emerge. On the political scene, in May 1987, Gary Hart, the Democratic frontrunner candidate for the 1988 Presidential election, was blind-sided by the press when his extramarital affair with Donna Rice was revealed. Hart was confused about why his sexual dalliances mattered, given that nobody covered President Kennedy's famous affairs (see further). Nonetheless, in a time frame where careless sex was being scrutinized, Hart dropped out. In hindsight, were it not for the Donna Rice controversy, Hart would have likely won the election, taken the country into progressive directions years earlier, changing the political fortunes for Bill and Hillary Clinton (see Chap. 4).

Reagan era films were also about a growing infertility epidemic due to women delaying childbirth in the wake of reproductive options and birth control, and also post-Sexual Revolution pelvic inflammatory disease (see under History of Medicine). Thus, creeping pronatalism films dotted the 1980s landscape, such as *The Big Chill* (1983), which features the character, Meg who is single, childless, had an abortion in college, now wants to have a baby before it's too late, and is borrowing one of her married college friends. Diane Keaton in *Baby Boom* (1987), a single, successful woman who—through freak circumstances—winds up with an orphaned baby, discovers her biological clock and suppressed maternal instincts. *Baby Boom* was, in essence, a pronatalist film about the fertility crisis. In *Another Woman* (1988), Gena Rowlands plays a childless 50-year-old philosophy professor who comes to regret an abortion she had when she was in graduate school. Ultimately, as the first generation of women came of age with real reproductive choices from the Pill and safe pregnancy termination, Reagan era popular culture was pushing motherhood. A new conservatism was dawning, which stood in stark contrast to the independent women's liberation themes of the 1970s, also spawning feminist works such as The Handmaid's Tale (Atwood 1986), in which war and environmental ruin render most of the women infertile, leading to a second right-winged revolution in which fertile women are enslaved as handmaids to the ruling class. The book would become a film in 1990, and a mini-series in 2017.

But more than anything, social class wars dominated Reagan era films; the working poor pitted against the privileged and wealthy is the definitive "Reaganite" plot (Baron and Bernard 2013), as the gap between working class and professional classes grew larger due to the breakup of unions and deregulation. In the 1980s, the rich got richer, and the poor got poorer. Another iconic 1987 film, *Wall Street*—touting the line by Michael Douglas' Gordon Geoko: "greed is good"—defined the era's sentiment. In *Wall Street*, Bud Fox, played by Charlie Sheen, will do anything to climb out of his working class roots. As noted by Gruner (2013), Charlie Sheen and Jennifer Grey also share a brilliant five-minute scene together in a police station in *Ferris Bueller's Day Off* (1986), just before each was catapulted into fame in memoir films about the 1960s. Sheen starred in the Vietnam memoir, *Platoon* (1986), based on Oliver Stone's experiences, while Grey's *Dirty Dancing* was a female memoir about the same era. However, *Ferris Bueller's Day Off*'s Reaganite plot plays out when Cameron (played by Alan Ruck) destroys his father's collector car in a fit of rage, screaming and kicking the car: "Who do you love? (kick!) Who do you love? (kick!) You love a car!"

Although set in 1963, *Dirty Dancing* is also a Reaganite film about class wars; the dance instructors are the working poor in servitude to the rich guests, creating the same class tension-based romances seen in *Some Kind of Wonderful* (1987) and *Pretty in Pink* (1986). *Jezebel* observes (Carmon 2010):

> Told her whole life that she could do anything and change the world, [Baby's] faced with the hypocrisy of a long-shunned minority enacting its own unexamined exclusion, this time on class grounds. The guests at Kellerman's look comfortable, but they were raised in the Depression and traumatized by World War II.

Meanwhile, *Less Than Zero* (1987) explored the unintended consequences of obscene wealth and materialism in the cocaine-tainted worlds of spoiled young adults with trust funds, absolutely no purposeful direction, and endless access to money. It was a film that particularly spoke to tail-end Baby Boomers born in 1963–64, the last years of that generational cohort (Sheehy 1995), and the starting years of Generation X, a newly named cohort to define a "lost generation" that had no real defining cultural events or values other than it had "missed the Sixties", and no living memory of the JFK assassination (Watergate and the Nixon resignation had certainly defined that generation, but it would not be evident until much later.) And so, *Dirty Dancing* was a film reflecting its own times, complete with 1980s music and an MTV influence (Tzioumakis and Lincoln 2013); but it was also about the 1960s, which tail-end Baby Boomers, in particular, had romanticized (see further).

And that gets us to one of the most defining features of Reagan era culture: a new processing and re-examination of the 1960s. By 1983, the country had reached the twentieth anniversary of the JFK assassination, but had changed beyond recognition.

With each successive year of the 1980s was the twentieth anniversary of some other major defining event in the 1960s. Consequently, Vietnam memoir films exploded into theaters. In addition to *Platoon* (1986) *Dirty Dancing* came out the same year as *Full Metal Jacket* (1987); *Good Morning Vietnam* (1987); and *Hamburger Hill* (1987), followed two years later by the Ron Kovic memoir, *Born on the Fourth of July* (1989). *Dirty Dancing* is full of Vietnam war references as the war is in its early stages in the Kennedy era, before it's escalated by Lyndon Johnson. While *Dirty Dancing* was in production in 1986, and then released in August 1987, a few defining political events were resonating with audiences at the time, which would have far reaching effects for future fans of *Dirty Dancing*.

1986: The Challenger Disaster and Chernobyl

January 28, 1986 marked the Challenger Space Shuttle disaster, in which seven crew members were killed when the Challenger exploded 73 seconds after it was launched. However, the crew member who captured the country's attention was 38-year-old Christa McAuliffe, a New Hampshire teacher who volunteered for the mission as a citizen astronaut. McAuliffe's family and students watched in horror as Challenger exploded. McAuliffe had planned to keep detailed journals on her observations and teach her students from space (NASA 2018). The disaster put the shuttle program on pause for two years, but McAuliffe's profile as a selfless woman who volunteered for her country to promote peace recalled the Peace Corps spirit of the Kennedy era (see further), which had also committed to the moon landing mission by end of the decade. In a 1962 speech, Kennedy stated: "We choose to go to the moon not because it is easy, but because it is hard." (Kennedy 1962). Challenger launched from the Kennedy Space Center in Cape Canaveral. It was a defining moment that linked the Kennedy and Reagan eras together, just as *Dirty Dancing* would.

On April 26, 1986, news reports of the Chernobyl nuclear disaster made headlines, and ultimately led to the fall of the Soviet Union, which would be complete by 1989; Reagan had begun warming relations with Gorbechev, and fears of nuclear war—despite Chernobyl—began to dissipate.

1987: The Robert Bork Nomination

The summer of 1987, when *Dirty Dancing* was released, was also the twentieth anniversary of the Summer of Love (1967), which on the heels of the Pill, launched the Sexual Revolution. But abortion remained inaccessible until 1973 (see under History of Medicine). However, the summer of 1987 would become a defining moment for women's reproductive lives that would extend far into the next century. On July 1, 1987, Ronald Reagan nominated Robert Bork to replace Justice Lewis Powell, a moderate, on the Supreme Court. Bork was considered a conservative extremist whose positions on civil rights and abortion were troubling to the Democrats; Bork also acceded to Richard Nixon's demands during Watergate (promised a Supreme

Court nomination in return). In 1981, Reagan had nominated a conservative moderate and first woman to the Supreme Court, Sandra Day O'Connor, who was confirmed unanimously by the Senate, and retired in 2006. He next nominated conservative Antonin Scalia in 1986, who later would be responsible for the contentious *Bush v. Gore* in 2000 (see Chap. 4). Reagan now was on his third Supreme Court nomination. In 1980, Reagan ran on appointing conservative judges who were also opposed to *Roe v. Wade* (National Journal 1980). Senator Ted Kennedy vehemently opposed Bork's nomination. The *New York Times* reported it this way on July 5, 1987 (Reston 1987):

> Mr. Kennedy asserted that "Bork's rigid ideology will tip the scales of justice against the kind of country America is and ought to be."
>
> He said that Judge Bork's firing of Archibald Cox as special prosecutor during the Watergate hearings was enough in itself to disqualify him for the Supreme Court, and he added:
>
> Robert Bork's America is a land in which women would be forced into back-alley abortions, blacks would sit at segregated lunch counters, rogue police could break down citizens' doors in midnight raids, schoolchildren could not be taught about evolution, writers and artists could be censored at the whim of the Government, and the doors of the Federal courts would be shut on the fingers of millions of citizens.

Bork was considered an "originalist" and did not recognize the right to privacy, upon which the *Roe v. Wade* decision is based (see under History of Medicine). The Bork nomination was also playing out alongside the Iran-Contra scandal (see further), and so the Bork vote was delayed until the Iran-Contra hearings were concluded. *Dirty Dancing* audiences in 1987 were concerned that Bork could be confirmed, which would be a vote for repealing *Roe v. Wade*. In October 1987, Bork was not confirmed by the Democratic-controlled Senate. At the time, it was considered a "partisan" vote, but, six Republicans voted against him, and two Democrats had voted for him. In November, Reagan ultimately nominated Anthony Kennedy, who wound up being a reliable swing vote on the court until 2018, protecting *Roe v. Wade* in many decisions.

But the Bork incident led to "pay back" when the Republicans controlled the Senate confirmations, and a "long game" of installing conservative judges to remake the courts with a singular focus to overturn *Roe v. Wade*. In 2016, when Justice Scalia died, President Obama nominated moderate Merrick Garland. But the Republican-led Senate still invoked the bitter Bork nomination process, and refused to hold hearings to confirm Garland claiming it was an election year. It worked. Scalia was replaced by Neil Gorsuch, a presumed opponent of *Roe v. Wade*. When Reagan's Bork replacement, Justice Kennedy retired, President Trump nominated Brett Kavanaugh for Justice Kennedy's vacancy. The Kavanaugh nomination led to the most bitter and partisan Supreme Court confirmation hearing in U.S. history, and he remains a wild card on *Roe v. Wade* protections. Kavanaugh was a privileged teenager in the 1980s, and in what could have easily been a Reaganite film plot, routinely drank to excess while "partying hard"—never with any consequences due to his wealth and class. In 2018, he was credibly accused of having sexually assaulted Christine Blasey Ford while drunk in 1982. A national "He Says/She Says" hearing was held in the Senate in which Ford gave credible testimony. Ford said she came forward to be a "good citizen" and wanted "to be helpful". In rehashing the Reagan era assault incident, the country got a Reagan era ending, too. Just as Baby risks her relationship with her family to do the right thing, so does Ford. In the Reaganite Kavanaugh hearing, Ford's life is disrupted while Kavanaugh gets confirmed. Ford could have delivered Baby's angry rant: "So I did it for nothing! I hurt my family…I did it for nothing!…You're right…you can't win no matter what you do!". Ultimately, the reach of the Reagan judiciary and the concerns over a repeal of *Roe v. Wade* was on the radar for 1987 audiences watching *Dirty Dancing*. But the concerns over the repeal *of Roe v. Wade* reached a fever pitch by 2018, when the last of Reagan's appointees retired, while *Dirty Dancing* only got more popular.

The Iran-Contra Scandal

In 2018, women in Iran who were fans of *Dirty Dancing* would begin openly "dirty dancing" in protest of the country's misogynistic policies. Iran had been a Westernized country until 1979, when after a revolution, it had morphed into an extremist Islamic regime enforcing Sharia law, especially known for its brutal treatment of women. Two events involving Iran coincided with the opening of *Dirty Dancing* in 1987, when the new Iran regime was less than a decade old. First, Betty Mahmoody, an American woman married to an Iranian had just published Not Without My Daughter (Mahmoody 1987), later made into a film in 1991. The book tracked her experiences of agreeing to visit Iran with her husband to see his family, and then being held hostage in the country with her small daughter when he wanted to repatriate. Betty lost all of her rights to her autonomy and her daughter, in keeping with Sharia law, and eventually fled the country in a harrowing escape.

Second, a complex foreign policy scandal involving the Reagan administration, Iran and Central America would dominate headlines, in the bizarre Iran-Contra scandal, or "Irangate" that nobody could quite follow. In brief, this was an odd "money laundering" scheme in which the United States illegally sold weapons to Iran as a way to funnel money to Central American rebels (the "Contras") trying to overthrow the socialist government in Nicaragua (Pincus and Woodward 1987). The arms arrangement was spearheaded by Oliver North, who began testimony before Congress in July 1987. The complicated scheme had such little impact on the average American, Oliver North's testimony played on televisions like a muted music video backdrop. But we would later learn that Reagan's covert dealings with Iran began long before—during the 1980 election, when Reagan negotiated as a private citizen with the Iranian regime to delay the release of American hostages taken during the Carter Administration until he was sworn in. Notwithstanding, the Iran-Contra scandal had future implications for women's rights—particularly reproductive rights—in both regions. In Iran and other Islamic countries, women's rights continued to erode, and as the violence throughout Central America raged, weaponized sexual assault would lead to mass migration of girls and women seeking asylum in the United States by the next century. In 2017, one such Central American girl who was pregnant due to rape was prevented by the U.S. government from getting an abortion while in its custody, waiting for an asylum hearing (Sacchetti and Somashekhar 2017).

The Summer of 1963

As the opening sequence of *Dirty Dancing* clearly established, setting the film in the summer of 1963 puts us squarely into the Kennedy administration before his assassination, which Americans equate with a time of innocence before the chaos of the Sixties had truly begun, and the hopeful and progressive March on Washington was in the works. Civil rights issues during this time frame are discussed more in Chaps. 1 and 9, but Kennedy's Civil Rights bill was being debated in Congress in October 1963. Kennedy had also given his famous West Berlin speech in protest of the newly constructed Berlin Wall on June 26, 1963, in which he said: "Ich Bin Ein Berliner".

With respect to Kennedy's policy agenda, perhaps the most important lasting legacy is that Kennedy took us to the moon by committing a budget to the NASA program—funds many thought were risky in light of the many domestic problems facing the country. But reaching for the moon became a metaphor of the Kennedy administration overall: it was about the future and reaching for a higher ideal, which united the country in optimism. By the Summer of 1963, Gordon Cooper had completed the last solo manned Mercury program flight in May 1963 with 22 orbits around the Earth, while the Gemini and the Apollo programs were gearing up.

By 1963—as in 1987—fears of nuclear annihilation were dissipating because of improved diplomacy with Russia. In the wake of the 1962 Cuban Missile Crisis, the U.S. Senate approved Kennedy's Atmospheric Test Ban Treaty with the Soviet Union, which banned above-ground testing of nuclear weapons. In June 1963, Kennedy had also signed into law the *Equal Pay Act*, which would become relevant in the 1970s, a prescient law in light of the reproductive autonomy that would come a decade later (see under History of Medicine).

The film makes references to "Southeast Asia", as Vietnam was just heating up, but to a 1987 audience, the Vietnam war was very familiar to them as five major Vietnam films were currently in theaters at the same time (see earlier). Baby's plans to join the Peace Corps in the summer of 1963 also reflected the idealism of the Kennedy administration, and was hugely popular at the time. Thousands of young adults—especially after college—were joining the two-year program, which had been established March 1, 1961. But with escalation of the Vietnam War, Peace Corps participation completely eroded with a loss of idealism until the Reagan era, when President Reagan renewed interest in it, again linking the Kennedy and Reagan periods. But on July 24, 1963, the Kennedy and Clinton era would also be joined: Bill Clinton, then 16, was in Washington D.C. as part of a Boys Nation program, and shook hands with President Kennedy, which became an iconic photograph in his 1992 campaign for President, and was considered the most impactful moment of

Clinton's life. In fact, when the events of *Dirty Dancing* are taking place, Kennedy and Clinton are both at the White House at the same moment.

By the summer of 1963, social changes for women had already begun with the February publication of Betty Friedan's *The Feminine Mystique*, which discusses the "problem that had no name"—boredom and depression in unsatisfying lives.

Friedan's book is essentially "ground zero" of the second wave feminist movement, which laid bare pronatalism and unfair social arrangements for women, which included the absence of reproductive control. Additionally, by 1963, while the Pill was available, it was only available to married women, which would not change until the *Griswold v. Connecticut* decision (1965) (see under History of Medicine). Single women had no way to protect themselves from pregnancy without a condom, which relied on male participation. When Penny tells Baby: "I don't sleep around…I thought he loved me…I thought I was something special", she is making clear that she was willing to risk pregnancy with the knowledge that marriage would likely result, since she thought she was involved with a loving partner with a future. In 1963, the Pill was also illegal in several states even for married women.

Kennedy's Affairs: "Have You Had Many Women?"

While Penny didn't sleep around, President Kennedy famously did; and things were coming to a head during the summer of 1963. One of his biographers noted in 2013 (Sabato 2013):

> In July 1963 FBI director J. Edgar Hoover informed Bobby Kennedy that he knew about the president's past relationship with an alleged East German spy…In late August 1963 [she] was flown back to Germany on a U.S. Air Force transport plane at the behest of the State Department… Kennedy had affairs with scores of other women, including two White House interns.

Kennedy had essentially engaged in sexual assault of one young intern who lost her virginity during the episode and was constantly shuttled to him (Sabato 2013). In 1963, however, Kennedy's sexually predatory behavior was condoned, but some believed it was getting to the point that summer, where it could have jeopardized his entire presidency. For example, his Secret Service agents noted "morale problems" over Kennedy's behaviors. According to one former agent (Sabato 2013):

> "You were on the most elite assignment in the Secret Service, and you were there watching an elevator or a door because the president was inside with two hookers…It just didn't compute. Your neighbors and everybody thought you were risking your life, and you were actually out there to see that he's not disturbed while he's having an interlude in the shower with two gals from Twelfth Avenue.

Dirty Dancing paints an accurate picture of some men sleeping around with abandon, without suffering any consequences at all. Robbie's attitude, and his imitation of Kennedy when he flirts with the older Houseman sister, Lisa—"Ask not what your waiter can do for you, ask what you can do for your waiter"—was likely a reference to Kennedy's womanizing. Yet as mentioned above, these behaviors were no longer acceptable by 1987, when Gary Hart's womanizing ended his campaign for President, while Arkansas Governor, Bill Clinton was eyeing a 1992 bid. It is indeed historically possible that at the same moment Bill Clinton shook Kennedy's hand at the Rose Garden, there was a White House intern inside who was having an affair with the President, something Clinton would repeat 30 years later (see Chap. 4).

The Third First Child: Patrick Kennedy

It was well documented that the First Lady Jacqueline Kennedy was aware of the affairs (Leaming 2014), but like other married women of her time, she tolerated it because the alternatives were not appealing. The summer of 1963 marked the last time a sitting First Lady was pregnant and would give birth. Jacqueline Kennedy, who had a history of miscarriages, including a stillborn daughter, delivered Patrick Kennedy prematurely on August 7, 1963. Patrick died two days later, on August 9, 1963 from infant respiratory distress syndrome, a common problem routinely treated in neonatal intensive care units today. It was Jackie Kennedy's fifth pregnancy, and third child. She and President Kennedy, in fact, had still been in mourning over Patrick when they went to Dallas November 22, 1963.

The History of Medicine Context: "I Carried a Watermelon"

Baby's first entry into Penny's and Johnny's world is helping to carry huge watermelons—classic symbols for the womb—into a cabin dance party. As Baby's holding a watermelon the size of a 39-week fetus, she is awestruck by fornication moves on the dance floor, which was known as "dirty dancing"—sexualized dancing that simulates sexual intercourse. Bergstein wants us to know that the film is about sex and "Baby carrying..." from the very beginning; and if we forget by the end, she cautions: "Nobody puts Baby in the corner!" When teaching the history of medicine relevant to *Dirty Dancing*, the focus is on the womb, which means discussing the history of abortion and the history of hormonal contraception, or the "birth control pill". Here, I will mostly focus on the history of abortion and will refer readers to other works for the history of birth control further on.

The first laws to criminalize abortion were passed in 1867, and by 1880 abortion was illegal across the country, including the sale of popular abortifacients through retailers or pharmacies. However, all such laws granted physicians the option to perform therapeutic abortions to save a mother's life, essentially reserving the practice for their medical judgments, including prescribing abortifacients. This led to the practice of widespread abortions performed by physicians for much of the later nineteenth century through the Depression era (Reagan 1997). Contraception (and abortion) was also made illegal by the Comstock Laws (1873), which forbid promotion or sale of anything immoral or obscene (see further).

The fight for legalized abortion has a different history than the fight for legalized "birth control", which was seen by its proponents as a public health crusade and antithetical to abortion, even though average women had difficulty with the distinction (Reagan 1997).

The Pill was approved in 1956 by the FDA, and was developed by a small team of advocates and researchers who began their work when abortion was a crime, and a "birth control pill" was seen as the ultimate tool for reproductive control (Eig 2014). But the Pill was not widely available to all women until 1969 (see further).

While these two reproductive justice histories are cousins, they are not twins and ran on different timelines. Yet access to the Pill ultimately led to *Roe v. Wade*, and unsafe abortions ultimately led to the Pill. Ethical and moral questions are discussed under Health Ethics Issues.

Penny for Your Thoughts: Pre-Roe America

Dirty Dancing depicts the last vestiges of illegal abortion in pre-Roe America. To a 1987 audience, Penny's plight resonated less than it did in 1997, on its re-release, when Targeted Restrictions of Abortion Providers, or TRAP laws, began to be passed (see further), which by 2017, had virtually eliminated abortion access in many states. Ironically, American women living in 1863 had better access to abortion than in 1963; and women in 1987 had better access to abortion than in 2017.

In the United States , abortion was not always illegal, and was the first method of reproductive control for American women until after the Civil War, when in 1867—for mainly business reasons—the white male medical establishment wanted to usurp control of the abortion trade (Reagan 1997). Prior to this, private abortion clinics were all over the place and were considered a private transaction between the client and abortionist. Other factors influencing criminalization of abortion were immigration trends, and fears that babies born to immigrants would outpace babies born to U.S. citizens. This led to state laws that criminalized nontherapeutic abortion but left "therapeutic abortion" legal and up to the discretion of physicians and their medical judgment. What constituted "therapeutic abortion" was quite broad (see under Healthcare Ethics Issues), but there remained a large network of physicians who made abortion a specialty practice, while physicians who did not perform abortions would refer patients to the ones who did.

Despite the illegalization of abortion, the practice was prolific nonetheless. Women shared information about self-abort techniques, which sometimes worked, while a lot of doctors in private practice performed abortions routinely, although there were still risks, and many women died of sepsis and peritonitis—particular prior to penicillin. It was married women who most frequently sought out abortion due to the cycle of misery and poverty that most women endured from involuntary motherhood, difficult childbirths, and far too many children than they could afford to feed or raise. In the first decades of illegal abortion (1880–1930), it was practiced as an "open secret" (Reagan 1997), and nobody looked upon abortion as "immoral" or anything other than a woman's right to do. In fact, an expose of the ease with which women could access abortion was published in the *Chicago Tribune* in 1888 by two investigative journalists who posed as an unwed couple seeking an abortion, exposing the world of underground "infanticide". Essentially, illegal abortion was treated like prohibition; only a small minority thought it was immoral in the early stages; few thought it should be illegal, few took the laws seriously, and several clinics paid for police to look the other way (Reagan 1997).

The prevailing attitude in the 1920s, for example, was that Americans approved of abortion as a social necessity, and plenty of physicians performed the procedure. This had to do with limited understanding of the timing of menstruation and conception (see under Healthcare Ethics Issues). Generally, women's desperate poverty, overly large families, and health circumstances led to a large network of helpers—just like in Penny's case—who would help get her "fixed up". Historians point out, too, that many men actively participated in helping to get an abortion for their wives or girlfriends (Reagan 1997).

Margaret Sanger, a birth control advocate, would receive letters from desperate pregnant women such as these (Eig 2014: 29, 48):

> Dear Mrs. Sanger…I am 30 years old have been married 14 years and have 11 children, the oldest 13 and the youngest a year…Can you help me? I have missed a few weeks and don't know to bring myself around…I would rather die than have another one.

> I am today the mother of six living children and have had two miscarriages. My oldest son is now twelve years old and he has been helpless from his birth. The rest of my children are very pale…One of my daughters has her left eye blind. I have tried to keep myself away from my husband since my last baby was born, but it causes quarrels, and once he left me saying I wasn't doing my duty as a wife.

Or (Reagan 1997:41):

> I am a poor married woman in great trouble…I'm in the family way again and I'm nearly crazy for when my husband finds out that I'm going to have another baby he will beat the life out of me.

Typically (Eig 2014:35):

> Sanger saw poor women resorting to "turpentine water, rolling down stairs, and…inserting slipper-elm sticks, or knitting needles or shoe hooks in the uterus" to end their pregnancies.

Sanger wrote in 1919: "[Women] suffer the long burden of carrying, bearing, and rearing the unwanted child. It is she who must watch beside the beds of pain where the babies who suffer because they have come into overcrowded homes." (Eig 2014:50).

By the 1930s, the Great Depression created enormous issues for women with unplanned/unwanted pregnancies, and abortion services peaked, as many physicians provided therapeutic abortions with broad definitions of what that meant. It was easier to get an abortion in 1933 than in 1963. In 1933, women could access safe abortions from a number of physician-abortionists in urban centers; obstetricians and gynecologists who specialized in abortions were widely considered part of the traditional medical community. Women were also fired from their jobs if they were pregnant, which meant that women who needed to work with other children at home, were forced to make a choice between employment or pregnancy. In cities like Chicago, safe abortions by physicians were easily accessed for about $67.00, a large amount at the time, but cheaper than having a baby in a hospital (Reagan 1997).

Things began to change around 1940, when various States began to enforce abortion laws; they began raids on various practices, which also entailed "outing" patients. Abortion clinics were treated like illegal gambling or prostitution rings, and began to close down. Several abortion providers were arrested, patients put on trial, and State courts upheld that referring physicians were legally complicit (Reagan 1997). This coincided with more women entering the workforce during World War II, and birth rates dropping precipitously. Pronatalism and an attempt to confine women to traditional roles and motherhood, fueled a crackdown on abortion. During the McCarthy era, abortion was called "subversive" and linked to "communism", making it even harder. By the 1950s, every accredited hospital began to form "therapeutic abortion committees" as a way to create its own guidelines and criteria for legal protections, which served to seriously limit access. These committees are discussed further on under Healthcare Ethics Issues.

Between the mid-1950s until *Roe v. Wade*, abortion went underground, became cost prohibitive ($250–500), and providers were frequently untrained. While hospitals declined to do many therapeutic abortions, at the same time, they needed to set up specific wards to treat complications such as sepsis from the illegal and botched abortions. Typical experiences were described like this (Reagan 1997):

> When Rose S. got pregnant at 19 … she "confirmed the pregnancy by seeing, under a false name and equipped with a Woolworth-bought wedding ring, a gynecologist…She acted under a false name because, she explained, "What seemed of paramount importance to me…was the secrecy"… she arranged for an abortion with "an expert" doctor in Chicago…She travelled by train …[and] had a 6 AM appointment, paid $250….

One woman recalling a college student she knew said (Reagan 1997:197–198):

> …she was "too frightened to tell anyone what she had done…so when she developed complications, she tried to take care of herself. She locked herself in the bathroom and quietly bled to death" …When another woman met her connection in Baltimore, he blindfolded her and walked her up the stairs, down the halls and down the stairs to thoroughly confuse her before taking her into a hotel room where the abortionist introduced a catheter. A woman who went to Tampa for an abortion in 1963 recalled being examined by the doctor, then being put in a van with several other women, the entire group blindfolded, and then driven to an unknown location where the abortions were performed.

Penny's abortion in *Dirty Dancing* was a typical experience, and if we presume that Dr. Jake Houseman had privileges at any hospital in that time frame, he had treated multiple cases of sepsis and likely peritonitis from complications of abortion. When he saw Penny, he was presented with a common medical problem due to unsafe abortions, and knew what to do. In fact, he had a duty to treat (see under Healthcare Ethics Issues). Historians also note that understanding abortion decisions pre-Roe had to do with understanding women's lives in context (Reagan 1997). Penny's pregnancy would have also led to her being unemployed, as she would not have been able to work as a pregnant dance instructor, which was her livelihood.

As the medical profession was confronted with the consequences of unsafe abortions and thousands of preventable deaths from sepsis, it reversed course from the previous century and began to lobby for legalizing abortion. Meanwhile, women began to organize and provide abortions on their own. The Jane Service performed thousands of abortions for women between 1969–73 (Bart 1981; Reagan 1997). Women would call a voicemail service and ask for "Jane" and would be contacted and taken to a secret location for the procedure. Laywomen ran Jane and learned how to perform abortions safely and properly. Another such service operated in California. By 1970 the Illinois state abortion law was overturned for being unconstitutional and then reversed on appeal, but it eventually led to *Roe v. Wade* (1973)—the reproductive equivalent of *Brown v. Board of Education* (1954).

Roe v. Wade (1973)

In 1969, Norma McCorvey was 21 and could not get an abortion in the state of Texas unless her life was endangered. She was unable to find any illegal abortion facility because it had shut down in her state. She went to attorneys Linda Coffee and Sarah Weddington, who renamed her Jane Roe, and challenged the Dallas District Attorney, Henry Wade. The following year, in 1970, the District court ruled in favor of Roe, stating the Texas law was unconstitutional, but wouldn't grant an injunction against the law. By then, Norma had already given birth and put the baby up for adoption.

Roe v. Wade next went to the Supreme Court, and in December 1971, along with *Doe v. Bolten* (a similar case surrounding Georgia's restrictive abortion law), arguments were heard. In a January 22, 1973 landmark decision, the court ruled in *Roe* and *Doe* that the restrictive state laws were wholly unconstitutional. The original 1973 decision included the following statement, which comprises the constitutional and medico-legal basis for legalizing abortion. At the time, roughly 10,000 women were dying each year from illegal abortions.

> This right of privacy, whether it be founded in the Fourteenth Amendment's concept of personal liberty and restrictions upon state action, as we feel it is, or, as the District Court determined, in the Ninth Amendment's reservation of rights to the people, is broad enough to encompass a woman's decision whether or not to terminate her pregnancy. The detriment that the State would impose upon the pregnant woman by denying this choice altogether is apparent. Specific and direct harm medically diagnosable even in early pregnancy may be involved. Maternity, or additional offspring, may force upon the woman a distressful life and future. Psychological harm may be imminent. Mental and physical health may be taxed by child care. There is also the distress, for all concerned, associated with the unwanted child, and there is the problem of bringing a child into a family already unable, psychologically and otherwise, to care for it. In other cases, as in this one, the additional difficulties and continuing stigma of unwed motherhood may be involved. All these are factors the woman and her responsible physician necessarily will consider in consultation. … On the basis of elements such as these, appellant and some amici argue that the woman's right is absolute

and that she is entitled to terminate her pregnancy at whatever time, in whatever way, and for whatever reason she alone chooses.

Rules of the Roe

The *Roe* decision clarified that the constitution protects a woman's decision to terminate pregnancy because right to privacy includes the decision whether to terminate pregnancy; additionally, the state does not have a compelling interest justifying restrictions on abortion until the point of viability (Huberfeld 2016). The Center for Reproductive Rights (2007) summarized the four medico-legal rules of *Roe v. Wade* bold emphasis mine:

> The Roe opinion was grounded on four constitutional pillars: (1) the decision to have an abortion was accorded the highest level of constitutional protection like any other fundamental constitutional right, **(2) the government had to stay neutral, legislatures could not enact laws that pushed women to make one decision or another, (3) in the period before the fetus is viable, the government may restrict abortion only to protect a woman's health**, (4) after viability, the government may prohibit abortion, but laws must make exceptions that permit abortion when necessary to protect a woman's health or life.

The ethical issues raised in the *Roe v. Wade* decision are discussed further under Healthcare Ethics Issues.

Post-Roe America

Female audiences who saw *Dirty Dancing* when it opened in 1987 were a hybrid of middle-aged women with living memories of pre-Roe America, and younger women who grew up with complete access to contraception and legal abortion. However, creeping pronatalism in the Reagan era was making a comeback as the Sexual Revolution had come to an end in the wake of the AIDS epidemic, and a new reproductive autonomy problem was unveiling: infertility and biological clocks. In 1984, a cover story in *Time* magazine declared the end of the Sexual Revolution, a byproduct of second wave feminism (see under Healthcare Ethics Issues). A number of women had delayed childbirth to discover they couldn't get pregnant. Additionally, pelvic inflammatory disease (PID) was rampant due to unchecked sexually transmitted diseases. *Time* magazine declared "an obituary for one-night stands and swinging lifestyles… Courtship, marriage and family have staged a comeback…" (Leo 1984).

Meanwhile, NBC aired *Second Thoughts on Being Single,* a one-hour documentary that reported many formerly sexually active women had become "born again prudes…a swing toward marriage and away from casual sex." (NBC 1984). By 1984, Mary Guinan at the CDC declared: "We are in the midst of an unprecedented epidemic of sexually transmitted diseases," (CDC 1984). By that point, approximately 20 million Americans had herpes; 40,000 women become involuntarily infertile each year from PID due to sexually transmitted diseases (STDs); gonorrhea rates had tripled since the 1950s, and there was an increase in syphilis again. By the 1980s, the most common cause of female factor infertility was scarring of the fallopian tubes from PID. However, the social production of infertility from delaying childbirth was the principal problem. In 1978, England's Louise Brown was the first baby born though in vitro fertilization (IVF), known then as the "test tube baby". By 1987, thousands of women were undergoing fertility treatments with what was called at the time, the "new reproductive technologies". Many Baby Boomer women were now spending thousands of dollars to get pregnant; embryos were being created and frozen; and attitudes about abortion began to shift in light of both fertility treatments and prenatal technologies such as ultrasound that began to be routine by around 1985 (Campbell 2013). Fertility treatments and premature births, as well as prenatal images with ultrasound, raised more questions about the moral status of the fetus (see further), and provided openings for fetal rights arguments launched by religious groups and the Pro-Life movement, which did not recognize "Pro-Choice" or women's autonomy as legitimate. Fetal interest arguments were beyond the scope and framework of *Roe v. Wade*, which had been decided based on the right to privacy and bodily autonomy. Fetal rights arguments would begin to erode the impact of the constitutional right to abortion.

1992: Planned Parenthood v. Casey

Five years after *Dirty Dancing* opened, individual states began to pass a number of restrictions on abortion access. By 1992, Planned Parenthood challenged a set of restrictions in Pennsylvania in the Supreme Court and won a pyrrhic victory, thanks to two Reagan-appointed justices (Sandra Day O'Connor and Anthony Kennedy), and two conservative justices appointed by George H. W. Bush (Bush 41): Clarence Thomas, replacing Thurgood Marshall (Thomas had credibly been accused of sexual harassment by Anita Hill but got confirmed anyway); and David Souter. The *Casey* decision upheld parts of *Roe v. Wade*, but not the whole thing. The Court clarified that the states could place restrictions on abortion previability (see under Healthcare Ethics Issues) so long as it was not an "undue burden". This fundamentally altered the *Roe v. Wade* decision, upending the long tradition of respecting Supreme Court precedents, known as "stare decisis". Based on the *Casey* decision, the state may not prohibit abortion before viability, but can certainly prohibit abortion after viability, except where necessary to protect the woman's life or health (Huberfeld 2016). *Casey* also permitted the state to *regulate* abortion before viability so long as it does not place an "undue burden" on the right to access abortion—acts with purpose or effect of discouraging abortion (Huberfeld 2016). Therefore, states were also allowed to encourage childbirth over abortion. Thus, the following became standard for state restrictions on abortion after 1992: (a) 24 hour waiting periods; (b) state-scripted information delivered by a doctor about putative risks of abortion; (c) adolescent parental notification (with judicial bypass); and statistical reporting requirements (Huberfeld 2016). The impact of the *Casey* decision, according to the Center for Reproductive Rights (2007):

Only two of the four *Roe* pillars remain today as a result of the Supreme Court's 1992 decision in Planned Parenthood of Southeastern *Pennsylvania v. Casey*. A woman's right to choose is still constitutionally protected, however, the "strict scrutiny" standard was jettisoned in favor of a lesser standard of protection for reproductive choice called "undue burden." Under *Casey*, state and local laws that favor fetal rights and burden a woman's choice to have abortion are permitted, so long as the burden is not "undue." No longer does the state have to be neutral in the choice of abortion or childbearing. Now the government is free to pass laws restricting abortion based on "morality," a code word for religious anti-abortion views. States are now permitted to disfavor abortion and punish women seeking abortions, even those who are young and sick, with harassing laws.

Medical Abortion

After the 1992 *Casey* decision, as restrictions to abortion began to increase, so did interest in the "abortion pill", or RU-486 (mifepristone). This was a drug that had been approved for use in France in 1988—just a year after *Dirty Dancing* was released, which caused contractions and miscarriage, and created the option of medical abortion. However, an FDA import ban on the drug was placed on it, barring American women from its use. By 1996, almost 10 years after *Dirty Dancing* was released, an application for approval in the United States was filed with the FDA (Lewin 1996).

At the same time, methotrexate was being used in the United States off-label as an alternative to mifepristone. In April 1997, a study published in The *New England Journal of Medicine* "shows that RU-486, when used in combination with a type of drug called a prostaglandin (misoprostol), medically terminates 92% of pregnancies when taken within 49 days of conception" (WebMD 2000).

By 2000, RU-486 in combination with misoprostol was finally approved for use in the first seven weeks of pregnancy, which broadened access to abortion in the wake of *Casey*. The *New York Times* reported the story amidst the 2000 election like this (Kolata 2000):

> A woman first takes mifepristone, which blocks the action of progesterone, a hormone required to maintain a pregnancy, and 36 to 48 hours later takes a second drug, misoprostol, which makes the uterus contract, expelling the fetal tissue. In clinical trials, the drug combination failed to result in complete abortion in about 5 percent of cases....Mifepristone was first synthesized 20 years ago, in France, and came on the market in Europe more than a decade ago. But from the beginning its introduction into the United States was entangled in abortion politics.

As further restrictions rose throughout the twenty-first century (see further), medical abortion became known as "do-it-yourself" abortion, and women began buying the abortion pills themselves from online sources. A 2016 *Glamour* magazine story begins (Zerwick 2016):

> With access to safe, legal methods becoming increasingly limited, some women are taking matters into their own hands with black-market remedies. Phoebe Zerwick goes underground to get their stories...For four months I've been investigating why more and more women like Renée are opting for what's being called do-it-yourself abortions. Rebecca Gomperts, M.D., Ph.D., founder of Women on Web, which sends abortion drugs to women in countries

where the procedure is banned, said she received nearly 600 emails last year from Americans frantic to end pregnancies under hard circumstances:

"Please, I need info on how to get the pills to do [an abortion] at home," one woman wrote. Has it come to this? How did women end up so desperate—even willing to break the law to get an abortion? And what does the new landscape mean for our health? Our rights? Our power as women?

Ultimately, online distribution of abortion pills was dubbed Plan C as TRAP laws (see further) continued to diminish access (Adams 2017), while Plan B, or emergency contraception (see further) is only effective 72 hours after intercourse.

Partial Birth Abortion Bans: Gonzales v. Carhart (2007)

By the twentieth anniversary of *Dirty Dancing*, Penny's story was very close to current law. Abortion was far more restrictive, and reports of women seeking information about self-aborting abounded (Stephens-Davidowitz 2016). This is because abortion restrictions were further expanded in the George W. Bush (Bush 43) era with the case, *Gonzales v. Carhart* (2007), which held that under no circumstances—even to save a woman's life, is a "partial birth" abortion permitted, a procedure that refers to a modified dilation and evacuation (D&E) in which a partial delivery of a fetus occurred—typically in the second trimester; such a procedure is reserved only for certain criteria.

Gonzales differed from prior decisions by giving greater deference when federal law created a hurdle to obtaining abortion. But it also removed the exception for preserving the health of the woman (Huberfeld 2016). After the *Gonzales* decision, State legislatures aimed to pass restrictions on abortion that would push the pregnancy past viability, and hence remove the abortion option altogether. *Gonzales* had unintended consequences for many women with planned and wanted pregnancies who discovered terrible anomalies, who were now forced to carry their fetuses to term knowing they would be giving birth to a fetus who would not survive (often with malformed brains or other critical organs). As a result, a number of States passed longer waiting periods (72 h instead of 24); created ultrasound viewing mandates; physician "contact" requirements; additional reporting requirements and limits on when abortion pills (medical abortion) could be used. State laws also had exemptions from insurance covering contraception or abortion; expanded child endangerment laws to the fetus, including fetal pain; and many states established much earlier limits on fetal viability (Huberfeld 2016).

TRAP Laws

By the time the Obama era began in 2009, State legislatures began to pass another set of laws known as "Targeted Restrictions for Abortion Providers" (TRAP). These laws singled out abortion for calculatedly burdensome regulations (Huberfeld 2016).

These were restrictions that were not imposed on medical procedures of similar or greater risk or on similar types of other health care providers. TRAP laws comprised mandating special facility licensure rules; additional reporting requirements; admitting privileges at a nearby hospital, including meeting ambulatory surgical center requirements. Provider "conscience clause" (refusal to provide abortion services) were also required. Typical states with TRAP laws mandated that abortion providers at any stage were subject to special licensure requirements; employee health testing; specific facility rules (e.g. written contract with hospital and ambulance service wherein they agree to accept and treat the abortion provider's patients if complications arise); unannounced inspections. These laws essentially sentence vulnerable or impoverished women to parenting by removing access to pregnancy termination services. Typically, these laws are particularly punishing to women in poverty who were raped or abused, who are unable to travel far. A fuller exploration of these laws from the perspective of women's healthcare providers can be seen in the documentary film, *Trapped* (2016), which featured calls with desperate women wanting information on how to self-abort. TRAP laws required many women having to drive hundreds of miles to obtain abortions. NPR reported on one state's new restrictions (i.e. TRAP law) this way (Domonoske 2016):

> First, it requires that all doctors who perform abortions have admitting privileges at a hospital within 30 miles of where the abortion takes place. But because the complication rate from abortions is so minuscule, most abortion providers cannot meet the minimum number of admittances that hospitals require before granting privileges.

> Second, the law requires that abortion clinics be retrofitted to meet elaborate statutory hospital-grade standards, including wide corridors, large rooms and other expensive construction and equipment standards that do not apply to all other outpatient facilities where other surgical procedures like liposuction and colonoscopies take place.

> The provisions also apply to doctors who prescribe medication-induced abortions; such procedures involve giving the patient two pills and sending her home.

The American Medical Association and the American College of Obstetricians and Gynecologists have made clear that TRAP laws do not make women safer.

In 2016, Indiana passed a law that made it illegal to terminate any pregnancy due to a genetic abnormality (Chicago Tribune 2016) as well as illegal to dispose of fetal remains without a proper burial. The federal courts declared only the first part of the law unconstitutional because it forces girls and women without means to care for a potentially *profoundly* disabled child that requires lifelong, complex medical care. But the second part of the law regarding mandated burial held (Higgins 2019).

In 2016, another landmark Supreme Court case, *Whole Woman's Health v. Hellerstedt*, challenged TRAP laws. In this case Texas TRAP laws required doctors who performed abortions to have admitting privileges to send patients to a hospital no further than thirty miles from the clinic, while each clinic was to have the same facilities as ambulatory surgical centers. Ultimately, the Supreme Court ruled that such TRAP laws were indeed unconstitutional and place an undue burden on women seeking an abortion. In *Whole Women's Health v. Hellerstedt*, the TRAP laws being challenged had nothing to do with making abortion safer, or making women's healthcare safer, and everything to do with making abortion access more difficult, and hence, posed an "undue burden" on a constitutionally protected right for all women (Denniston 2016).

TRAP laws helped to force postponement of termination procedures past viability. These laws also could force women's health clinics offering termination services to women to close by imposing onerous requirements that generally have nothing to do with patient safety. The Whole Women's Health decision validated constitutional protections for women's reproductive rights and health. Because it was decided in 2016, when there were only eight justices (due to the Merrick Garland situation), the decision was a 5-3 ruling. The *New York Times* reported on the decision this way (Liptak 2016):

> The 5-to-3 decision was the court's most sweeping statement on abortion since Planned Parenthood v. *Casey* in 1992, which reaffirmed the constitutional right to abortion established in 1973 in *Roe v. Wade*. It found that Texas' restrictions — requiring doctors to have admitting privileges at nearby hospitals and clinics to meet the standards of ambulatory surgical centers — violated *Casey*'s prohibition on placing an "undue burden" on the ability to obtain an abortion.

In this case, President Obama tweeted: "Pleased to see the Supreme Court reaffirm" that "every woman has a constitutional right to make her own reproductive choices." (Liptak 2016).

In *Whole Women's Health*, the court stated the TRAP laws "vastly increase the obstacles confronting women seeking abortions in Texas without providing any benefit to women's health capable of withstanding any meaningful scrutiny."

More Extreme Abortion Laws

The most restrictive abortion law in American history, known as the *Human Life Protection Act*, was passed in Alabama in 2019, which led to several similar laws in other states (Tavernise 2019). Missouri also passed a controversial bill in 2019 that bans abortion beyond 8 weeks (Smith 2019).

The 2019 Alabama law makes it a crime for doctors to perform abortions *at any stage* of a pregnancy unless a woman's life is threatened or there is a lethal fetal anomaly (Alabama.com 2019). Doctors in Alabama who violate the new law may be imprisoned up to 99 years if convicted, but the woman seeking or having an abortion would not be held criminally liable. Thus, Alabama's law severely punishes healthcare providers for terminating pregnancies for any reason unless the patient's life is threatened, effectively forcing them into wrenching ethical dilemmas and moral distress, and upending common practices in fertility treatment care. The American Society for Reproductive Medicine has made clear that it opposes the Alabama fetal "personhood" law (ASRM 2019).

In Kentucky, one 2019 bill prohibits abortion after six weeks (most women don't discover they are pregnant prior to six weeks); another bill prohibits abortion if it's related to results of fetal diagnosis. The ACLU challenged both laws shortly after passage in federal court arguing it is unconstitutional (Amiri 2019). On March 27, 2019 a judge ordered the laws be suspended indefinitely until the court issues a final ruling on whether they are constitutional.

The History of the Birth Control Pill

The history of abortion, the main focus of this section, is indeed a different history than the contraceptive pill, which was spearheaded by feminist Margaret Sanger (1879–1966), philanthropist Katharine McCormick, biologist Gregory Pincus and obstetrician and gynecologist, John Rock. Sanger, who published in the early twentieth century, *Birth Control Review*, championed the concept of a "birth control pill" as an alternative to abortion, and in recognition of how many women were desperate to control reproduction and terminate their unplanned pregnancies. The research and development story of the first oral contraceptive pill for women, and the first controversial drug trials in Puerto Rico, is beautifully covered in Jonathan Eig's The Birth of the Pill (2012), in which Sanger's entangled history with the Eugenics Movement is also explored. For the purposes of teaching *Dirty Dancing*, it's important to note that the Pill, when it was first approved for use in 1960, was not legal in all states, and then when it was by 1965, it was only prescribed to married women who needed the consent of their husbands. The legal battle for women to access the Pill autonomously involved two landmark decisions. *Griswold v. Connecticut* (1965), granted all married couples access to the Pill in all states in 1965, while *Eisenstaat v. Baird* (1972) granted all women access to the Pill regardless of marital status. Thus, in the summer of 1963, Penny had no access to the Pill as an unmarried woman.

After the *Griswold* case, as thousands of married women began to use the Pill, questions arose surrounding informed consent over risks of the pill, and a major book by journalist and women's health advocate, Barbara Seaman, called The Doctor's Case Against the Pill (Seaman 1969) excoriated the FDA, the American College of Physicians and Surgeons, and the Planned Parenthood Federation for failing to adequately warn women of side effects and adverse events associated with the Pill. After the *Eisenstadt* ruling, as single women began to access the Pill, side-effects were often not clear to users, and risks of the Pill began to be communicated in the late 1970s. It would take until 1980 before the FDA required package inserts for users of the Pill (Planned Parenthood 2015).

Chara summarized where things stood in 1991, just a few years after *Dirty Dancing* was released (Chara 1991):

> Yet only a relatively short while ago in this country, contraceptives were considered obscene. They were also illegal, and there were several vigorous campaigns in the 1950s and 1960s to keep them that way. The battle in the United States was lost, however, when the Supreme Court ruled—first in *Griswold v. Connecticut* in 1965 and then in *Eisenstadt v. Baird* in 1972—that there is a constitutionally protected zone of privacy that extends to the purchase and use of contraceptives.

It would be a mistake, however, to view that battle as simply one centered on contraception or even on sexual morality. Rather, it was part of a larger debate about the power of women to control their reproductive capabilities and their lives. It began with nineteenth-century "voluntary motherhood" organizations fighting for contraception. This movement did not reject the idealization of motherhood; it fought merely to make the timing one of discretion rather than chance. The twentieth-century family planning movement went further, supporting a broader effort to ensure equal opportunity and independence for women. Personal control of reproduction was a crucial first step toward women's rights.

As abortion debates heated up throughout the 1990s, misinformation about the Pill abounded, leading to "conscientious objectors" (see further) who refused to either prescribe or fill prescriptions, even though the Pill does not have any abortifacient effect whatsoever, and prevents pregnancy by thickening cervical mucus and preventing ovulation. The claims by conscientious objectors that the Pill stops a fertilized egg from implantation is not accurate (Planned Parenthood 2015). By 1999, a new use of the Pill was made available through Emergency Contraception, also called Plan B, in which high doses of the Pill (high doses of a progestin) taken within 72 hours after unprotected sex could disrupt fertilization; it was also known as the "morning after pill". Plan B was made available as an option over-the-counter during the Obama administration, coinciding with TRAP laws that were restricting abortion access.

Healthcare Ethics Issues: Reproductive Justice

Dirty Dancing is a film about Reproductive Justice, which presumes that women have autonomy over their bodies, and have the right to self-determination and procreative liberty with respect to their reproductive lives. Reproductive Justice evolved through the "Women's Health Movement", which came of age through the book, Our Bodies, Ourselves (Boston Women's Health Collective 1969). The Women's Health Movement focused on creating access to reproductive care and services, strived to eliminate health disparities and immoral healthcare delivery frameworks based on gender, poverty and race. In the post-Roe period, Reproductive Justice issues also concern healthcare provider duties of reproductive care including adequate prenatal and postpartum care; ethical frameworks for "therapeutic abortion"; "fetal patienthood", and conscientious objection. Finally, Reproductive Justice frameworks necessarily involve limits to autonomy post-viability, which can create wrenching moral dilemmas balancing the rights of legal existing persons with the unborn, who have no common law legal status as potential persons.

The Women's Autonomy Movement: Second Wave Feminism

Reproductive Justice and the Women's Health Movement are subsets of the women's autonomy movement, which can be viewed in two historical timelines. "First wave feminism" tracks with the suffragette movement and the right to vote in the late nineteenth and earlier twentieth centuries. First wave feminism was about securing political equality and voice. This was the time frame in which Margaret Sanger began the fight for birth control (see earlier), and the oral contraceptive pill (the Pill), launching the second historical time frame: the "Second wave feminism" movement (1960–1981). By the 1980s, a "feminist backlash" period began in the Reagan era, which tracks with AIDS and the Pro-Life/Anti-Choice movements. Reproductive Justice was a second wave feminism fight, which led to *Roe v. Wade* (see earlier).

The Pill marks the first victory for the Reproductive Justice movement, but severe restrictions on its use continued until 1965: *Griswold v. Connecticut* granted married women access to the Pill in the United States without their spouse's permission. In 1972, *Eisenstadt v. Baird* granted unmarried women access to the Pill. *Eisenstadt* is really the decision that sparked the Sexual Revolution, which would not be in full swing until the 1970s, after legalized abortion was granted through *Roe v. Wade* (see earlier).

Social Role Reform: Moral Agency and Autonomous Decision-Making

Second wave feminism also involved a reordering of women's repressive social roles that had removed their decision-making authority as moral agents. For example, it was typical in patriarchal and paternalistic frameworks for male physicians to bypass women for consent and go to a male spouse or father. (Similarly, women could not get their own credit cards or bank accounts without male co-signers.) Reclaiming both healthcare and reproductive decisions was a major social role reform restoring procreative liberty—which included the right not to procreate. Two significant works that impacted women's reclaiming of their decision-making were The Feminine Mystique (Friedan 1963) and Women and Madness (Chesler 1972). Friedan identified "the problem that had no name," which was situational depression based on societal repression and oppression of women. Chesler identified the mass "incarceration" of women by the psychiatric community, and identified widespread labelling of "autonomy" and "feminism" as mental illness in women who did not want to be confined to restrictive social roles. But Chesler also clarified that restrictive social roles, poverty, and sexual abuse without reproductive rights, leads to the natural outcome of clinical depression.

In reclaiming control of their social roles and lives, many women began to question pronatalism, "compulsory motherhood", and some more radically suggested that science should be used for artificial wombs to remove women from the burden of childbearing and motherhood altogether (Giminez 1980). In other words, rigid social constructs and unfair social arrangements, which led to poor mental health outcomes and quality of lives, were no longer tolerated. Women questioned institutionalized sexism, and a number of famous adages emerged: "We are living the unlived lives of our mothers" (Gloria Steinem 1970s); "Biology is not destiny"; "The personal is political". Although second wave feminists raised the same questions as first wave feminists regarding "compulsory" or "voluntary" motherhood, women also confronted the underlying political goals of a pronatalist agenda in their newfound autonomy. Giminez (1980) noted: "The notion that all women should be and desire to be mothers has been used to keep women in a subordinate position...".

A third feature of second wave feminism was the women's sexual liberation movement, which was different than reproductive autonomy. Not only were women questioning reproductive choices, but they also wanted better control over their sex lives. Three critical works also defined the Sexual Revolution: Human Sexual Response (Masters and Johnson 1966); The Female Eunich (Greer 1970); and The Hite Report (Hite 1976). As women begin to question the quality of their sex lives, many women reclaim their lives as they leave bad relationships and bad marriages, leading to soaring divorce rates in the 1970s and 1980s, as part of the women's autonomy movement.

By reclaiming bodily autonomy and social moral agency, women's demand for reproductive justice included wider and greater access to contraception; abortion; and by the mid-1980s, reproductive technologies and fertility treatments.

Women's Healthcare Provider Frameworks

Women's healthcare issues comprise all healthcare services for women, and in the United States , routine discriminatory practices in the health insurance industry— known as "healthism" and "wealthcare"—frequently barred women from access to routine primary care (see Chap. 4), which included routine prenatal or postpartum care, or accessing pediatric care for their children. But sexism was frequently layered onto healthism in which simply being female could be a "pre-existing condition". Past pregnancy complications, or even yeast infections could count as pre-existing conditions that could deny coverage for any reproductive care.

Complicating matters is moral imperialism regarding the moral status of the fetus, which has led to state regulations surrounding access to contraception, and post-*Casey* restrictions on therapeutic abortion. Therapeutic abortion ranges from ending a pregnancy due to catastrophic fetal anomalies and a never-viable fetus in desired pregnancies, to termination of an unwanted pregnancy pre-viability. Post-viability, states have forced women to carry catastrophic and even never-viable fetuses to term.

Regardless of what the laws state, the bioethics principles governing reproductive justice do not change: Healthcare providers have an ethical and legal duty to provide reproductive care, and respect women's reproductive autonomy and decisions. As a result, the fetal patienthood framework emerged in the late twentieth century to guide reproductive care and prenatal care previability. Meanwhile, the principles of beneficence and non-maleficence guide decisions surrounding therapeutic abortions, as well as prenatal care post-viability.

Fetal Patienthood

This prenatal ethics framework was developed by McCullough and Chervenak (McCullough and Chervenak 2008; Chervenak and McCullough 1995, 2003; Chervenak et al. 2001) in which the mother's values and preferences establish whether the previable embryo or fetus is a "patient" in which dependent (rather than independent) moral status is granted. Within this ethics framework, when the mother presents the previable embryo or fetus for moral consideration as a "patient", it is a "patient" requiring a beneficent approach, where clinical harms are minimized, and clinical benefits are maximized.

The concept of embryo/fetal "patienthood" has been misread by critics for implying moral equivalency between the autonomous mother, and the unborn child (Strong 2008; Brown 2008; Lyerly et al. 2008). But McCullough and Chervenak correct their critics (McCullough and Chervenak 2008) and reassert that in any discussion of embryo/fetal patienthood, we are first dealing with "dependent" moral status granted by the mother. This not the same as independent moral status granted by society irrespective of the mother. In the "patienthood" framework, the mother decides whether the embryo/fetus is to be morally considered. "Patienthood" status is thus an extension of the mother's autonomous wishes, and merely clarifies in a perinatal setting whether the physician has an obligation of beneficence to the embryo(s)/fetus or not. When the fetus/embryo is a patient, this does not mean granting independent moral status or moral equivalence to the embyo/fetus, where the embryo/fetus's interests are equal to the mother's. Rather, patienthood status respects the mother's wish for the doctor to maximize benefits and minimize harms to her embryo/fetus, and not automatically assume the embryo/fetus is morally inconsequential. When the mother does not grant "patienthood" status, as in cases where she wishes to terminate pregnancy, the embryo/fetus is not morally considered until viability.

Withdrawal of Life Support on Pregnant Patients

Worse, incapacitated pregnant women are maintained as fetal containers against their documented wishes in advance directives, or against their surrogates' wishes, when in a persistent vegetative state, or even when pronounced dead by neurological criteria (aka brain dead), as in the Munoz case (Fernandez 2014), where a Texas woman who was dead by neurological criteria (see Chap. 7) was kept a fetal container against her family's wishes, misinterpreting the law. The majority of states have laws that mandate incapacitated women to carry to term (Doran 2019) even if it is not in their medical interests, and no family member wants the pregnancy to continue.

Half of the states in this country make it illegal to withdraw life support on a pregnant woman—even if there is an advance directive stating to do so. It is not uncommon for women who are severe intravenous drug users, for example, to go into cardiac arrest after an overdose, and go into a coma or persistent vegetative state after an anoxic brain injury; just when we are ready to withdraw life support, however, we discover the woman is 3–6 weeks pregnant. Many states force hospitals to keep these women as fetal containers, while the fetuses develop in their comatose mothers. Such babies are frequently born without parents, and become wards of the state.

Therapeutic Abortion

In the pre-Roe era, therapeutic abortion was always legal, and was left to the individual practitioner's medical judgement to define. Before there were clear bioethics principles to guide clinical ethics decisions, medical historians noted that both the social circumstances as well as the medical and psychological health of the patient defined whether an abortion was "therapeutic" (Reagan 1997). Poverty and over-crowded families were commonly considered to be reason enough for therapeutic abortion in which "lifeboat ethics" considerations, from a utilitarian justice framework, were part of the rationale. In situations of dire poverty, and an inability to care or feed a large family, abortion until the 1940s was considered to be a socio-economic necessity, and many physicians advocated to improve the social misery of patients they saw sentenced to compulsory childbirth strictly due to their second-class status as women (Reagan 1997). Additionally, the terrible stigma of unwed motherhood was also taken into consideration, while sometimes "eugenics" was provided as a rationale in cases where it was thought that the patient was "feeble minded" or from "undesirable" roots.

In other cases, pre-Cesarean section era physicians recognized the need for therapeutic abortion for women deemed at high risk of dying in childbirth because of poor health. Such cases truly met "beneficent" criteria for therapeutic abortion in the best interests of the patient, as the risks of continuing the pregnancy and delivering were greater. In the 1950s until the *Roe* decision, hospitals set up "therapeutic abortion committees" to determine criteria. In these cases, social implications of compulsory pregnancies and motherhood were re-interpreted to meet psychiatric criteria of "mental anguish" and other mental health sequela of compulsory motherhood and childbirth (Reagan 1997). Based on Penny's anguish, in *Dirty Dancing*, she probably would have met this criteria had she had access to a psychiatrist.

However, psychiatrists began to resent having to classify women wanting abortions as having mental health issues when they thought they were perfectly rational and capacitated, and many began to lobby for legalizing abortion, which they thought should be the patient's decision previability. Ultimately, there was no uniform criteria for "therapeutic abortion" but hospitals began to narrow the criteria as a defensive medicine practice to remove state scrutiny (Reagan 1997), even though all state laws permitted physicians to define "therapeutic abortion" even pre-Roe. As the criteria narrowed, and access started to shut down, caring for patients from unsafe or botched abortions became ethically imperative, and the numbers of women dying from unsafe abortions demanded different ethical frameworks. There was always a clear duty to treat abortion complications, even in situations where the practitioner was morally opposed, as in Dr. Houseman's situation. See Chap 8. for an expanded discussion of healthcare provider duties, professionalism and humanism.

Post-Roe, therapeutic abortions pre-viability are performed even in hospitals that are legally barred from terminating pregnancies due to certain federal restrictions, such as receiving federal funding. In such cases, therapeutic abortions can still be done to save a mother's life. Ad hoc therapeutic abortion committees typically review post-viability requests on a case by case basis. In such cases, premature delivery to save a mother's life, or compassionate delivery if the fetus is determined to be nonviable and incompatible with life, but many women who could not access abortion pre-viability are forced to go to term by law without compelling medical justification. Additionally, "healthism" and "wealthcare" (see Chap. 4) that deny prenatal care to women may lead to higher risk catastrophic pregnancy outcomes.

Pediatric Frameworks

Therapeutic abortion decisions also consider the best interests of the potential child, and the existing children, which, in the pre-Roe era, was frequently a "lifeboat ethics" decision for women who barely had the resources to care for the children they had, while childbirth could put their lives at risk, making their existing children orphans, with unintended consequences for society and the state. Post-Roe, prenatal testing can prevent the birth of a severely disabled child. The concept of preventing a "harmed life" and the notion that "sometimes disability is a disaster" (Purdy 1996) was also a factor. *Roe v. Wade* took this into account:

> The detriment that the State would impose upon the pregnant woman by denying this choice altogether is apparent. Specific and direct harm medically diagnosable even in early pregnancy may be involved. Maternity, or additional offspring, may force upon the woman a distressful life and future. Psychological harm may be imminent. Mental and physical health may be taxed by child care. There is also the distress, for all concerned, associated with the unwanted child, and there is the problem of bringing a child into a family already unable, psychologically and otherwise, to care for it. In other cases, as in this one, the additional difficulties and continuing stigma of unwed motherhood may be involved. All these are factors the woman and her responsible physician necessarily will consider in consultation. … On the basis of elements such as these, appellant and some amici argue that the woman's right is absolute and that she is entitled to terminate her pregnancy at whatever time, in whatever way, and for whatever reason she alone chooses…In addition, population growth, pollution, poverty, and racial overtones tend to complicate and not to simplify the problem."

Duty to Intervene

In *Dirty Dancing*, when his daughter wakes him up to come save a life, Dr. Jake Houseman doesn't hesitate. He treats Penny's sepsis with professionalism and empathy, even though it is a complication from an unsafe abortion he does not condone. Even in 1963 in the pre-Roe era, he had a clear duty to treat without prejudice, discussed more in Chap. 8. However, Dr. Houseman also unwittingly provided the funds for Penny's unsafe abortion, which is why his involvement could have had professional consequences. But at the same time, the doctor's actions are heroic because he does not abandon the patient; in fact, it seems he holds the male impregnator more responsible than Penny, making a false assumption that Johnny, and not Robbie is responsible, when he asks "who is responsible for this girl?" which can also mean who really "cares for this girl"? According to Bergstein (Jones 2013):

> One of the reasons Baby's father was a hero is because he could have lost his medical license [for being complicit in an illegal abortion]… I wanted to show what would have happened if we had illegal abortions…These battles go round and round. They're never won.

But the duty to intervene can also revolve around saving a fetus if the fetus is a patient (see earlier). The prenatal ethics literature clearly establishes that in circumstances where the mother is placing the viable fetus(es) in certain harm's way, the physician has a beneficence-based duty to intervene and protect the fetus(es) (Bessner 1994; Chervenak and McCullough 1991; Mahowald 2006; Strong 1997). Cases in which Cesarean section is refused, or substance abuse is putting a viable fetus in harm's way, may ethically permit intervention, too.

Conscientious Objection

Culturally diverse populations of patients and practitioners will generate countless situations in which providers may find themselves in morally compromising situations because their views may differ from patients/surrogates; or their views may differ from secular frameworks of healthcare. Different healthcare systems will produce different moral conflicts. Abortion does create valid moral distress for practitioners opposed to it, and as a result, practitioners can recuse themselves from participating. (Similarly, physician-assisted suicide, discussed in Chap. 3, may require recusal.) Women's reproductive healthcare has been the target of conscientious objectors with respect to both contraception and abortion. Moral diversity necessitates a broad understanding of what healthcare providers may find morally compromising, but there may be some positions that are incompatible with human rights, evidenced-based medicine and the standard of care, or secular frameworks. There are both legitimate, and illegitimate claims of conscientious objection (e.g. refusing to treat based on race or gender); illegitimate claims need no accommodation. Voicing a legitimate conscientious objection is known as moral courage, in which one takes a stand for one's ethical values (e.g. speaking up, or speaking out), even if there are negative consequences (Murray 2010). Most healthcare institutions and professional organizations recognize legitimate claims of conscientious objection, and generally have similar policies to accommodate such objections. Referral, as was done in the pre-Roe era (Reagan 1997), accommodates both patients and providers in recognition of a morally diverse population. Brock (2008), in his core work on conscientious objection, discusses referral as part of the "conventional compromise", in which he distinguishes between an individual healthcare provider's moral objections from the wider Medical Profession's obligation to the patient. Brock states (p. 194): "According to the conventional compromise, a [healthcare provider] who has a serious moral objection to providing a service/product to a patient is not required to do so only if the following three conditions are satisfied: (a) The [health-care provider] informs the patient/customer about the medical option/service if it is medically relevant and within the standard of care; (b) The [healthcare provider] refers the patient/customer to another professional willing and able to provide the medical option/service; (c) The referral does not impose an unreasonable burden on the patient.

Despite these conditions, it is accepted that there can be a "complicity objection to the conventional compromise". Brock concedes that these "three conditions will still leave [healthcare providers] unacceptably complicit in the practices to which they have serious moral objections" (Brock 2008), which may include treating complications of unsafe abortion as with Dr. Houseman in *Dirty Dancing*.

However, it is also widely accepted that moral diversity exists with respect to referral; some will find referral to be a morally neutral action that is a reasonable accommodation and accept the "conventional compromise", while others will find that referral is a morally significant action that does not accommodate their moral objections. Brock asserts: "Society can legitimately enforce requirements of social justice [also reproductive justice] that require persons to act in a way that may violate their moral integrity, and can use coercive measures to enforce its judgment…and…a particular profession can likewise legitimately decide that a specific social or professional role may require actions that might violate a particular individual's integrity, and so can in turn exclude individuals from that role who are unwilling to perform those actions…" (Brock 2008). Finally, societal debates over the moral status of the fetus are complicated by greater advances in neonatology in which peri-viable fetuses are now patients.

Conclusions

Dirty Dancing tells a common story of pre-*Roe* America, which is resonating much more in a post-*Casey* America. The future is pointing toward individual states regulating abortion with the likelihood of some states banning the practice completely, and some states offering complete and comprehensive access. Ultimately the umbilical cord tying individual states to *Roe v. Wade* may be cut, even if the law is not completely overturned. The original intention of *Roe v. Wade* was to grant a pregnant woman autonomous decision-making when it came to her own pregnancy, and the law placed few restrictions on the timing of the abortion. Post-*Casey*, restricted access, restricted education and counseling, restricted training, and anti-choice terrorism activities have made Penny's story from 1963 current news, as many women, once again, have no access to abortion. Regardless of whether abortion is legal or illegal, reproductive justice demands that women have self-determination over their bodies, and history teaches that they will ultimately take control of their pregnancies anyway (Reagan 1997). To paraphrase from Baby: There are a lot of things about women's bodies that aren't what we thought, but if healthcare providers are going to take care of them, they have to take care of the whole person—and that includes respecting a person's decisions about her own body.

Theatrical Poster

Dirty Dancing (1987)

Directed by: Emile Ardolino
Produced by: Linda Gottlieb
Written by: Eleanor Bergstein
Starring: Patrick Swayze, Jennifer Grey, Jerry Orbach, Cynthia Rhodes
Music by: John Morris, Erich Bulling, Jon Barns
Production: Great American Films Limited Partnership
Distributed by: Vestron Pictures
Release date: August 21, 1987.

References

Adams, P. (2017). Spreading Plan C to end pregnancy. *New York Times*, April 27, 2017. https://www.nytimes.com/2017/04/27/opinion/spreading-plan-c-to-end-pregnancy.html.
Alabama.com. (2019). *Alabama abortion pill passes*, May 15, 2019.   https://www.al.com/news/2019/05/alabama-abortion-ban-passes-read-the-bill.html.
Amiri, B. (2019). Kentucky just banned abortion. *ACLU*, March 14, 2019. https://www.aclu.org/blog/reproductive-freedom/abortion/kentucky-just-banned-abortion?fbclid=IwAR17Uypgea fdMW-_mYCgPxpt1RkalH1y-xGhJJkjgWQgAs7yu99VxnwtPGA.
Aroesty, S. (2017). A conversation with Eleanor Bergstein. *Tablet Magazine*, August 28, 2017. https://www.tabletmag.com/scroll/244045/jewcy-dirty-dancing.
Aronowitz, A. (2002). The strange case of Max Cantor. *The Blacklisted Journalist*, June 1, 2002. http://www.blacklistedjournalist.com/column72.html.
ASRM. *Statement on personhood measures*. https://www.asrm.org/about-us/media-and-public-aff airs/public-affairs/asrm-position-statement-on-personhood-measures/.
Atwood, M. (1986). *The Handmaid's Tale*. New York: Houghlin Miflin.
Baron, C., & Bernard, M. (2013). Dirty Dancing as Reagan-era cinema and "Reaganite Entertainment". In Y. Tzioumakis & S. Lincoln (Eds.), *The time of our lives: Dirty Dancing and popular culture*. Detroit: Wayne State University Press.
Bart, P. (1981). Seizing the means of reproduction. In H. Roberts (Ed.), *Women, health and reproduction*. London: RKP.
Bessner, R. (1994). State intervention in pregnancy. In G. E. M. Basen & A. Lippman (Eds.), *Misconceptions: The social construction of choice and the new reproductive technologies* (Vol. 2). Hull, Quebec: Sean Fordyce.
Brock, D. W. (2008). Conscientious refusal by physicians and pharmacists: Who is obligated to do what, and why? *Theoretical Medicine and Bioethics, 29*(3), 187–200.
Brown, S. D. (2008). The "Fetus as Patient": A critique. *The American Journal of Bioethics, 8*, 47–49.
Bryant, T. (2012). Still having the time of my life 25 years on: Dirty Dancing star Jennifer Grey on Patrick Swayze, dancing and her "nose job from hell". *Mirror UK*, August 23, 2012.
Butnick, S. (2011). Is Dirty Dancing the most Jewish film ever? *Tablet Magazine*, August 16, 2011. https://www.tabletmag.com/scroll/74789/is-dirty-dancing-the-most-jewish-film-ever.
Campbell, S. (2013). A short history of sonography. *Facts, Views and Visions in Obgyn, 5*(3), 213–229. https://www.ncbi.nlm.nih.gov/pmc/articles/PMC3987368/.

Carmon, I. (2010). Dirty Dancing is the greatest movie of all Time. *Jezebel*, April 29, 2010. https://jezebel.com/dirty-dancing-is-the-greatest-movie-of-all-time-5527079.

Center for Reproductive Rights. (2007). Roe v. Wade: Then and now. https://reproductiverights.org/document/roe-v-wade-then-and-now.

Chara, A. (1991). A political history of RU-486. In K. E. Hanna (Ed.), Institute of Medicine (US) Committee to study decision making. *Biomedical politics*. Washington (DC): National Academies Press (US). https://www.ncbi.nlm.nih.gov/books/NBK234199/.

Chervenak, F. A., & McCullough, L. B. (1991). Legal intervention during pregnancy. *JAMA, 265,* 1953.

Chervenak, F. A., & McCullough, L. B. (1995). Ethical implications for early pregnancy of the fetus as patient: A basic ethical concept in fetal medicine. *Early Pregnancy, 1,* 253–257.

Chervenak, F. A., & McCullough, L. B. (2003). The fetus as a patient: An essential concept for the ethics of perinatal medicine. *American Journal of Perinatology, 20,* 399–404.

Chervenak, F. A., McCullough, L. B., & Rosenwaks, Z. (2001). Ethical dimensions of the number of embryos to be transferred in in vitro fertilization. *Journal of Assisted Reproduction and Genetics, 18,* 583–587.

Chesler, P. (1972). *Women and madness*. New York: Avon Books.

Chicago Tribune. (2016). Indiana governor Oks fetal defects abortion ban. *Chicago Tribune*, March 24, 2016. http://www.chicagotribune.com/news/nationworld/midwest/ct-indiana-abortion-ban-genetic-abnormalities-20160324-story.html.

Dawn, R. (2017). Dirty Dancing turns 30. *Online*, January 25, 2017. https://www.today.com/series/family-entertainment/dirty-dancing-turns-30-here-are-6-things-know-about-t107375.

Denniston, L. (2016). Abortion rights re-emerge strongly. *Scotus blog*, June 27, 2016. http://www.scotusblog.com/2016/06/opinion-analysis-abortion-rights-reemerge-strongly/.

Domonoske, C. (2016). Supreme court strikes down abortion restrictions in Texas. *NPR*, June 27, 2016. https://www.npr.org/sections/thetwo-way/2016/06/27/483686616/supreme-court-strikes-down-abortion-restrictions-in-texas.

Doran, C. (2019). A pregnant pause. *Mayo Clinic Blog*, May 21, 2019. https://advancingthescience.mayo.edu/2019/05/21/a-pregnant-pause/.

Eig, J. (2014). *The birth of the pill*. New York: W.W. Norton and Company.

Eisenstadt v. Baird. (1972). 405 U.S. 438.

Fernandez, M. (2014). Texas woman is taken off life support. *New York Times*, January 26, 2014. https://www.nytimes.com/2014/01/27/us/texas-hospital-to-end-life-support-for-pregnant-brain-dead-woman.html.

Friedan, B. (1963). *The Feminine Mystique*. New York: Dell Publishing.

Giminez, M. E. (1980). Feminism, pronatalism and motherhood. *International Journal of Women's Studies, 3*(3), 215–240.

Griswold v. Connecticut. (1965). 381 U.S. 479.

Gonzales v. Carhart. (2007). 550 U.S. 124.

Gruner, O. (2013). "There are a lot of things…The Politics of Dirty Dancing. In Y. Tzioumakis & S. Lincoln (Eds.), *The time of our lives: Dirty Dancing and popular culture*. Detroit: Wayne State University Press.

Hamilton, K. (1999). "It's Like, Uh … Jennifer Grey". *Newsweek*, March 22, 1999.

Higgins, T. (2019). Supreme court allows Indiana abortion law. *CNBC*, May 28, 2019. https://www.cnbc.com/2019/05/28/supreme-court-allows-indiana-abortion-law-governing-disposal-of-fetal-remains-but-wont-revive-discriminatory-abortion-bansupreme-court-allows-indiana-abortion-law-governing-disposal-of-fetal-remains-b.html.

Huberfeld, N. (2016). *Pregnancy and the law*. Presentation, University of Kentucky Program for Bioethics Grand Rounds.

IMDB. (2019). *Max Cantor biography*. https://www.imdb.com/name/nm0134674/bio?ref_=nm_ov_bio_sm.

Jones, A. (2013). The not-so-dirty Dirty Dancing story. *The Forward*, September 22, 2013. https://forward.com/culture/184141/the-not-so-dirty-dirty-dancing-story/.

Kennedy, J. F. (1962). Houston, Texas: Speech at Rice University Stadium, September 12, 1962.

Kolata, G. (2000). U.S. Approves abortion pill. *New York Times*, September 29, 2000. https://www.nytimes.com/2000/09/29/us/us-approves-abortion-pill-drug-offers-more-privacy-and-could-reshape-debate.html.

Kolson, A. (1997). August 17, 1997. https://www.nytimes.com/1997/08/17/movies/fairy-tale-without-an-ending.html.

Laurie, A. (2010). Jennifer Grey never recovered from Matthew Broderick car crash. *San Francisco Chronicle*, September 22, 2010.

Leaming, B. (2014). *Jacqueine Bouvier Kennedy Onassis: The untold story*. New York: St. Martin's Press.

Leo, J. (1984). The revolution is over. *Time, 123*(15), April 9, 1984.

Lewin, T. (1996). FDA approval sought for French abortion pill. *New York Times*, April 1, 1996. https://www.nytimes.com/1996/04/01/us/fda-approval-sought-for-french-abortion-pill.html.

Liptak, A. (2016). Supreme court strikes down Texas abortion restrictions. *New York Times*, June 27, 2016. https://www.nytimes.com/2016/06/28/us/supreme-court-texas-abortion.html.

Lowry, M. (2015). Real life baby having time of her life. The *Star Telegram*, July 2, 2015. https://www.star-telegram.com/entertainment/arts-culture/article26065504.html#storylink=cpy.

Lyerly, A. D., Little, M. O., & Faden, R. R. (2008). A critique of the 'Fetus as Patient'. *The American Journal of Bioethics, 8,* 42–44.

Mahowald, M. B. (2006). Noncompliance during pregnancy. In M. B. Mahowald (Ed.), *Bioethics and women across the lifespan*. New York: Oxford University Press.

Matthews, J. (1987). September 9, 1987. https://www.latimes.com/archives/la-xpm-1987-09-09-ca-4383-story.html.

McCullough, L. B., & Chervenak, F. A. (2008). Response to commentaries on "a critical analysis of the concept and discourse of 'unborn child'". *The American Journal of Bioethics, 8,* W4–W6.

McLaughlin, K. (2013). Life is still a Cabaret. *CNN*, February 13, 2013. https://www.cnn.com/2013/02/13/showbiz/cabaret-40th-anniversary/index.html.

Murray, J. S. (2010). Moral courage in healthcare: Acting ethically even in the presence of risk. *OJIN: The Online Journal of Issues in Nursing, 15*(3). https://doi.org/10.3912/ojin.vol15no03man02.

NASA. (2018). January 23, 2018. https://www.nasa.gov/feature/nasa-challenger-center-collaborate-to-perform-christa-mcauliffe-s-legacy-experiments.

NBC. (1984). Documentary. Second thoughts on being single. Airdate. Reported in *Washington Post*, April 25, 1984. https://www.washingtonpost.com/archive/lifestyle/1984/04/25/the-loneliest-number/1cda73d3-001f-431e-ae1b-84214722253f/?utm_term=.cb72ecab3873.

Nikkhah, R. (2009). Dirty Dancing: The classic story. *The Telegraph*, February 14, 2009. https://www.telegraph.co.uk/culture/4621585/Dirty-Dancing-The-classic-story.html.

Pincus, W., & B. Woodward. (1987). Iran Contra report. *Washington Post*, November 5, 1987. https://www.washingtonpost.com/archive/politics/1987/11/05/iran-contra-report-wont-tell-whole-story/a40c20f8-c291-4b56-a127-bf79764fff83/?utm_term=.7fef0d5f6f21.

Planned, P. (2015). *History of the birth control pill*. https://www.plannedparenthood.org/files/1514/3518/7100/Pill_History_FactSheet.pdf.

Purdy, L. M. (1996). Genetics and reproductive risk: Can having children be immoral? In L. M. Purdy (Ed.), *Reproducing persons: Issues in feminist bioethics*. Ithaca: Cornell University Press.

Reagan, L. J. (1997). *When abortion was a crime: Women, medicine, and law in the United States, 1867–1973*. Berkeley: University of California Press.

Reston, J. (1987). Kennedy and Bork. *New York Times*, July 5, 1987. https://www.nytimes.com/1987/07/05/opinion/washington-kennedy-and-bork.html.

Roe v. Wade. (1973). 410 U.S. 113.

Sabato, L. (2013). John F. Kennedy's final days. *Forbes*, October 16, 2013. https://www.forbes.com/sites/realspin/2013/10/16/john-f-kennedys-final-days-reveal-a-man-who-craved-excitement/#15a3c4a71a9c.

Sacchetti, M., & Somashekhar, S. (2017). An undocumented teen is pregnant. Can the U.S. stop her from getting an abortion? *Washington Post*, October 17, 2017. https://www.washingtonpost.com/local/immigration/an-undocumented-teen-is-pregnant-and-in-custody-can-the-us-stop-her-from-getting-an-abortion/2017/10/17/6b548cda-b34b-11e7-9e58-e6288544af98_story.html?utm_term=.359b6d914e96.

Seaman, B. (1969). *The doctor's case against the pill*. New York: Peter H. Wyden Inc.

Sheehy, G. (1995). *New passages: Mapping your life across time*. New York: Ballantine Books.

Shilts, R. (1987). *And the band played on*. New York: St. Martin's Press.

Smith, M. (2019). Missouri governor signs bill outlawing abortion after 8 weeks. *New York Times*, May 24, 2019. https://www.nytimes.com/2019/05/24/us/missouri-abortion-law.html.

Stephens-Davidowitz, S. (2016). The Return of the DIY abortion. *New York Times*, March 6, 2016. http://www.nytimes.com/2016/03/06/opinion/sunday/the-return-of-the-diy-abortion.html?_r=0.

Strong, C. (1997). *Ethics in reproductive and perinatal medicine: A new framework*. New Haven: Yale University Press.

Strong, C. (2008). Do Embryonic "patients" have moral interests? *The American Journal of Bioethics, 8*, 40–52.

Tabashnik, S. (2011). Like Dirty Dancing. *Blogspot*, October 16, 2011. http://likedirtydancing.blogspot.com/2011/10/dirty-dancing-character-penny-was-based.html.

Tauber, M. (2008). Fighting together. *People*, March 24, 2008. https://people.com/archive/cover-story-fighting-together-vol-69-no-11/.

Tavernise, S. (2019). States are rushing to restrict abortion, or protect it. *New York Times*, May 15, 2019. https://www.nytimes.com/2019/05/15/us/abortion-laws-2019.html?module=inline.

Tzioumakis, Y., & Lincoln, S. (Eds.). (2013). *The time of our lives: Dirty Dancing and popular culture*. Detroit: Wayne State University Press.

WebMD. (2000). *Brief history of the abortion pill in the U.S*, September 28, 2000. https://www.webmd.com/women/news/20000928/brief-history-of-abortion-pill-in-us#1.

Wenn, A. (2006). Dirty Dancing—Dirty Dancer Grey's Nightmare Nose job. *Contactmusic*, October 29, 2006.

Whole Woman's Health v. Hellerstedt. (2016). 136 S. Ct. 2292.

Zerwick, P. (2016). The rise of the DIY abortion. *Glamour*, May 31, 2016. https://www.glamour.com/story/the-rise-of-the-diy-abortion.

Chapter 6
"The Wrong Bat Met the Wrong Pig": Pandemic Ethics in *Contagion* (2011)

It's hard to find great films about pandemics that are not "zombie" films, such as *28 Days* (2000), its sequel, *28 Days Later* (2002); or *World War Z* (2013). *Contagion* is a fast-paced thriller with an ensemble cast, brilliantly directed by Steven Soderbergh. It opened in theaters on September 9, 2011—two days before the tenth anniversary of 9/11. *Contagion* is an original screenplay for which its source material was science and medical news about recent pandemic threats, such as Severe Acute Respiratory Syndrome (SARS) in 2003 and the 2009 H1N1 flu. It has fresh relevance in light of the COVID-19 coronovirus pandemic of 2020. The novel flu virus in the film is called MEV-1—"a deadly, highly communicable strain that strikes the brain and lung, triggering dramatic convulsions for the camera" (Hall 2011). MEV-1 devastates the body within days of infection, transmitting like the flu—droplets and fomites. Throughout the film, no one ever never really knows how it all starts except that "the wrong bat met the wrong pig". But in the final two brilliant minutes of the film, it all plays out for the viewer. Deforestation forces a fruit bat to leave a treetop; it flies over a nearby pig pen, dropping its feces onto the ground that one of the pigs ingests. That same pig is next sent to slaughter, shipped to a fine restaurant in Hong Kong, in which the chef is slicing it open raw to begin a fine pork dish; he is then interrupted mid-raw slice to be introduced to an American restaurant guest: the lovely Beth Emhoff (played by Gwyneth Paltrow). The chef wipes his hands on his apron without washing them and then goes into the dining room to shake hands with Beth. He goes back to prepare his pig. Beth then becomes the human superspreader—infecting several people in Hong Kong before she gets on a plane and brings it to the United States.

Experts from the World Health Organization (WHO) and professors of infectious disease were consultants on the script. *Contagion* is not just a great thriller, but praised by scientists for its accuracy. Three years after its release, *Contagion* resonated in light of the Ebola pandemic of 2014, and was absolutely prescient during the COVID-19 pandemic of 2020 (see Afterword). When you're looking at the myriad of healthcare ethics issues that arise in a pandemic, this is the film to view. This chapter discusses

© Springer Nature Switzerland AG 2020
M. S. Rosenthal, *Healthcare Ethics on Film*,
https://doi.org/10.1007/978-3-030-48818-5_6

the production aspects of the film, its social location, the history of medicine context and the justice and pandemic ethics issues raised in the film.

Making a Film Go Viral: Production of *Contagion*

Contagion is an original screenplay by Scott Burns (see further) and Director Steven Soderbergh (see further), who also was the cinematographer under the pseudonym Peter Andrews (Douglas 2011). The germ of the idea arose when the two collaborated on the 2009 film *The Informant*, starring Matt Damon, who was also in *Contagion*. *The Informant* was about an agribusiness whistleblower (based on the real story of Mark Whitacre), featuring a scene in which Whitacre is germaphobic (Douglas 2011). Soderbergh and Burns were planning to collaborate on a different film when they decided to rethink the project and do something else. Burns recalled in an interview (Douglas 2011) that when Soderbergh asked him:

> Do you have any other ideas?" I said, "Well, I always thought it would be cool to do a pandemic movie but a pandemic movie that was more rooted in reality." I was certainly aware there were other pandemic movies, but I wanted to do one that really felt like what could happen. There's a scene in "The Informant!" where Matt (Damon) is watching Scott Bakula's character talk on the phone and Scott coughs on the phone, and there's this whole ramp that Matt goes off on of "Oh, great, now what happens? He gets sick and then I'm going to get it, my kids are going to get it." I've always been fascinated by transmissibility, so I said to Steven, "I want to do an interesting thriller version of a pandemic movie" and he said, "Great! Let's do that instead."

Burns began the project about six months prior to the 2009 H1N1 virus news, and began thinking about ramifications of social distancing and social media. He noted (Douglas 2011):

> "Well, do you close the schools and if you close the schools, then who stays home with the kids? And will everyone keep their kids at home?" Things happening online, which is where the Jude Law character came from, that there's going to be information that comes out online where people want to be ahead of the curve, so some people will write things about anti-virals or different treatment protocols, and so there's always going to be information …[with] sort of a viral pulse. So it's not just the disease that you have to track, it's how the disease is interpreted by the population. It was great, because while I was writing this, it was playing out in real time.

Burns' script was informed by experts at the Centers for Disease Control (CDC), and the WHO; several scenes were shot at the CDC in Atlanta, with one scene shot at the WHO headquarters in Geneva. Experts included Lawrence Brilliant, a virologist and pandemic expert, W. Ian Lipkin, an expert in diagnostics, microbial discovery and outbreak response, as well as science journalist, Laurie Garrett, author of The Coming Plague (1995). The experts on the film praised the film for its realism. Says Burns (Douglas 2011):

> …Dr. Lipkin and Laurie Garrett were in Hong Kong for the SARS outbreak and they've seen the movie and they were like, "Yeah, this is what it felt like." You walk around and you

are paranoid and you look at social gatherings and you look at the world and you think as animals we are messy and we do eat from the same bowl and touch the same knobs and we hug each other and touch each other and that's part of the great thing about being a human being. It's also dangerous.

Burns' screenplays included several partnerships with Soderbergh. Prior to *Contagion*, he wrote *The Bourne Ultimatum* (2007), and also produced Al Gore's *An Inconvenient Truth* (2006), which won an Academy Award (see Chap. 4).

Each storyline in *Contagion* was shot in different locations, comprising the storyline of the index patient's saga; the WHO epidemiologist, played by Marion Cotillard; the CDC experts played by Laurence Fishburne (see Chap. 1) and Kate Winslet. There is also the storyline surrounding the virologists, played by Jennifer Ehle and Elliot Gould; and finally, the social media and the spreading of misinformation online by a menacing blogger played by Jude Law (a scenario far less severe in 2009 than it was to become by 2016 and beyond). The script was prescient surrounding the worst virus of all: the spread of misinformation in a pandemic event. Burns noted (Douglas 2011):

> Steven and I have talked about this for a long time and my view is… Look, that line that I wrote about blogging as being graffiti with punctuation, I feel like there's a lot of unfiltered content in the world now. It is both a great freedom and a huge danger. I don't think we spend enough time talking about that… But we're living in the Wild West on the internet…and again, like a virus, it's one of these things where it can get outta control really, really quickly. I think Steven and I share a concern. On the one hand, I think we applaud the openness of media and the exchange of ideas; on the other hand, if you're going to live in a world where everyone has a microphone and that big a loud speaker, it's a scary world. It can get pretty noisy.

With Burns script, Soderbergh was motivated to make a film about the response to a global pandemic informed by HIV/AIDS in the 1980s (see Chap. 2), as well as more recent events, such as 9/11, Hurricane Katrina, and what countries dealing with SARS experienced.

Film critic, David Denby in the *New Yorker* noted this (Denby 2011):

> Contagion confronts reality head on; it's a brief against magical thinking. Soderbergh and his screenwriter, Scott Z. Burns, may not have intended it, but their movie could become an event in an ongoing political debate over the nature of American life. In "Contagion," paranoia reaches its logical end point: the everyday streams of connection—the personal and professional meetings that make economic and social life and pleasure possible—become the vehicle of our destruction…. The speed and the range of the movie are analogous to the way the disease spreads. The Internet, another system of universal connection, becomes a kind of plague, too. [Jude Law's blogger character] Krumwiede's aggressions echo those of 9/11 conspiracy theorists and, perhaps, of Julian Assange at his most self-righteous. As the filmmakers tell it, Krumwiede, spreading distrust, is part of the disease…

The film's release in 2011 coincided with the tenth anniversary of 9/11, which was not lost on its audience; Denby also notes (Denby 2011):

> *Contagion* is, of course, a 9/11-anniversary movie, though probably not one that the public was expecting. Soderbergh appears to be saying, "I'll show you something far worse than a terrorist attack, and no fundamentalist fanatic planned it." The film suggests that, at any moment, our advanced civilization could be close to a breakdown exacerbated by precisely what is most advanced in it.

Contagion now stands the test of time with fresh relevance during the coronavirus COVID-19 pandemic (Castrodale 2020; Yang 2020). Ironically, the film did particularly well in its release weekend in Italy in 2011, which became the first European hot zone for the spread of Coronavirus and COVID-19 and went on lock-down in March 2020 (Mounk 2020; Newsweek 2020; Lintern 2020). Praised by scientists for its accuracy, *Contagion* also received accolades from Paul Offit, an expert in vaccines (Offit 2011). for a good explanation of the Basic Reproduction Number or R nought (R0)—the anticipated number of cases directly generated by one case, and fomites.

Steven Soderbergh

A tail-end Baby Boomer born in 1963, Soderbergh's debut film when he was 26 was the fascinating and hailed *Sex, Lies and Videotape* (1989), which explored the burgeoning changes technology interjected into sexual relationships, as many adults in the wake of HIV/AIDS looked to technology as an alternative to intimate relationships. *Sex, Lies and Videotape* also tapped into pre-social media behaviors decades before it was invented. The film was considered to have helped launch the Independent film industry in the 1990s. Known for his avant garde, French new wave and experimental style, reminiscent of a European filmmaker, Soderbergh is a master of the Independent film. His repertoire includes *Traffic* (2000), a tour de force about the Mexican-American drug wars; *Erin Brokovich* (2000), a film about the environmental tort law crusader; and the *Ocean's Eleven* through *Ocean's Thirteen* films (2001 through 2018). He also does his own cinematography under the pseudonym, Peter Andrews.

Ensemble Cast

Contagion boasts an impressive ensemble cast, not unlike *And the Band Played On* (see Chap. 2). Gwyneth Paltrow, as the index patient, has a more minor role and dies early in the film. Paltrow is also known as the CEO of Goop, a multi-million dollar company she started in her kitchen, which promotes healthy living and has a number of unique and highly priced products. At the same time, Paltrow has become a lightning rod for several in the medical and bioethics community for promoting pseudoscience through Goop's products and messages (Pruden 2018).

The cast also boasts Matt Damon, Jennifer Ehle, who was also known in the role of Elizabeth Bennett in the particularly compelling 1995 version of *Pride and Prejudice* by *Masterpiece Theater* on PBS, opposite Colin Firth. Laurence Fishburne (see Chap. 1), Jude Law, Kate Winslet, and Marion Cottilard fill out the cast.

Synopsis

Contagion's fast-paced plot focuses on multiple simultaneous narratives, but each actor's storyline was shot in one separate sequence, which included a large ensemble cast of characters. The first gripping scene begins with a black screen and then a cough. And then we meet coughing Beth Emhoff (played by Gwyneth Paltrow) at the airport. She's on the phone with the man she just made love to during her layover in Chicago, letting him know she's at the gate, waiting for her connecting flight home to Minneapolis, where she returns to her family—husband, Mitch Emhoff, played by Matt Damon, her young son from a previous marriage, and Mitch's teenage daughter from a previous marriage. Emhoff is the index case who becomes the super-spreader while she's in Hong Kong on business and then brings the novel virus into the United States while traveling home from Hong Kong. On her layover in Chicago, she meets up with an old flame and infects him before catching her connecting flight from Chicago to Minneapolis, next infecting her son in Minneapolis, and, of course, anyone else along the way she's been in contact with at airports, restaurants, and so forth. Beth Emhoff has flu symptoms, and then suddenly has a seizure and dies, leaving her widowed second husband and teenaged daughter behind, both of whom seem to be immune. As more cases begin to be seen in Hong Kong, and then around the world, Beth Emhoff's death sets off a detective story where infectious disease experts at the CDC and the WHO try to isolate the virus; epidemiologists at the CDC and WHO try to contact-trace and figure out etiology, echoing the various ground zero clinical, research and social theaters also seen in *And the Band Played On* (see Chap. 2). As social distancing and mandatory quarantine becomes the only tool of containment while a vaccine is being sped through development and trials, the United States becomes mired in shortages of supplies and healthcare workers, panic buying, looting, stopping of basic services such as garbage collection, with martial law declared in various states. Finally, as the vaccine becomes slowly available, ethically sound rationing frameworks are used to dispense the vaccine via "birthday lottery" to global populations, as life starts to return to more normalcy.

One reviewer summarizes it this way (Moviesonline 2011):

> Intelligently written, superbly directed and impeccably acted, Steven Soderbergh's new bio-threat thriller, "Contagion," examines what happens when our worst fears about a global pandemic become reality. An international traveler reaches into the snack bowl at an airport bar before passing her credit card to a waiter. A business meeting begins with a round of handshakes. A man coughs on a crowded bus. One contact. One instant. And a lethal virus is transmitted.

The Pro-science Obama Era: The Social Location of *Contagion*

When *Contagion* was being written and produced during the first term of the Obama Administration, scientists around the world were breathing a sigh of relief as science and technology appeared to be back on the agenda. In December 2008, before his inauguration (Tollefson 2012):

> [Obama] announced other members of his future staff, who would make up a star-studded science team: marine ecologist Jane Lubchenco would head the National Oceanographic and Atmospheric Administration in Washington DC and physicist John Holdren would be Obama's science adviser and head the Office of Science and Technology Policy, also in

Washington DC. They joined Lisa Jackson, a respected chemical engineer with political experience, who had been named to run the US Environmental Protection Agency (EPA) in Washington DC. After taking office, the president completed the team by appointing geneticist Francis Collins at the National Institutes of Health (NIH) in Bethesda, Maryland, and geophysicist Marcia McNutt at the US Geological Survey in Reston, Virginia. Never before had a president assembled such a strong crop of researchers to lead his science agencies.

"The truth is that promoting science isn't just about providing resources—it's about protecting free and open inquiry," Obama proclaimed as he made the initial appointments. "It's about listening to what our scientists have to say, even when it's inconvenient—especially when it's inconvenient."

Scientists and environmentalists swooned; they had spent 8 years complaining that the administration of President George W. Bush had overly politicized science. Climate researchers in government had charged that they were being muzzled and that their data were being manipulated. Pollution regulations were blocked or watered down. With Obama's election, scientists would finally have a president who not only said the right things but actually appointed the right people. Even journalists drooled. "Science Born Again in the White House, and Not a Moment Too Soon," read a headline in Wired magazine, endorsing Obama's appointments with a swipe at Bush's reputation as a born-again Christian.

Nature later recalled in 2016 (Monastersky 2016):

Many researchers who watched Barack Obama's inauguration in 2009 were thrilled by his pledge to "restore science to its rightful place".… In general, government researchers have enjoyed more freedom—and endured less political meddling—than they did under the previous president, George W. Bush. Bush's administration was accused of muzzling or ignoring scientists on subjects ranging from stem cells to climate change.

Under Obama, science was back on-track, climate change was now framed as a national security issue; while environmental protections and other policies were being informed by real scientists. There was even a White House Science Fair being introduced (Whitehouse.gov 2015). In medical news, the biggest Obama era change was healthcare reform with the *Affordable Care Act* (see Chap. 4), while a decade after 9/11, healthcare for those exposed to World Trade Center dust (see Chap. 4) would finally have coverage.

The timing for a scientifically accurate film about the next pandemic was ripe, especially since the next flu pandemic appeared to be pending with concerns over the H1N1 virus (see under History of Medicine).

But in 2009, the Obama Administration had other problems. Two post-9/11 wars were raging in Iraq and Afghanistan and the financial crisis meltdown, freshly inherited from the Bush Administration, was in full bloom. Obama's *American Recovery and Reinvestment Act* (ARRA) of 2009 was a huge fiscal stimulus bill that addressed the financial crisis (Amadeo 2019). But here is a picture of what the Obama Administration was dealing with (Amadeo 2019):

The unemployment rate rose to 10% in October 2009, the worst since the 1982 recession. Almost 6 million jobs were lost in the 12 months prior to that. Employers added temporary workers, too cautious about the economy to add full-time employees…Meanwhile, a

Federal Reserve report showed that lending was down 15% from the nation's four biggest banks…[which] cut their commercial and industrial lending by $100 billion, according to the Treasury Department data…Lending from all banks surveyed showed the number of loans made fell 9% between October 2008 and October 2009…Bank of America pledged to President Obama it would increase lending to small and medium-sized businesses by $5 billion in 2010. But that's after they slashed lending by 21% or by $58 billion in 2009.

Obama's ARRA did not leave science in the dust. It also stimulated science by providing $10 billion to modernize science facilities and fund research jobs that investigated disease cures; $4 billion to increase broadband internet access to rural and inner-city areas, as well as $4 billion for physics and science research (Amadeo 2020). Neil Lane, the former director of the National Science Foundation (NSF) was interviewed by *Chemistry World* in 2017 (Trager 2017):

[Lane said] Obama's science and technology record is "quite remarkable given the political circumstances in which he served"…Obama was responsible for this 'unprecedented increase' in funding for research through the one-time stimulus…Obama's economic recovery package gave a 'huge bump' of almost 50% to the NSF's budget….

Giving research a boost was a challenge given that Obama arrived as the financial crisis was in full swing, and roadblocks laid by his Republican opponents on Capitol Hill. 'It was made very clear by the Republicans in Congress that they were not going to work with President Obama, and they were going to do everything they could to ensure that he was not successful,' Lane recalls.

In 2009, the NIH made a public statement over the science funding (NIH Record 2009):

NIH is extremely grateful to President Obama and the Congress for recognizing both the economic and health impacts of biomedical and behavioral research…The science funded by this bill will stimulate the national economy and have an impact for many years to come.

In *Contagion*, there is a lot of emphasis regarding the science infrastructure needed in combatting a pandemic—containment facilities and equipment; under Bush, no such funding was available. Out of the $10.4 billion allocated by ARRA, $300 million was designated for "shared instrumentation and other capital equipment" and $500 million was for "NIH buildings and facilities" for "high-priority repair, construction and improvement projects" (NIH 2009). When teaching this film, it's important to emphasize that while science doesn't happen in a vacuum, infectious diseases can't be studied in under-resourced, unsafe facilities, either, a condition that would reappear in the Trump era, which included an under-resourced response to the COVID-19 pandemic (see Afterword).

In the wake of the financial crisis and ARRA, it was also clear that social fabric was delicate, and a pandemic health crisis layered onto a society in which few had any real social safety nets, would be an existential threat. Meanwhile, a war-weary United States, in which its all-volunteer military was stretched to the breaking point, required Obama to rethink military strategic planning, given that both wars were going badly and had led to chaos (Woodward 2010). By December 2009, Obama ordered a troop surge into Afghanistan, and began to plan for exiting Iraq. Two days after *Contagion* opened, so did the first 9/11 Memorial on the event's tenth anniversary while the first 9/11 museum was still under construction until 2014. By December 2011, three months after *Contagion* opened, American troops left Iraq.

Postmortem on Bush Era Policies and the "War on Terror"

When the Obama era began, the postmortem on the Bush Administration had begun, too, as Americans felt freer to exercise their civil rights and criticize Bush era policies that had to do with decreasing individual freedoms and liberties in the name of the "War on Terror". If Americans were willing to give up their rights in the wake of terrorism, what would they do in the midst of a pandemic?

Contagion is a post-SARS film in which many of the social conditions of a pandemic, as well as the balancing of individual liberties with pandemic conditions (see further) are depicted; however, to an Obama era audience, the slow erosion of their personal liberties was all-too familiar. The SARS pandemic is discussed under the History of Medicine section below, but the social location of SARS was in February 2003, just at the time that the Bush Administration was making an argument for the invasion of Iraq, based on the false premise that Iraq had weapons of mass destruction. *Contagion* also comments on the extent to which a public health or public safety threat—such as the post-9/11 environment—inspires Americans to give up their civil rights. When SARS hit, the Bush Administration had already expanded its war powers, and passed significant legislation that had encroached upon individual liberties, although the public's acceptance of this was perhaps more significant, as Bush enjoyed an approval rating of 90% after 9/11, and above 60% until January 2004 (Gallup 2018). *The Patriot Act* (2001) sailed through Congress, while the "War on Terror" began to hold individuals suspected of ties to Al Qaeda or terrorist cells without legal due process, in violation of their civil liberties. Meanwhile, American citizens were being held without due process, as extraordinary rendition, in which those suspected of terrorist activities were detained and sent to another country for interrogation, became much too common in the Bush era. According to the ACLU in 2007 (ACLU.org 2020):

> In the name of national security, the U.S. government is sponsoring torture programs that transcend the bounds of law and threaten our most treasured values. The ACLU has brought two lawsuits against the U.S. government and the airline that facilitates flight planning for C.I.A. renditions. The cases are El Masri v. Tenet, a lawsuit against former CIA director George Tenet on behalf of German citizen Khaled El-Masri, and *Mohamed v. Jeppesen*, a lawsuit against Boeing subsidiary Jeppesen Dataplan, Inc.

The Patriot Act gave away a number of hard-won civil liberties (CRF 2019)

> Some of the most controversial parts of the Patriot Act surround issues of privacy and government surveillance. The Fourth Amendment to the U.S. Constitution protects the "right of the people to be secure in their persons, houses, papers, and effects, against unreasonable searches and seizures…." It requires law-enforcement officers to obtain warrants before making most

searches…. The judge may only issue a search warrant if officers show "probable cause" that the person is engaged in criminal activity….

The Patriot Act now authorizes the court to issue search orders directed at any U.S. citizen who the FBI believes may be involved in terrorist activities. Such activities may, in part, even involve First Amendment protected acts such as participating in non-violent public protests.

The Patriot Act also expanded the property that could be searched (CRF 2019):

In Sect. 215, "any tangible things" may include almost any kind of property–such as books, documents, and computers. The FBI may also monitor or seize personal records held by public libraries, bookstores, medical offices, Internet providers, churches, political groups, universities, and other businesses and institutions.

But what enabled *The Patriot Act* most of all, was that public opinion had dramatically shifted. By 2002, 47% of Americans felt that civil liberties should be sacrificed to prevent terrorism; 60% were in favor of Bush policies in restricting civil liberties, while 25% thought it hadn't restricted them enough (Gallup 2018; CRF 2019).

By 2004, when the Abu Ghraib story had been outed, in which American soldiers were found to have tortured Iraqi citizens held in the Abu Ghraib prison (CBS 2004), accidentally replicating the Stanford Prison Study experiment (Haney et al. 1973), the country began to become disillusioned with the "War on Terror" and expanded practices of torture in the Bush Era (U.S. Senate 2014).

Hurricane Katrina

Also discussed in Chap. 4, Hurricane Katrina was considered one of the most disastrous events during the Bush era, not because of a natural disaster but because of the manmade disaster that followed. On August 31, 2005, Hurricane Katrina, a category 5 hurricane, made a bullseye for New Orleans and stranded its vulnerable population: citizens who were sick, elderly, poor, and largely African American. Levees breached as thousands were flooded out of their homes. The four-year old Department of Homeland Security (established in 2001), now responsible for the Federal Emergency Management Agency (FEMA) failed to respond in a timely and appropriate manner. For days, stranded victims of the hurricane went without proper resources, as Americans watched in horror how paralyzed the federal government had become in coordinating the allocation of abundant, not even scarce, resources to the victims. In fact, one of the federal employees responsible for the failed Katrina response was Kirstjen Nielsen, who would go on to even more notorious acts while becoming Director of Homeland Security in the Trump Administration. Aside from abandoning American citizens in the wake of disaster, what Steven Soderbergh noted (Hoffman 2011) was the complete breakdown in social order as citizens were left to their own devices to obtain resources to survive. Mass looting and citizens self-arming for defense became the norm very quickly—an environment that begins to take shape in *Contagion*. Ultimately, Hurricane Katrina would not have become as memorable were it not for the weeks-long social order breakdown that was completely preventable. Social order was only restored when the U.S. military went into the region and established martial law. It took years for New Orleans to bounce back, but the United States depicted in *Contagion* was not a stretch. In fact, in a particularly disturbing scene in which Matt Damon is trying to escape Minnesota's anarchy by trying to cross into Wisconsin, he is stopped as state borders are enforced. Similar local borders were enforced in the wake of Katrina, barring citizens from leaving New Orleans into other jurisdictions (Burnett 2005). However, Katrina also led to specific disaster ethics and pandemic ethics protocols that had contingency plans for vulnerable populations (see under Healthcare Ethics Issues).

Viral Information: Social Media

Contagion has a powerful subplot involving the role of social media in disrupting containment and social order, breeding disinformation, conspiracy theories, and misinformation by greedy and immoral actors looking to profit from the destabilized environment and panic. But Soderbergh was not depicting the social media of 2009–2010 when *Contagion* is in production. He is inadvertently depicting the social media of the future—which ultimately becomes its own social virus no one can contain, and which is only more dangerous now in the wake of foreign interference and message hijacking. In 2009, *Mashable* stated this (Parr 2009):

> To most observers, 2009 marked the year Twitter conquered the world. Yet it wasn't the only social media company that grew like wildfire. There's another that grew even more rapidly, adding over 200 million new users and raising $200 milllion dollars—double that of Twitter's most recent round.

> 2009 was a breakout year for Facebook, even if some of its successes were overshadowed by its emerging rival. In fact, the two have been locked in a new battle for the soul of the web, and the right to be the platform where the world converses. To that end, Facebook's 2009 has partly been about fighting back through the opening up of its data and profiles, a process we sometimes "Twitterfercation"

Indeed by 2016, just five years after *Contagion* is released, social media becomes the Frankenstein monster no one can control anymore—not even its creators (PBS 2018). But most of all, the "tool that Liberals built" (PBS 2018) became weaponized within the United States by Americans themselves serving anti-science groups, right wing conspirators and a variety of actors with the goal of spreading disinformation. *Contagion* does not tell us whether the U.S. government is functioning under a Democratic, Republican or Independent party, but instead, points to the global science communities as taking charge, while martial law appears to become the norm under a faceless and nameless administration. In a real pandemic, social media will likely become the critical messaging tool of local health authorities, as it has now displaced all other communication tools. By 2020 the COVID-19 virus was even called a "hoax" by the President of the United States (NBC News 2020).

The Zombie Context

When *Contagion* was in development, the "Zombie Apocalypse" genre for infectious diseases was well known. In fact, on May 18, 2011, the CDC had even published an article called "Preparedness 101: Zombie Apocalypse" providing tips on preparing to survive a zombie invasion. The CDC recognized that the zombie was a good allegory for preparedness messaging. The article by Khan (2011) starts:

> There are all kinds of emergencies out there that we can prepare for. Take a zombie apocalypse for example. That's right, I said z-o-m-b-i-e a-p-o-c-a-l-y-p-s-e. You may laugh now, but when it happens you'll be happy you read this, and hey, maybe you'll even learn a thing or two about how to prepare for a real emergency.

The explosion of the cultural allegory in media led the CDC to make use of it as an educational opportunity (CDC 2018). Scholars have noted that the zombie is merely a personification of how infectious disease spreads (Verran and Reyes 2018):

> In the zombie, internal damage to the host becomes externalized, and contagion patterns among populations are demonstrated as the zombie hordes rampage. With no subclinical manifestation, the zombie makes the apocalypse visible, enabling us to physically map the spread of infection. In other words, the zombie becomes an "allegory of infectious disease" and a "metaphor of ubiquitous contagion…In their hordelike structure, zombies also operate metonymically, standing in for large swaths of the population (the infected), or viruses (the infection). The mathematics of zombie outbreaks has therefore also been explored as an education tool to represent contagion patterns and containment strategies…. Our innate knowledge of real disease epidemiology is thus illustrated in much zombie literature by the behavior of the humans who are under threat. In the absence of any treatment strategy, options are restricted to quarantine; immunization strategies (protection of the uninfected and control. As zombies become the manifestation of virulent infection, they do not just address our fear of pandemic disease and apocalypse; they also allow us to explore coping strategies.

If *Contagion* is being shown in a more specific course about infectious disease, pairing it with *28 Days Later* (2002) and *World War Z* (2013) is recommended.

History of Medicine Context

It's important to note that in the infectious disease genre of science fiction and films, some are based on the Plague (called Yersinia pestis) as in Stephen King's The Stand (1978), some are based on AIDS (see Chap. 2), while the "rage virus" in *28 Days* is based on Ebola. From a history of medicine context, *Contagion*'s story is based on early twenty-first century global pandemics: Severe Acute Respiratory Syndrome (SARS), which emerged in 2003, and concerns over a novel flu virus, H1N1 in 2009. However, the virus' etiology is based on the Nipah virus, which was identified as a virus in a type of fruit bat in Malaysia jumping to a pig, and occurred because of an overlap between fruit bat habitats and piggeries in Malaysia. At a farm where the "wrong bat met the wrong pig", fruit orchards were too close in proximity to a piggery, "allowing the spillage of urine, feces and partially eaten fruit onto the pigs" (Chua et al. 2002).

The film was released just three years before the 2014 Ebola pandemic, which indeed travelled to the United States and could have threatened large population centers if it were not for aggressive contact tracing and quarantine, which helped to contain it (see under History of Medicine). This section briefly reviews the 1918 flu pandemic, but focuses on SARS, H1N1 and the 2014 Ebola virus. As *Contagion* also deals with vaccine misinformation, this section also deals with the Andrew Wakefield incident, which influenced the script.

The 1918 Flu Pandemic

Contagion references the 1918 Great Influenza pandemic (aka Spanish Flu), which was a global pandemic that probably started in a small Kansas town in the United States in the early part of 1918. According to clinicians at the time (Barry 2005):

> The flu 'was ushered in by two groups of symptoms: in the first place the constitutional reactions of an acute febrile disease—headache, general aching, chills, fever, malaise, prostration, anorexia, nausea or vomiting; and in the second place, symptoms referable to an intense congestion of the mucous membranes of the nose, pharynx, larynx, trachea and upper respiratory tract in general and of the conjunctivae…[causing] absolute exhaustion and chill, fever, extreme pain in back and limbs…cough was often constant. Upper air passages were clogged.' (Pg. 232)

There were also Ebola-like symptoms: 15% suffered from bleeding from the nose (epistaxis), and women bled from their vaginas due to uterine mucosa, not menstruation. "Blood poured from noses, ears, eye sockets" and when they died, "bloody liquid was seeping from the nostrils or mouths" (Barry 2005). For some flu victims, symptoms were very sudden and violent: sudden, extreme joint pain, chills and a high fever—almost an instantaneous reaction with no gradual buildup. In many flu victims, they could recall the exact moment they became sick, and many died within hours. According to Barry (2005): "JAMA reported 'One robust person showed the first symptoms at 4 pm and died by 10 am'."

The first victims of the flu were soldiers, who spread it through crowded conditions of various military bases and hospitals during World War I. For example, on October 4, 1918, over 100 men at Camp Grant died in a single day; in less than a month, 2800 troops reported ill in a single day (Barry 2005, p. 216):.

> The hospital staff could not keep pace. Endless rows of men coughing, lying in bloodstained linen, surrounded by flies—orders were issued that 'formalin should be added to each sputum cup to keep the flies away'—the grotesque smells of vomit, urine and feces made the relatives in some ways more desperate then the patients.

Anyone with a living memory of 1918 in the United States—particularly if they lived in an urban center such as New York or Philadelphia, experienced conditions similar to the plagues of the middle ages. Overcrowding and slums helped to spread the flu in cities, which were the same social conditions in Europe. As Barry (2005) notes, although the plague in the middle ages killed a much larger *proportion* of the population, in sheer body count, the 1918 pandemic killed more people in a year than the Black Death killed in a century. As another parallel, the 1918 flu killed more people in 24 weeks than AIDS killed in 24 years (see Chap. 2). In Louisville, Kentucky, 40% of those who died were aged 20–24. The Great Influenza pandemic of 1918 was different than other flu epidemics because it seemed to target adults in the prime of life, rather than the very old or young, which was more typical. Half the victims were men and women in their 20s and 30s, or roughly 10% of all young adults alive in that time frame. The death toll so overwhelmed cities, there were coffin shortages; piles of dead bodies; shortages of doctors and nurses. (In Philadelphia, when the flu exploded in the population, all 31 of its hospitals were filled with flu patients.) Relatives were either burying their own dead or wrapping them in cloth until "body wagons" (in trucks or by horse-drawn carriage) could pick them up. The deaths increased in 10 days from 2 deaths per day to hundreds of deaths per day. During this time, as in *Contagion*, people isolated themselves, and children would see one neighbor after the other die and wonder who would die next. Everyone was wearing masks. Everyone had someone in the family who was dying or sick, and many households had several sick and dying. Common public health signs from 1918–20 would read "Spit spreads death" while ordinances were passed that made not wearing a mask a felony. Public events were cancelled, and city streets became empty and "ghost town" like. The pandemic didn't fade away until about 1920, but its impact lingered into the next century (Barry 2005).

Severe Acute Respiratory Syndrome (SARS)

SARS was a novel coronavirus that was first identified in China's Guangdong province and quickly spread to Hong Kong, which borders Guangdong province. It began to spread throughout Asia from there, and then travelled by plane to Europe and Canada, brought by passengers infected with SARS, just like Beth Emhoff, played by Gwyneth Paltrow, who flies from Hong Kong to Chicago and then Minneapolis.

SARS spread like the flu—contact with another person, droplets which can spread through the air, which makes air travel a particularly potent carrier. Crowded conditions, in fact, are known as "super spreader events". A Toronto resident who had been to Hong Kong in Spring of 2003 brought SARS to Toronto, Ontario Canada—a city of roughly 6 million people, creating a preview of coming attractions for other large global cities. The Toronto Index case was described by Low (2004):

The index case and her husband had vacationed in Hong Kong and had stayed at a hotel in Kowloon from February 18 to 21, 2003. The index case began to experience symptoms after her return on February 23 and died at home on March 5. During her illness, family members, including her son (case A), provided care at home. Case A became ill on February 27 and presented to the index hospital on March 7 [2003].... Toronto's experience with SARS illustrated how quickly the disease can spread in hospitals and highlighted the dangerous phenomenon of SARS superspreaders. The absence of rapid tests to distinguish this new disease from pneumonia, influenza, or other common diseases bodes ill for future outbreaks.

In Toronto, which is heralded for having one of the best healthcare systems in the world, the SARS epidemic spread through the hospital environment. According to the *Canadian Medical Association Journal* (Borgundvaag et al. 2004):

> In Toronto, there were several "super-spreading" events, instances when a few individuals were responsible for infecting a large number of others. At least 1 of these events occurred in an emergency department, 6 where overcrowding, open observation "wards" for patients with respiratory complaints, aerosol treatments, poor compliance with hand-washing procedures among health care workers and largely unrestricted access by visitors may have contributed to disease transmission.

SARS hit several cities in Asia, South America, Europe and Canada, and could have been the ultimate nightmare scenario had it hit New York City or another major center in the United States aggravated by its unjust healthcare system, which surely would have denied care to patients without insurance (see Chap. 4).

Contagion should be taught as a post-SARS film as it depicts scenarios and decision-making based on that particular outbreak. Canadian social science scholars noted (Ries 2004).

> The outbreak took 44 lives in our country, threatened many others and created numerous challenges for public health officials and the acute health care system. In particular, SARS highlighted serious deficiencies in public health infrastructure and preparedness. As in other countries, officials in Canada were required to weigh the legalities and ethics of various interventions to control the spread of the disease, including quarantine.

H1N1: 2009

In 2009, when *Contagion* is in development, the United States was bracing for the H1N1 influenza (flu) pandemic. It was not clear at the time how serious it was, but it hit the United States in April 2009, infecting a 10-year-old boy (CDC 2010). The concern was that the virus "was a unique combination of influenza virus genes never previously identified in either animals or people. The virus genes were a combination of genes most closely related to North American swine-lineage H1N1 and Eurasian lineage swine-origin H1N1 influenza viruses" (CDC 2010). The virus originated from pigs but was spreading among humans only. Two days after the first child was infected, and 8-year-old was infected but there was no contact between the two children, suggesting it was likely in multiple places. H1N1 was a novel flu virus, which had all the ingredients of another 1918 situation—even an ongoing war, which in 1918 was World War I. On April 18, 2009, the United States reported the 2009 H1N1 influenza cases to the World Health Organization (WHO 2010). By April 25, 2009, after more cases were identified in other states, the WHO declared the outbreak a Public Health Emergency and by April 26, 2009, 25% of the U.S. government's stockpiled supplies were released, which "included 11 million regimens of antiviral drugs, and personal protective equipment including over 39 million respiratory protection devices (masks and respirators), gowns, gloves and face shields, to states (allocations were based on each state's population)." After several reports of severe symptoms and deaths in Mexico and other countries, the threat level went to a phase 5:

> As the outbreak spread, CDC began receiving reports of school closures and implementation of community-level social distancing measures meant to slow the spread of disease. School administrators and public health officials were following their pandemic plans and doing everything they could to slow the spread of illness. (Social distancing measures are meant to increase distance between people. Measures include staying home when ill unless to seek medical care, avoiding large gatherings, telecommuting, and implementing school closures).

By June 11, 2009, a worldwide H1N1 pandemic had been declared, and the threat level was raised to phase 6. However, most of the H1N1 cases were not fatal. By August 8, 2009, after 1 million infections, 477 Americans died from the 2009 H1N1 flu, including 36 who were under 18. Sixty-seven percent of the children who died had at least one high-risk medical condition. By September, H1NI vaccines were developed, and by October, they were available in very limited supplies. By December 2009, major campaigns were initiated to encourage vaccination, however the anti-vax movement was beginning. The WHO described the 2009 H1N1pandemic this way (WHO 2010):

> After early outbreaks in North America in April 2009 the new influenza virus spread rapidly around the world. By the time WHO declared a pandemic in June 2009, a total of 74 countries and territories had reported laboratory confirmed infections. To date, most countries in the world have confirmed infections from the new virus…. The new virus has also led to patterns of death and illness not normally seen in influenza infections. Most of the deaths caused by the pandemic influenza have occurred among younger people, including those who were otherwise healthy. Pregnant women, younger children and people of any age with certain chronic lung or other medical conditions appear to be at higher risk of more complicated or severe illness. Many of the severe cases have been due to viral pneumonia, which is harder to treat than bacterial pneumonias usually associated with seasonal influenza. Many of these patients have required intensive care.

In 2009, in the setting of the financial crisis and economic collapse, the H1N1 pandemic could have been much more disastrous given that it was transmitted through "respiratory droplets…can survive for 6 hours on a dry surface…[and] when touched by hands it can be retransmitted. Transmissibility of the H1N1 2009 strain is higher than that of seasonal influenza strains" (Al Muharrmi 2010).

The Wakefield Case and Anti-vaccination

Contagion was released two years after a major scientific misconduct case surfaced, which surrounded a falsely reported connection between autism and vaccinations. In 1998, researcher Andrew Wakefield published a major paper in *The Lancet*, connecting autism spectrum disorder to the MMR vaccine (Wakefield 1998). Wakefield claimed that the families of eight out of 12 children attending a routine clinic at the hospital had blamed the MMR vaccine for their child's autism within days of receiving the vaccine. Wakefield also claimed that a new inflammatory bowel disease was also discovered in these children. This led to widespread mistrust of vaccination, and rates of vaccination dramatically dropped (DeStefano and Chen 1999). Several epidemiological studies began to refute Wakefield's claims (Taylor et al. 1999).

According to (Walton 2009):

> Researchers owe it to the public to do their best to ensure that their findings and interpretations of their data are presented accurately and simply through the mass media. Despite the fact that the original article had a sample size of only 12 children (and Wakefield's claims are modest at best about the correlation between autism and immunization), once the media got hold of the story, many viewed the small study as a landmark article. According to the media, rates of immunization dropped from ~92 to 80% in the UK and many journalists claim that this is a direct result of the Wakefield article. First of all, that is a difficult claim to make. Second of all, it is quite likely due to the way that journalists actually interpreted and reported the Wakefield data rather than the data as presented in the Lancet article.

Wakefield et al. retracted their paper in a stunning admission that they committed fraud. Rao and Andrade (2011) note this:

> According to the retraction, "no causal link was established between MMR vaccine and autism as the data were insufficient". This was accompanied by an admission by the *Lancet* that Wakefield et al. had failed to disclose financial interests (e.g. Wakefield had been funded by lawyers who had been engaged by parents in lawsuits against vaccine-producing companies). However, the *Lancet* exonerated Wakefield and his colleagues from charges of ethical violations and scientific misconduct.

Ultimately, *The Lancet* completely retracted the paper by February 2010, while the studies on the children did not follow proper ethical guidelines; the *British Medical Journal* revealed that Wakefield et al. had actually committed fraud (Rao and Andrade 2011). The long-term consequences of the Wakefield incident led to unending distrust in vaccines by parents. It was the catalyst to the entire anti-vaccine movement that persists.

The 2014 Ebola Pandemic

Although *Contagion* was released in 2011, the film was prescient when the Ebola outbreak in 2014 threatened to become as out of control in the United States as it had become in parts of Africa, when an index patient presented at a Texas hospital and was sent home to expose potentially hundreds of unsuspecting people. Global health ethics issues also interfered with priority-setting (see further).

Ebola is a blood borne virus that spreads easily through contact with bodily fluids once symptoms present, and it is usually lethal. In March 2014, there were 49 cases of Ebola in Guinea, and 29 deaths. Doctors Without Borders had set up treatment centers. Two cases were identified in Liberia in March, but concern

mounted in Guinea, as Doctors Without Borders stated it was facing an "unprecedented epidemic". Kent Brantly was a 33-year-old healthcare provider working in
Liberia at a missionary hospital with Samaritan's Purse and had downloaded a 1998
guide on "controlling Ebola". His hospital opened an Ebola ward.

Between April-June, the hospital couldn't find any staff to work in the Ebola ward,
and scrambled to find people. In a news conference in Geneva, the WHO stated the
outbreak was "relatively small". By late May, two cases and two deaths were in
Sierra Leone, and seven cases were now in Monrovia. There was also uncertainty
between whether the CDC or the WHO had jurisdiction over trying to contain and
control the outbreak. By June 20, 2014, Doctors Without Borders warned that the
virus is "totally out of control"; WHO ramped up its response. At this point, the
Samaritan Purse Hospital in Liberia had become overwhelmed, and Melvin Korlkor,
a healthcare provider who contracted Ebola along with 5 nurses and 4 others, was the
only one to survive; the other nine died. On July 20, Liberian doctor Patrick Sawyer
collapsed in Nigeria and died of Ebola, and by July 23, Brantly became infected
and was flown to the United States and was the first to receive the experimental
drug, ZMAPP. Brantly had begun to vomit blood and had diarrhea when he got to
the United States. Another American, Nancy Writebol, also working at Samaritan
Purse, contracted Ebola and was also flown to the United States. By July 24, 2014,
the WHO upgraded the crisis to a 3. On July 29th, the lead virologist in Sierra Leone
died from Ebola. At this point, Dr. Joanne Liu, president of Doctors Without Borders,
requests a meeting with the WHO director in Geneva, imploring her to declare an
emergency, which finally occured on August 8, 2014. On August 31, Tom Frieden,
then Director of the CDC stated that the outbreak was: "a plague on a medieval
scale…a scene out of Dante."

By September 2014, among the major global health organizations and governments, no one can exactly determine how to respond or who's in charge. Doctors
Without Borders implored countries to utilize their military personnel to help with
control; by this time, Liberia had thousands of cases. Frieden noted: "We and the
world failed [the Ebola] test"; Lui noted: "We're losing the battle to contain it." The
U.S. government warned that there would be about 1.4 million infected by January
2015 without a robust response (Sun et al. 2014).

By September 30, 2014, Ebola arrived in the United States when Thomas Duncan,
traveling from Liberia to Dallas, is visiting family, and staying in a home with two
children. He began to display symptoms of fever and vomiting, and went to a Dallas
hospital, which sent him home, as the practitioners there did not suspect Ebola, even
though he did state he was from an Ebola-infected region. Duncan left the hospital
and vomited on a public sidewalk; he potentially exposed two children as well,
and the children then went to school, potentially exposing many more. Ultimately,
Duncan returned two days later to the hospital by ambulance, where an uprepared staff
admitted him, determined it was Ebola, and without proper precautions, instructed
two nurses to take care of him; both of the nurses then got infected with Ebola, too.
Duncan died within a few days. Once the U.S. index case was revealed, a race to do
contact tracing, isolation and quarantine ensued, and it was contained. However, it
was a close call, and could have had disastrous consequences (Rosenthal 2014). It

revealed, however, that the index case hospital, Texas Health Presbyterian in Dallas, was grossly unprepared (Spigner 2014; Berman and Brown 2014).

In a few other cases of Americans who had been potentially exposed to Ebola in their travels or charity medical work—including one physician who began to show symptoms while traveling on a New York City subway—various quarantine measures were implemented while public health, politicians and infectious disease experts argued over quarantine criteria. Fortunately, the Ebola epidemic was contained, but no one could predict that by 2018, the robust White House pandemic office would be disbanded in the Trump era, leaving the country completely unprepared for the 2020 COVID-19 pandemic that had brought the United States to its knees by the spring of 2020 (Cameron 2020).

The Coronovirus COVID-19 Pandemic

Ultimately, the nightmare scenario came true when the COVID-19 pandemic upended life across the globe, caused by a novel coronavirus that was highly infectious, with estimates of roughly 2.2 million deaths in the United States alone without extreme mitigation measures of physical and social distancing, shelter in place, and shutting down all non-essential business. In the absence of sufficient testing, shortages in hospital supplies, and no means to quarantine the infected from the non-infected, a stunning pandemic plot that is eerily close to *Contagion*, began to play out in China in 2019, and the rest of the world in 2020. Unpreparedness, delays and missteps in the United States led to a mass casualty death toll of over 100,000 by Memorial Day of that year (see Afterword).

I'm Testing My Vaccine

Contagion makes reference to scientists using themselves as research subjects, and discusses the ulcer-causing H. pylori discovery in 1985 (Ahmed 2005). When the character played by Jennifer Ehle tests her vaccine on herself, she makes reference to the 1985 research team that discovered that the bacteria H. pylori was the principal cause of ulcers. To prove that ulcers were not caused by stress, researcher Barry Marshall presented data suggesting that H. pylori caused between 80 and 90% of ulcers. Marshall ultimately used himself as a research subject and underwent a gastric biopsy as a baseline, and then deliberately infected himself with H. pylori to see if he developed an ulcer, which he did, and was able to prove was caused by H. pylori. He published his findings in the *Medical Journal of Australia* (Marshall 1985). Noted Ahmed (2005): "This extraordinary act of Marshall demonstrated extreme dedication and commitment to his research that generated one of the most radical and important impacts on the last 50 year's perception of gastroduodenal pathology. Their research made H. pylori infection one of the best-studied paradigms of pathogen biology, paving way for intense and hectic basic and clinical research activity leading to about 25,000 scientific publications till date" (Ahmed 2005).

In any *Contagion* scenario, including COVID-19, fast-tracking a vaccine by self-testing a "rough draft" may indeed become an ethical standard.

Healthcare Ethics Issues: Pandemic Ethics

Pandemic ethics frameworks require implementation of utilitarian justice principles in which a consequentialist framework is used, whereby the greatest good for the greatest number of people is the goal. This is an autonomy-limiting framework in which priority setting may restrict individual liberties, privacy and confidentiality, and ignore patient autonomy. This section covers the duties of healthcare providers in a pandemic; ethical frameworks for allocation of resources and priority setting; ethical issues in triage; ethical considerations regarding vaccination refusal and mandated vaccination; and individual liberty restrictions. It is this specific framework that *Contagion* presents accurately.

The field of Pandemic Ethics emerged as a legitimate subspecialty of bioethics in the wake of SARS in 2003 (Singer et al. 2003), and amidst anticipation of a bird flu pandemic (H1N1), which the SARS episode exposed as a potentially more problematic pandemic without more focused planning (see above). In 2005, a novel Canadian Report emerged from the SARS experience titled Stand On Guard For Thee: Ethical Considerations In Preparedness Planning For Pandemic Influenza (University of Toronto 2005).

This report outlined an ethical framework to guide planning efforts in the health sector, and led to a major research effort at the University of Toronto to establish an internationally focused Program of Research on Ethics in a Pandemic. This program published a White Paper Series in 2009, titled Ethics and Pandemic Influenza. In 2008, the WHO published its own White Paper series titled Addressing Ethical Issues In Pandemic Influenza Planning. These documents informed the substantive literature worldwide with respect to pandemic ethics. The core pandemic ethical issues include: duty to care of health professionals; priority setting (a.k.a. resource allocation) of limited health resources; restrictive measures (e.g. quarantine); and global governance. The need for an ethical framework to guide local pandemic planning has been reinforced in multiple disciplines engaged in pandemic planning. Ultimately, the public health goals identified by Pandemic Ethics researchers comprised: (a) building and maintaining public trust; (b) protection of vulnerable populations; (c) establishing the obligations of health care workers in a pandemic; (d) establishing the reciprocal obligations of the health care system to health care workers; (e) establishing a framework to allocate strained resources such as ventilators, antiviral medication, or community health services and; (f) establishing a framework for communicating information to the public. These resources were intended for policy makers to deal with local pandemic planning based on a "reverse engineering" of what went wrong, and what went right, with SARS.

Ethical Duties of Healthcare Providers in a Pandemic

Health care providers—both clinical and nonclinical—face disproportionate health risks in a pandemic situation. As shown in *Contagion*, they may face competing personal and professional obligations to their patients, colleagues, employers, family members, and to their own health. Research indicates that 25–85% of healthcare providers report being unwilling to show up for work in a pandemic. Pandemic Ethics researchers have raised the following questions (CanPrep 2009): Do healthcare providers have an obligation to treat patients despite risk of infection? What limits, if any, are there to health care workers' duty to care? What institutional supports are owed to health care providers in a pandemic? Health care providers' ethical duty to care (distinct from the legal duty to treat) is both a professional duty and societal duty, but professional codes are typically insufficient in addressing duty to care in a pandemic. Communicating duty to care to healthcare providers is best done within the context of societal obligations (a social contract framework) rather than professional obligations, but that reciprocity should be a consideration in priority setting. For example, in *Contagion*, healthcare providers were first to receive limited supplies of vaccine. Research indicates that the public's perception of healthcare providers is that they have special obligations to care because of the profession they entered, but that their institutions or government must ensure they have reciprocity, meaning they are safe, and fairly compensated for their risk, and given priority for resources. Guidelines were cited by Toronto experts as follows (CanPrep 2009):

1. Pandemic planners should ensure the right of healthcare workers (HCWs) to safe working conditions is maximized to ensure the discharge of duties and that HCWs receive sufficient support throughout a period of extraordinary demands, which will include training on hygienic measures to reduce infection risk.
2. Consideration should be given to needs of health care providers to ensure care to their families.
3. Professional associations should provide, by way of their codes of ethics, clear guidance tomembers in advance of an influenza pandemic. This may include information regarding existing mechanisms to inform members as to expectations and obligations regarding the duty to provide care during a communicable disease outbreak.
4. Pandemic planners should ensure that processes be in place to accommodate legitimate exceptions to the provision of clinical care (e.g. pregnancy, immunodeficiency).
5. Pandemic planners should assess local circumstances and ensure the participation of the community sector in planning of formal and informal care networks and engage clinical and non-clinical, professional and non-professional HCWs.

Ethical Frameworks for Allocation of Resources and Priority Setting

"Priority Setting" is the dominant term used by pandemic ethics researchers in the discussion of resource allocation in a pandemic setting, in which ordinary healthcare resources and services are expected to exceed demand.

Access to ventilators, vaccines, antivirals, and other necessary resources in hospitals and in the community will need to be prioritized, and typical clinical criteria is insufficient in priority setting. Value-based decisions in a pandemic setting will need to be made, but how? Should we give priority to the sickest or should those most likely to survive be the benchmark? The following questions have been raised by pandemic ethics researchers (CanPrep 2009): Should resources be allocated to save the most lives or to give everyone a fair chance at survival? Should special consideration be given to vulnerable populations in determining access to resources? Who should make these allocation decisions?

The ethical goals of resource allocation or priority setting are legitimacy, fairness, and equity. Research indicates the following parameters are acceptable to the public in resource allocation decisions: need, survivability, and social value. Need takes into consideration not just the sickest person; persons who are responsible for caring for others may take priority. Social utility of individuals (healthcare workers, critical infrastructure workers, etc.) who are sick is a key concept in prioritizing. Establishing transparent priority setting criteria in advance of a crisis is another key concept, to enforce fairness and public trust in priority setting. There is public consensus that priority should be given to healthcare workers, whose social utility value is high; and whose risk assumption is high. Research indicates there is public consensus that children should be given second priority after healthcare workers.

The WHO (2008) emphasized that priority setting is typically based on the principle of efficiency (saving most lives), which prioritizes protecting individuals responsible for caring for the sick, and is not necessarily based on prioritizing resources for the "sickest". The principle of equity is typically a failed principle in priority setting because equitable distribution of resources may not achieve the goals of public safety in pandemic situations. The WHO White Paper on priority setting provides a detailed and thorough discussion of the strengths and weaknesses of various moral frameworks for establishing priority setting guidelines, however the 2009 University of Toronto report distills much of this information into practical guidelines (CanPrep 2009):

1. Governments and health sector officials should engage the public actively in transparent, inclusive, and accountable deliberations about priority-setting issues related to the use of limited resources for treatment and prevention.
2. Governments and health care sector officials should engage stakeholders (including health care workers and administrators, and the public) in determining what criteria should be used to make resource allocation decisions (e.g. access to ventilators, vaccines, antivirals).

3. Governments and health care sector officials should provide an explicit rationale for resource allocation decisions, including priority groups for access to limited health care resources and services. The rationale should be publicly accessible, justified in relation to the defined criteria, and include a reasonable explanation for any deviation from the pre-determined criteria.
4. Governments and health care sector officials should ensure that there are formal mechanisms in place for stakeholders to bring forward new information, to appeal or raise concerns about particular allocation decisions, and to resolve disputes.

Ethical Issues in Triage

The WHO (2008) emphasized the following with respect to triage:

> Similar to judgments about medical futility, triage decisions should be based upon professional standards that are publicly justifiable. In this way, controversial and deeply troubling decisions are not left to the discretion or subjective assessment of individual caregivers. Priorities should be based upon general triage criteria that are reasonably acceptable to everyone. On the one hand this involves appeal to the basic normative principles discussed previously; maximization of health benefits (notably saving lives) and equity. On the other hand, criteria should be defined and specified on the basis of medical evidence about health needs and factors that determine the chance of recovery.

In critical care, the primary focus is on saving lives by responding to acute health crises. Triage decisions aimed at saving the most lives with limited resources will give less priority to patients who are expected to recover less easily. Although the implications of such decisions will be harsh and controversial, the basic principle to save the greatest number of lives possible can be reasonably justified to anyone.

Ethical Considerations Regarding Vaccination Refusal and Mandated Vaccination

The main issue identified with a vaccine for a novel virus is time; it is typically not expected that an appropriate vaccine will be available to the public at least for the first six months after the start of any pandemic, and for large numbers of people this will be far too late. Yet even when a vaccine has finally been developed and approved, deployment will be incremental and there will be insufficient production capacity to accommodate the enormous demand worldwide (see further). Priority Setting guidelines can help to get vaccine to the critical populations. However, there is a considerable ethical issue on the rise regarding vaccine refusal. Vaccination refusal (see earlier) is linked to two issues in public health: (1) a flawed vaccine that was distributed in 1976, in anticipation of a flu pandemic, which produced a number of side effects; (2) parental distrust of vaccines, and their unproven association with autism due to the scientific misconduct of the Andrew Wakefield research (see under

History of Medicine). There is an increasing distrust by the public regarding the safety of vaccines. Refusal of vaccines is also now seen among healthcare providers themselves. Vaccination refusal has been dealt with by Diekema (2005) in the context of the Harm Principle, originally outlined by J. S. Mill in his On Liberty treatise (1859). The Harm Principle states:

> That the only purpose for which power can be rightfully exercised over any member of a civilised community, against his will, is to prevent harm to others… The only part of the conduct of any one, for which [an individual] is amenable to society, is that which concerns others. In the part which merely concerns himself, his independence is, of right, absolute. Over himself, over his own body and mind, the individual is sovereign.

This principle makes it clear that when parents exercise their right to refuse to vaccinate a child, that right infringes on another's right, and may harm another child. Vaccines in a pandemic situation are not only for the benefit of the individual receiving the vaccine but also for the benefit of the public. The process of creating "herd immunity" allows for individuals who do not get vaccinated, cannot get vaccinated, or do not develop sufficient immunity from vaccination to derive some measure of protection from others in the population being successfully immunized. Determining the purpose of the vaccination program is a key concept: is it to protect the public or individual? If it's to protect the individual, then the individual's autonomy to refuse vaccination should be honored. However, if the program is designed to protect the public, then the principle of "solidarity" and protecting the public from harm justifies coercive policies in mandating vaccination, and infringing upon individual liberty. Pandemic ethics researchers assert that in order for public health officials to justify the use of more coercive measures, they need to have scientific evidence that supports the population health benefits of the vaccination program. Coercive policies can include consequences for HCWs who refuse to get the vaccine. Some work places have introduced laws that require health care workers to go home without pay when an influenza outbreak occurs if they refuse vaccination.

Mandated school vaccination programs are also common. Coercive policies could be justified, such as not permitting school attendance during an outbreak if the child is not vaccinated. While there may be a reluctance to use and justify coercion, public health officials also have a responsibility to justify the lack of use of coercive policies for vaccination, particularly if there is evidence for the population health benefits of such policies. The failure to do so would violate the principle of solidarity and protecting the public from harm, resulting in avoidable illness and death. In making this decision, officials will have to balance the potential risks and benefits of the vaccination program taking into account the strength of evidence for both of these. Officials will also have to be guided by the "precautionary principle", which advocates a lower evidentiary standard for taking action to protect against a large scale risk than what is traditionally used in evaluating the benefit of health technologies at the individual level.

In any mandated vaccine program, there are reciprocal responsibilities of the state to vaccine recipients: ensuring the safety and effectiveness of the vaccine, and providing just compensation to those who suffer an adverse event following vaccination.

Individual Liberty Restrictions

Research indicates that roughly 85% of the population supports states and governments to suspend some individual rights (e.g. traveling, right to assemble) during an influenza pandemic. However, such rights can only be suspended in the public's view, with reciprocity: reciprocal obligation of governments to provide for the basic needs of restricted individuals, as well as support services after the restrictive measures end. For example, restricted individuals should not be penalized by an employer for following a quarantine order (e.g. losing a job). Pandemic Ethics researchers have summarized guidelines regarding individual liberty restrictions as follows (CanPrep 2009):

1. Public health officials should ensure that pandemic response plans include a comprehensive and transparent protocol for the implementation of restrictive measures. The protocol should be founded upon the principles of proportionality and least restrictive means, should balance individual liberties with protection of public from harm, and should build in safeguards such as the right to appeal.
2. Governments and the health care sector should ensure that the public is aware of the rationale for restrictive measures, the benefits of compliance, and the consequences of non-compliance.
3. All pandemic influenza plans should include measures to protect against stigmatization and to safeguard the privacy of individuals and/or communities affected by quarantine or other restrictive measures.
4. Measures and processes ought to be implemented in order to guarantee provisions and support services to individuals and/or communities affected by restrictive measures during a pandemic emergency. Plans should state in advance what backup support will be available to help those affected by restrictive measures (e.g. food, bills, loss of income).
   Government should have public discussions of appropriate levels of compensation, including who is responsible for compensation.
5. In order to get the public "on board" with decisions regarding restrictive measures, policymakers need to include the public in deliberations about public policy with respect to a pandemic.

Global Health Ethics Considerations

Contagion raises serious global health ethics questions surrounding fair distribution of resources in a pandemic. Should wealthy countries, which may ultimately be responsible for creating any vaccines or treatments, usurp the resources and have greater access than poorer countries? In the film, healthcare workers from wealthy countries are taken hostage to ensure equitable distribution of resources. In the Ebola epidemic, Americans who got infected were flown to the United States and provided with experimental treatments over patients in West Africa who were dying. In August 2014, for example, a WHO Ethics Task force on Ebola was severely criticized for failing to address distribution of resources and cultural ethics issues that led to the spread of the disease (e.g. burial practices). At the same time, dependence on aid from wealthy countries led to a failure of local strategies to contain Ebola. Meanwhile, viruses ravaging poor countries are taken less seriously until they reach wealthy countries. Oyewale Tomori, a Nigerian virologist noted in 2014: "Ebola is swimming in an ocean of national apathy, denial and unpreparedness...After the first cases occurred, it took three months for WHO to know....[Africa] should take the lead of Ebola control efforts—not Geneva, not Washington, not New York" (Kupferschmidt 2014). He noted that one major problem was that when the international medical charity groups leave at the end of any given epidemic, the countries dependent on aid will remain in the dark about how to effectively respond to the next one. Ebola left in its wake "squandered millions in international aid from unstable leadership and governments" (Farmer 2014; Kupferschmidt 2014).

Global health ethics issues expose a disconnect between global health officials and infectious disease control (Sun et al. 2014). In the Ebola case, West Africa was not equipped, while other sociopolitical issues such as civil war, chronic poverty, poor healthcare, and less than 50 doctors in Liberia, turned it into a manmade disaster as well. Other global health ethics issues in a pandemic occur when foreign aid disappears because the workers get sick or flee, while contact tracing cannot be done effectively. Meanwhile the conditions, criteria and ethical basis for quarantine and travel bans can still divide experts.

Conclusions

Discussions surrounding pandemic ethics issues, and global health disparities are situated within the broader theme of Justice and Healthcare Access in the context of allocation of scarce resources as healthcare delivery may become overwhelmed and vaccine production and distribution may be limited. Some of these discussions may overlap with discussion of the global HIV/AIDS crisis as well as the mismanagement of HIV/AIDS in the early years of the epidemic (see Chap. 2). Addtionally, emerging analyses of the COVID-19 pandemic will put these issues into greater context (see Afterword). When teaching *Contagion* to healthcare trainees, it's important to note that they will now have a living memory of being on the front lines during the 2020 pandemic. However, in light of the growing epidemic of science denialism and vaccine refusal, it may be that the pandemics will recycle, or a familiar virus from the past will revisit. Ultimately, *Contagion* enhances any curriculum dealing with pandemic ethics, scarce resources, or infectious disease.

Theatrical Poster

Contagion (2011)

> Directed by: Steven Soderbergh
> Produced by: Michael Shamberg, Stacey Sher, Gregory Jacobs
> Written by: Scott Z. Burns
> Starring: Marion Cotillard, Matt Damon, Laurence Fishburne, Jude Law, Gwyneth Paltrow, Kate Winslet
> Music by: Cliff Martinez
> Production Company: Participant Media, Imagenation Abu Dhabi
> Distributed by: Warner Bros. Pictures
> Release Date: September 9, 2011.

References

ACLU. (2020). *Rendition, the Movie*. https://www.aclu.org/other/rendition-movie.

Ahmed, N. (2005). 23 years of the discovery of helicobacter pylori: Is the debate over? *Annals of Clinical Microbiology and Antimicrobials, 4*, 17. https://doi.org/10.1186/1476-0711-4-17. https://www.ncbi.nlm.nih.gov/pmc/articles/PMC1283743/.

Al Muharrmi, Z. (2010). Understanding the influenza a H1N1 2009 pandemic. *Sultan Qaboos University Medical Journal, 10*(2), 187–195. https://www.ncbi.nlm.nih.gov/pmc/articles/PMC3074714/.

Amadeo, K. (2019). Financial crisis bailouts. *The Balance*, June 25, 2019. https://www.thebalance.com/2009-financial-crisis-bailouts-3305539.

Amadeo, K. (2020). ARRA: Its details, with pros and cons. *The Balance*, April 11, 2020. https://www.thebalance.com/arra-details-3306299.

Anonymous. Young and unafraid of the coronavirus pandemic? Good for you. Now stop killing people. *Newsweek*, March 11, 2020. https://www.newsweek.com/young-unafraid-coronavirus-pandemic-good-you-now-stop-killing-people-opinion-1491797.

Barry, John. (2005). *The great influenza*. New York: Penguin Books.

Berman, M., & Brown, D. L. (2014). Texas ebola patient has died from ebola. *Washington Post*, October 8, 2014. https://www.washingtonpost.com/news/post-nation/wp/2014/10/08/texas-ebola-patient-has-died-from-ebola/?utm_term=.214758893a0b.

Burnett, J. (2005). Evacuees were turned away. *NPR*, September 20, 2005. https://www.npr.org/templates/story/story.php?storyId=4855611.

Borgundvaag, B., et al. (2004). SARS outbreak in the Greater Toronto Area. *CMAJ, 171*(11), 1342–1344, November 23, 2004. https://doi.org/10.1503/cmaj.1031580.

Cameron, B. (2020). I ran the White House pandemic office. Trump closed it. *Washington Post*, March 13, 2020. https://www.washingtonpost.com/outlook/nsc-pandemic-office-trump-closed/2020/03/13/a70de09c-6491-11ea-acca-80c22bbee96f_story.html.

Canadian Program of Research on Ethics in a Pandemic (CanPrep). (2009). *Ethics and pandemic influenza white paper series.* University of Toronto.

Castrodale, J. (2020). *Coronavirus has led to a surge in popularity for the 2011 movie, Contagion.* January 31, 2020. https://www.vice.com/en_us/article/qjdv7m/coronavirus-has-led-to-a-surge-in-popularity-for-the-2011-movie-contagion.

CBS. (2004). Prison abuse at Abu Ghraib. *60 minutes II.* April 28, 2004.

CDC. (2010). *The 2009 H1N1 pandemic: Summary highlights,* June 16, 2010. https://www.cdc.gov/h1n1flu/cdcresponse.htm.

CDC. (2018). Zombie preparedness. Center for preparedness and response, *CDC,* October 11, 2018. https://www.cdc.gov/cpr/zombie/index.htm.

Chua, K. B., Chua, B. H., & Wang, C. W. (2002). Anthropogenic deforestation, El Niño and the emergence of Nipah virus in Malaysia. *The Malaysian Journal of Pathology, 24*(1), 15–21. PMID 16329551.

Constitutional Rights Foundation (CRF). (2019). *America responds to terrorism.* https://www.crf-usa.org/america-responds-to-terrorism/the-patriot-act.html.

Denby, D. (2011). Contagion. *The New Yorker,* September 9, 2011. https://www.newyorker.com/magazine/2011/09/19/call-the-doctor-david-denby.

DeStefano, F., & Chen, R. T. (1999). Negative association between MMR and autism. *Lancet, 353,* 1987–1988.

Diekema, D. S. (2005). Responding to parental refusals of immunization of children. *Pediatrics, 115*(5), 1428–1431.

Douglas, E. (2011). Interview with Contagion writer Scott Burns. *Coming soon,* December 6, 2011. http://www.comingsoon.net/news/movienews.php?id=81596.

Farmer, P. (2014). Diary: Ebola. *London review of books* (Vol. 36). http://www.lrb.co.uk/v36/n20/paul-farmer/diary.

Gallup. (2018). *Presidential approval ratings—George W. Bush.* https://news.gallup.com/poll/116500/presidential-approval-rat305ings-george-bush.aspx.

Hall, C. (2011). How the "Contagion" virus was born. *Reuters,* September 13, 2011. https://advancingthescience.mayo.edu/2019/05/21/a-pregnant-pause/.

Haney, C., Banks, W. C., Zimbardo, P. G., et al. (1973). A study of prisoners and guards in a simulated prison. *Naval Research Review, 30,* 4–17.

Hoffman, J. (2011). Interview with Steven Soderbergh. *UGO.com,* September 7, 2011. https://web.archive.org/web/20120129093342/ https://www.ugo.com/movies/steven-soderbergh-interview.

Interview with Steven Soderbergh. (2011). *Movies Online,* September, 2011. http://www.moviesonline.ca/2011/09/steven-soderbergh-interview-contagion/.

Khan, A. (2011). Preparedness 101: Zombie apocalypse. *Public Health Matters Blog, CDC,* May 16, 2011. https://blogs.cdc.gov/publichealthmatters/2011/05/preparedness-101-zombie-apocalypse/.

Kupferschmidt, K. (2014). Nigerian virologist delivers scathing analysis of Africa's Ebola response. *Science,* November 3, 2014. http://news.sciencemag.org/africa/2014/11/nigerian-virologist-delivers-scathing-analysis-africas-response-ebola.

Lintern, S. (2020). 'We are making difficult choices': Italian doctor tells of struggle against coronavirus. *The Independent,* March 13, 2020. https://www.independent.co.uk/news/health/coronavirus-italy-hospitals-doctor-lockdown-quarantine-intensive-care-a9401186.html.

Low, D. (2004). SARS: Lessons from Toronto. In S. Knobler, A. Mahmoud, & S. Lemon (Eds.), *Institute of medicine (US) Forum on microbial threats.* Washington (DC): National Academies Press (US). https://www.ncbi.nlm.nih.gov/books/NBK92467/.

Marshall, B. J., Armstrong, J. A., McGechie, D. B., & Glancy, R. J. (1985). Attempt to fulfill Koch's postulates for pyloric campylobacter. *Medical Journal Australia, 142,* 436–439.

Monastersky, R. (2016). Obama's science legacy: Uneven progress on scientific integrity. *Nature, 536,* August 23, 2016. Posted to:https://www.nature.com/news/obama-s-science-legacy-uneven-progress-on-scientific-integrity-1.20467.

Mounk, Y. (2020). The extraordinary decisions facing Italian doctors. There are now simply too many patients for each one of them to receive adequate care. *The Atlantic*, March 11, 2020. https://www.theatlantic.com/ideas/archive/2020/03/who-gets-hospital-bed/607807/.

NBC News. (2020). Trump calls coronavirus new "hoax". *NBC News*, February 28, 2020. https://www.nbcnews.com/politics/donald-trump/trump-calls-coronavirus-democrats-new-hoax-n11 45721.

NIH Record. (2009). *ARRA results in unprecedented boost for NIH*. https://nihrecord.nih.gov/sites/recordNIH/files/pdf/2009/NIH-Record-2009-03-20.pdf.

Offit, P. (2011). *Contagion, the movie: An expert medical review*, September 13, 2011. https://www.medscape.com/viewarticle/749482.

Parr, B. (2009). How Facebook Dominated in 2009. *Mashable*, December 30, 2009. https://mashable.com/2009/12/30/facebook-2009/

PBS. (2018). *The Facebook Dilemma*. October 29–30, 2018. https://www.pbs.org/wgbh/frontline/film/facebook-dilemma/.

Pruden, J. (2018). Has Tim Caulfield become the Canadian nemesis of pseudoscience? *The Globe and Mail*, December 31, 2018. https://www.theglobeandmail.com/amp/life/article-has-tim-caulfi eld-become-the-canadian-nemesis-of-pseudoscience/

Ries, N. (2004). Public health law and ethics: Lessons from SARS and quarantine. *Health Law Review*, *13*(1), 2004. https://pdfs.semanticscholar.org/b3f2/5e98cbf81576261dcdf15488101c0f afdd3f.pdf.

Rosenthal, M. S. (2014). *What in the World?* The Ebola Epidemic: Presentation, University of Kentucky Program for Bioethics.

Sathyanarayana Rao, T. S., & Andrade, C. (April-June 2011) The MMR vaccine and autism: Sensation, refutation, retraction, and fraud. *Indian Journal Psychiatry*, *53*(2), 95–96. https://www.ncbi.nlm.nih.gov/pmc/articles/PMC3136032/.

Singer, P. A. et al. (2003). Ethics and SARS: Lessons from Toronto. *British Medical Journal*, *327*(7427), 1342–1344. https://www.ncbi.nlm.nih.gov/pmc/articles/PMC286332/.

Spigner, C. (2014). *Patient Zero: Thomas Eric Duncan and the Ebola Crisis in West Africa and the United States*. October 29, 2014. https://www.blackpast.org/global-african-history/perspecti ves-global-african-history/patient-zero-thomas-eric-duncan-and-ebola-crisis-west-africa-and-united-states/).

Sun, L., Brady, D., Lenny, B., & Joel, A. (2014). Out of control: How the world's health organizations failed to stop the Ebola disaster. *Washington Post*, October 4, 2014. https://www.washingtonpost.com/sf/national/2014/10/04/how-ebola-sped-out-of-control/?utm_term=.89916d45c456.

Taylor, B., Miller, E., Farrington, C. P., Petropoulos, M. C., Favot-Mayaud, I., Li, J., et al. (1999). Autism and measles, mumps, and rubella vaccine: No epidemiologic evidence for a causal association. *Lancet, 353,* 2026–2029.

Tollefson, J. (2012). US science: The Obama experiment. *Nature, 489,* September 26, 2012. https://www.nature.com/news/us-science-the-obama-experiment-1.11481.

Trager, R. (2017). Obama's science legacy. *Chemistry World*, January 10, 2017. https://www.chemistryworld.com/news/obamas-science-legacy-rhetoric-versus-reality/2500217.article.

United States Senate. (2014). *Senate intelligence committee study on CIA detention and interrogation*. https://www.intelligence.senate.gov/sites/default/files/publications/CRPT-113srpt288.pdf.

University of Toronto Joint Centre for Bioethics. (2005). *Stand on guard for thee: Ethical considerations in preparedness planning for pandemic influenza*. http://www.utoronto.ca/jcb/home/news_pandemic.htm.

Verran, J., & Reyes, X. (2018). Emerging infectious literatures and the zombie condition. *Emerging Infectious Diseases*, *24*(9), 1774–1778. https://doi.org/10.3201/eid2409.170658. https://wwwnc.cdc.gov/eid/article/24/9/17-0658_article.

Wakefield, A. J., Murch, S. H., Anthony, A., Linnell, J., Casson, D. M., Malik, M., et al. (1998). Ileal-lymphoid-nodular hyperplasia, non-specific colitis, and pervasive developmental disorder in children. *Lancet, 351,* 637–641.

Walton, N. (2009). The Wakefield story and the need for clarity. *Research Ethics Blog*, February 15, 2009. https://researchethicsblog.com/2009/02/15/the-wakefield-story-and-the-need-for-clarity/.

Whitehouse.gov. (2015). *Obama's remarks at White House Science Fair*, March 23, 2015. (https://obamawhitehouse.archives.gov/the-press-office/2015/03/23/remarks-president-white-house-science-fair).

WHO. (2010). What is the pandemic H1N1 2009 virus? *WHO*, February 24, 2009. https://www.who.int/csr/disease/swineflu/frequently_asked_questions/about_disease/en/.

Woodward, B. (2010). *Obama's Wars*. Simon and Schuster.

World Health Organization. (2008). Addressing ethical issues in pandemic influenza planning. *Discussion Papers*. Available at: http://www.who.int/ethics/publications/en/.

Yang, et al. (2020). Clinical course and outcomes of critically ill patients with SARS-COV2-2 pneumonia in Wuhan. *The Lancet*, February 24, 2020. https://www.thelancet.com/journals/lanres/article/PIIS2213-2600(20)30079-5/fulltext.2020 https://doi.org/10.1016/S2213-2600(20)30079-5.

Chapter 7
Solid Organ Transplantation: Donors and Recipients in *21 Grams* (2003)

For several years, finding good, credible films about solid organ transplantation resulted in the "some good parts" problem. A major contender historically has been *John Q* (2002), a pediatric ethics film about health disparities in solid organ transplantation. However, the film derails into an improbable subplot and "thriller" by its second half, leaving it as a "clips" film in which there are some good parts. Several medical educators take the surreal, comic or science fiction route, and use *Frankenstein* (1931) or various iterations of it, *Repo Men* (2010), or even the absurd, demonstrated by the "Live Organ Transplants" skit in Monty Python's *The Meaning of Life* (1983), in which the State sends government representatives to homes of citizens who have opted in as donors to remove their organs against their wills while they're still alive and well (Monty Python 2020).

With respect to organ selling/trafficking and the exploitation of vulnerable populations, *Dirty Pretty Things* (2002) can be used, which is set in pre-Brexit London. The film revolves around exploitation of immigrants, who trade organs for the coveted British passport in a freshly post-9/11 world that is becoming hostile to them. However, the pace of the film and its various subplots make it suboptimal as a teaching film.

The top bone marrow transplant film for bioethicists is *Marvin's Room* (1996), which examines family systems and stands out as an excellent clinical ethics film. *My Sister's Keeper* (2009), another pediatric ethics film, is about the "heir and a spare" system of the donor parts child, but like *John Q*, it also devolves into improbability (much of it is based on a poor script), and has never been considered a credible teaching film from my perspective.

If you're looking for the quintessential film about solid organ transplantation, *21 Grams* (2003) is the film for you. This is a messy ethics story of heart transplantation and how the donor, the recipient and the family members become entangled. It's difficult to tell which character is made more miserable by the successful transplant. The story is told in non-chronological order, and bioethicists seem to conclude that among the many flawed "transplant films" out there, this one is… well, solid. *21 Grams* depicts troubling organ procurement discussions with grief-stricken family

© Springer Nature Switzerland AG 2020
M. S. Rosenthal, *Healthcare Ethics on Film*,
https://doi.org/10.1007/978-3-030-48818-5_7

members; the recipient's guilt; and the suffering of the surviving third party whose actions created a "donor" by causing the car accident. Ultimately, the film is about a flawed system, flawed people, and one good heart. This chapter discusses the solid organ transplant ecosystem of scarce resource allocation, increasing the donor pool, recipient listing, and trauma medicine as a key "first responder" in identifying potential donors resulting from catastrophic accidents. The non-linear film is also a riveting, highly acclaimed film by visionary director, Alejandro González Iñárritu, with an outstanding cast of character actors including Naomi Watts, Sean Penn and Benicio Del Toro.

Origins of *21 Grams*: A "Stop and Gawk" Moment

21 Grams is an original screenplay by Guillermo Arriaga, who had partnered with Iñárritu on previous films. The title of the film refers to a bizarre 1907 study by Dr. Duncan MacDougall (New York Times 1907) in which he tried to weigh the "soul" by weighing dying patients prior to death and after, and claimed the bodies after death were 21 grams lighter (see under History of Medicine). The script arose from Arriaga's witnessing of a car accident in 2000 on his birthday (on his way to his own party). As he watched all the people involved in the accident, he began to weave an organ donor narrative surrounding an everyday occurrence. Driving is an everyday occurrence; but so is death, dying and organ donation. Arriaga wondered how his life would change if he were to be involved in such an accident, and began to think about each person's role in the trauma (Focus Features 2018). Iñárritu liked the idea, which led to the film's original screenplay. *Focus Features* noted on the film's 15th anniversary (Focus Features 2018):

> For *21 Grams*, Iñárritu constructed a poetic puzzle with his screenwriter Guillermo Arriaga about how fate, coincidence, and hope connect the destinies of three strangers. Paul (Sean Penn) is a math professor in need of a heart transplant. Cristina (Naomi Watts) is a recovering addict whose future is shattered when her husband and children die in a hit-and-run crash, a terrible twist of fate that provides the heart Paul needs. Jack (Benicio Del Toro) is a born-again ex-con whose out-of-control behavior leads to the traffic accident that sets the film's harrowing events into motion.

The Director

Alejandro González Iñárritu had made a previous successful film with Arriaga entitled *Amores Perros* (2000). *Amores Perros* is also about a car crash and parallel lives and was nominated for Best Foreign film. He would go on to make *Babel* (2006) with Arriaga as well.

Iñárritu studied at Mexico's Iberoamaricana University, and worked as a disc jockey for Mexican radio and as a concert promoter. He studied theater with Polish-Mexican director Ludvik Margule, and began his film career by making commercials for the network that owned his radio station, and then began making commercials in Mexico for multiple companies. He also composed music for Mexican films, and eventually founded his own production company (Curiel 2003).

Iñárritu also explored other healthcare issues in film; in 2010, he made *Biutiful* (2010), about a man diagnosed with terminal prostate cancer. And then Iñárritu made the 2014 film that won for best picture, best director, best cinematography, and best original screenplay: *Birdman (Or The Unexpected Virtue of Ignorance)*, which looks like it was filmed in a single take.

Married to an editor and graphic designer, María Eladia Hagerman, Iñárritu has two children, but lost his first son two months after he was born due to medical complications in 1996. He noted (Curiel 2003):

> I felt the doctors didn't really talk to me in advance, and they didn't take the necessary precautions to prevent the death...The reason my kid died is very complicated. When I began to investigate it, I began to plan things – and how I (could) damage those guys. Suddenly I realized, by a very tough process, that nothing will bring my kid back. It's a very subjective world. And I just let it go. I had to find a way to let go. If not, you begin to get crazy.

Iñárritu's first film, *Amores Perros,* is dedicated to his late son, Luciano, while *21 Grams* is dedicated to Maria. In Spanish, the dedication reads: "A María Eladia, pues cuando ardió la pérdida reverdecieron sus maizales" which means "To Maria Eladia for when the loss burned, their cornfields became green again." (Rediff 2016).

Iñárritu's cinematic style is unique. According to the *New York Times* "plumbs here are so rarely touched by filmmakers that *21 Grams* is tantamount to the discovery of a new country" (Mitchell 2003). When screening *21 Grams*, the viewer will note a completely original and raw style. First, the film was shot mostly with hand-held cameras, which gives a grittier and reality-based feel. This was a technique used in Steven Spielberg's *Saving Private Ryan* (1998) to simulate wartime coverage as well as the chaos of the D-Day Omaha Beach landing scenes. Second, although the film is in non-chronological order, it was shot in chronological order. One scholar notes (McCormack 2016):

> Non-linear narrative forms are staples of both independent and Hollywood cinematic produc-tions. The analepses (flashbacks) and the prolepses (flashforwards), while interrupting the linear form, often complement the diegetic flow... Indeed, Iñárritu creates a significant stac-cato effect that, until the linear ending of the film, disrupts time and thus the coherence of the story being portrayed. Our sense of what is happening in the film is, especially at the outset, constantly uncertain as scenes change quickly and give little indication of where they occur in or how they are relevant to the story. Random scenes leak into each other, connected by what Iñárritu has described as 'emotional time'.

Cast

21 Grams has an extraordinary cast. Iñárritu first approached Naomi Watts for the role of Cristina while she was filming a horror movie, *The Ring* (2002). Watts was familiar with the director's work, and joined the project without auditioning. As for Sean Penn, who plays Paul Rivers, he was very impressed with Iñárritu's films, and was eager to work with him. Similarly, Benicio Del Toro was "awestruck by *Amores Perros* and wanted to work with Iñárritu, who went from making relatively small-budget films in Mexico to a U.S. studio project whose budget approached $20 million" (Curiel 2003). Benicio Del Toro had been introduced as a tour de force actor in Steven Soderbergh's riveting film about the drug cartel wars, *Traffic* (2000), featuring an ensemble cast; Del Toro played a Mexican police detective who spoke mainly in Spanish throughout the film. (Soderbergh is discussed in Chap. 6). Melissa Leo, who plays Del Toro's wife in *21 Grams*, had mainly a television drama career prior to this film, most known as a detective in *Homicide: Life on the Street* (1993–1997).

Charlotte Gainsbourg, who plays Paul's estranged wife, is an English actress who had famous parents: English singer, Jane Birkin, and French actor and singer, Serge Gainsbourg, whose marriage was always reported on in the British tabloids. Her father was an alcoholic and ultimately died of his alcoholism. Gainsbourg had won notoriety for her role in the 1996 film, *Jane Eyre*, and won an award for her work in a 1999 film about a family drama, *La Buche*. Gainsbourg helped to shape some critical scenes in *21 Grams*. For example, when her father was dying, he apparently hid cigarette butts in his pill bottles; Paul, played by Sean Penn, also hides his cigarettes in a pill bottle in their bathroom at home (IMDB 2020). The issue of smoking and transplant candidacy is discussed under Healthcare Ethics Issues.

Sean Penn had a very long and distinguished career prior to *21 Grams*, and along with Iñárritu, had participated in a post-9/11 film project (see further), and so was very familiar with Iñárritu. Penn was the most well-known actor in the cast at the time, and had been introduced to American audiences through a teen cult classic, *Fast Times at Ridgmont High* (1982) which launched the careers of several actors; Penn was the "surfer dude" in the film. Penn had also won an academy award for his role in *Mystic River* (2003), which was released the same year as *21 Grams*. After *21 Grams* was released, the cast went on to significant roles, awards and accolades. Of note, Penn and Watts would later star together in the film *Fair Game* (2010), which was about the intelligence failures at the start of the Iraq War (March 21 2003), and the saga of outed CIA agent, Valerie Plame and her husband (see further). Penn also played the character of Harvey Milk (see Chap. 2), in the film, *Milk* (2008). Penn was almost equally known for his tabloid appearances and fights with the paparazzi due to his interesting personal life, which included highly publicized marriages to Madonna, Robin Wright, divorces from each, and relationships with Jewel and Charlize Theron. Penn was the son of blacklisted actor and director, Leo Penn (1921–1998), who refused to name names during the McCarthy era and the House UnAmerican Activities Committee (HUAC) (Burns 2002).

According to the *Washington Post* (Hunter 2003):

> One can certainly see why such a trio of powerhouse performers was attracted to this project. For actors, the movie is pure candy, a succession of gigantic actorly moments to be painted in the brightest of primary colors: grief (Cristina gets the news), remorse (Jack realizes what he has done), anguish (Paul attempts to understand why he has been spared a sure death

from heart failure at the expense of three innocent lives), as well as such other fun things as drug addiction (Cristina turns junkie in the aftermath), vengeance (Paul gets that gun and goes hunting to settle the score and establish justice) and bitterness (having given his life to Jesus, Jack cannot understand why Jesus punished him by putting him behind the wheel of a truck that struck three innocent people).So the movie is right on that thin membrane between pathos and bathos. It has so many Big Emotional Moments…[and] raw honesty of the production.

Synopsis

In this case, sharing the synopsis in advance likely ruins some of the viewing experience, as the film challenges the viewer to a temporal jigsaw puzzle. Nonetheless, here is the plot summary unraveled by different reviewers. The *San Francisco Chronicle* describes it this way (Curiel 2003):

> "21 Grams" links various characters to a single, random car crash. One of the principal figures is a brutish ex-con (Del Toro) who tattoos religion onto his skin and threatens others who don't adhere to his dogmatic principles. Another central figure (Watts), traumatized by the deaths of loved ones, seeks comfort in murdering their killer. Others are also swept up by raw emotions. It's a template for drama and disaster.

The *Washington Post* summarized it like this (Hunter 2003):

> One night, driving home in a hurry after having lost his day job – this is the inciting event, but by the film's narrative strategy it is not revealed until the halfway point – [Del Toro] strikes and kills three pedestrians, a man and his two daughters in a suburb, then flees. Cristina [played by Watts] is the mother and wife of those who died; Paul [played by Penn], terminally ill, becomes the recipient of the heart from her husband's chest.

And the *New York Times* tells the story like this (Mitchell 2003):

> Prof. Paul Rivers (Mr. Penn) is suffering from a damaged heart, which is about to give out on him, and the transplant he receives so consumes him with guilt that he's like a death-row prisoner with a horrible secret who's been pardoned. He still believes he deserves to die. Cristina (Ms. Watts), a reformed party girl, has returned to her fallen ways after taking up with Paul. She's lost her family in an auto accident. And Jack (Mr. Del Toro) is a shaggy, trembling mountain of anger he can barely contain. His already tenuous grasp on sobriety is slipping away even faster since being involved in a terrible incident.

Film scholar Todd McGowan, in his book, Out of Time: Desire in Atemporal Cinema (McGowan 2011), notes: "Iñárritu edits the film in an atemporal fashion …and forces the spectator to infer the order in which they take place." McGowan describes that the editing style of the film was "to stress the ubiquity of loss, and to emphasize how loss outstrips every gain" in a "traumatic narration".

The United Nations Educational, Scientific and Cultural Organization (UNESCO) and the UNESCO Chair in Bioethics recommends this film from its "Teaching Bioethics" website, and synopsizes it this way (UNESCO 2018):

21 grams is a disquieting story concerning the value of life and death. Paul Rivers suffers a serious heart condition and must have an urgent transplant. He is not expected to live long so he goes with his wife Mary to a fertility center in search of help to conceive a child despite the difficult circumstances. He finally receives the heart of a man who died, together with his two daughters, in a car accident. The widow, devastated by the tragedy, later receives the unexpected visit of Paul, the recipient of the organ donated by her dead husband, thus opening a complex web of passion, guilt and vengeance.

Even the Scottish Council on Human Bioethics recommends it in their Human Transplantation Film Library (SCHB 2020):

Sean Penn and Benecio Del Toro, play wildly different men linked through a grieving woman (Naomi Watts) in 21 Grams. Del Toro delves deep into the role of an ex-con turned born-again Christian, a deeply conflicted man struggling to set right a terrible accident, even at the expense of his family. Penn captures a cynical, philandering professor in dire need of a heart transplant, which he gets from the death of Watts' husband. 21 Grams slips back in forth in time, creating an intricate emotional web out of the past and the present that slowly draws these three together; the result is remarkably fluid and compelling. The movie overreaches for metaphors towards the end, but that does not erase the power of the deeply felt performances.

Ultimately, what the three central characters have in common is an understanding of the cycle of life and death. For Jack Jordan (played by Del Torro), the driver of the car, he is a born-again Christian who lives in continuous repentance for the sins of his past life, and then is almost killed by Cristina out of revenge when she discovers him, only to be saved by Paul. For Paul Rivers (played by Penn), the organ recipient, he was in the last stages of dying before the accident, and had also agreed to fertility treatment with his estranged wife (Gainsbourg) so she could have his child posthumously if need be to make up for a regretted abortion she had when she and Paul were separating. As for Cristina (played by Watts), the widow of the organ donor, she had a history of substance abuse and suicide prior to her marriage, and returns down that path in her grief.

A literature scholar (McCormack 2016) provides this synopsis:

21 Grams tells the story of chance encounters where people are brought together through everyday and yet somewhat extraordinary circumstances, including a car crash, the death of a father and his two daughters, and a heart transplant. The characters collide with one another and seep into each other's lives. Random events bring together their previously disconnected lives. Jack accidentally kills Cristina's husband and children in a hit-and-run car crash. Because Michael (Cristina's husband) dies in the car accident, Paul receives the much-awaited heart transplant necessary to prevent him from dying. This accident shatters the lives of these three characters, breaking apart their worlds in very different ways: Cristina cannot cope with the loss of her family and therefore returns to what is represented as an abusive use of drugs and alcohol; Jack cannot cope with the guilt of having killed these three people and therefore, unable to look his children in the eye, leaves the family home to punish himself through hard labour and alcohol and drug abuse; and Paul regains a temporary state of improved health and in so doing decides both to terminate his dying relationship with his partner Mary and to try to understand who he is now that he has another person's organ inside himself. Although prohibited to do so, Paul seeks contact with the donor's spouse Cristina.

The Social Location of *21 Grams*: Grief, Loss and Revenge in Post-9/11

21 Grams was very much a product of its time in the wake of post-9/11 (September 11, 2001) as the country was coming to grips with the war on terror, and thousands of families were coping with grief and loss either directly, or indirectly. The war in Afghanistan had begun in October 2001, which as of this writing, is still ongoing, while the war in Iraq had begun in March 2003, while *21 Grams* was in production. In fact, both Iñárritu and Penn had been involved in a 2002 international film project called *September 11* (also known by its European title *11'09'01*) in which 11 film-makers from around the world made 11-minute and nine second long films about their perspectives and experiences of September 11th. The *San Francisco Chronicle* noted in 2003 (Curiel 2003):

> Now that he has time to reflect on "21 Grams," Alejandro González Iñárritu understands that he made a movie influenced by the terrorist attacks of Sept. 11, 2001. It's subtle, he says. It might have even been an unconscious decision to incorporate themes of violence, revenge and God into his highly anticipated sequel to "Amores Perros," but Iñárritu can't deny the connection. "I didn't think about it when I was doing it, but '21 Grams' has a very Sept. 11 (theme)…"What I discovered is that the journey of Cristina (Watts' character) –which is about loss, then grief and confusion and the need of revenge – works on a human scale and on a country scale, which is exactly what happened on Sept. 11," Iñárritu says. "The sense of loss, then confusion and grief, then the need for revenge – that's what happened in Afghanistan and now Iraq."

Iñárritu moved to Los Angeles from Mexico four days before 9/11. Because of his success with *Amores Perros*, Iñárritu was invited as one of 11 international filmmakers to contribute to the *September 11* film project. In Iñárritu's short film, *Episode 7: Mexico*, the film is described like this on the Project's website (Wikipedia 2020):

> Black screen. Background noises and rumors of everyday life, suddenly interrupted by the screams of the witnesses of the crash of American 11 against the North Tower of the World Trade Center. While the black screen is occasionally interrupted by the repertoire images of the attacks, the voices of the announcements on television, the screams of the victims, the explosions of the planes, the calls made by the victims and their relatives overlap. The sound stops and you can see the two towers collapse without sound. The background voices start again on a background of violins, while the screen gradually changes from black to white. Two writings appear in both Arabic and Latin characters of the same meaning: "Does God's light guide us or blind us?" The sentences finally disappear in a blinding light."

Ironically, Sean Penn also contributed to the September 11 project for the American film. Episode 10: *United States of America* by Penn is described like this (Wikipedia 2020):

> An elder spends his life alone in an apartment overshadowed by the Twin Towers. The widowed man vents his loneliness by talking to his late wife as if she were still alive and cultivating her flower pot, withered by the lack of light. The collapse of the Towers finally allows the light to flood the apartment and suddenly revitalizes the flowers. The elder, happy

for what happened, tries to show the vase to his wife, but the light "reveals" the illusion in which he lived until then. Between tears, he regrets that his wife is not there to finally see the vase bloom again.

Penn's State: An Antiwar Advocate

In the aftermath of 9/11, as the March to the Iraq war heated up, Sean Penn was a particularly strong anti-war voice. On October 19, 2002, he paid $56,000 to run a full-page Open Letter to President George W. Bush in the *Washington Post*, which read (Penn 2002):

> An Open Letter to the President of the United States of America.
>
> Mr. Bush:
>
> Good morning sir. Like you, I am a father and an American. Like you, I consider myself a patriot. Like you, I was horrified by the events of this past year, concerned for my family and my country. However, I do not believe in a simplistic and inflammatory view of good and evil. I believe this is a big world full of men, women, and children who struggle to eat, to love, to work, to protect their families, their beliefs, and their dreams. My father, like yours, was decorated for service in World War II. He raised me with a deep belief in the Constitution and the Bill of Rights, as they should apply to all Americans who would sacrifice to maintain them and to all human beings as a matter of principle.
>
> Many of your actions to date and those proposed seem to violate every defining principle of this country over which you preside: intolerance of debate ("with us or against us"), marginalization of your critics, the promoting of fear through unsubstantiated rhetoric, manipulation of a quick comfort media, and position of your administration's deconstruction of civil liberties all contradict the very core of the patriotism you claim. You lead, it seems, through a blood-lined sense of entitlement. Take a close look at your most vehement media supporters. See the fear in their eyes as their loud voices of support ring out with that historically disastrous undercurrent of rage and panic masked as "straight tough talk." How far have we come from understanding what it is to kill one man, one woman, or one child, much less the "collateral damage" of many hundreds of thousands. Your use of the words, "this is a new kind of war" is often accompanied by an odd smile. It concerns me that what you are asking of us is to abandon all previous lessons of history in favor of following you blindly into the future. It worries me because with all your best intentions, an enormous economic surplus has been squandered. Your administration has virtually dismissed the most fundamental environmental concerns and therefore, by implication, one gets the message that, as you seem to be willing to sacrifice the children of the world, would you also be willing to sacrifice ours. I know this cannot be your aim so, I beg you Mr. President, listen to Gershwin, read chapters of Stegner, of Saroyan, the speeches of Martin Luther King, Jr. Remind yourself of America. Remember the Iraqi children, our children, and your own.
>
> There can be no justification for the actions of Al Qaeda. Nor acceptance of the criminal viciousness of the tyrant, Saddam Hussein. Yet, that bombing is answered by bombing, mutilation by mutilation, killing by killing, is a pattern that only a great country like ours can stop. However, principles cannot be recklessly or greedily abandoned in the guise of preserving them.
>
> Avoiding war while accomplishing national security is no simple task. But you will recall that we Americans had a little missile problem down in CubaCuba once. Mr. Kennedy's restraint (and that of the nuclear submarine captain, Arkhipov) is to be aspired

to. Weapons of mass destruction are clearly a threat to the entire world in any hands. But as Americans, we must ask ourselves, since the potential for Mr. HusseinHussein, Saddam to possess them threatens not only our country, (and in fact, his technology to launch is likely not yet at that high a level of sophistication) therefore, many in his own region would have the greatest cause for concern. Why then, is the United States, as led by your administration, in the small minority of the world nations predisposed toward a preemptive military assault on Iraq?

Simply put, sir, let us re-introduce inspection teams, inhibiting offensive capability. We buy time, maintain our principles here and abroad and demand of ourselves the ingenuity to be the strongest diplomatic muscle on the planet, perhaps in the history of the planet. The answers will come. You are a man of faith, but your saber is rattling the faith of many Americans in you.

I do understand what a tremendously daunting task it must be to stand in your shoes at this moment. As a father of two young children who will live their lives in the world as it will be affected by critical choices today, I have no choice but to believe that you can ultimately stand as a great president. History has offered you such a destiny. So again, sir, I beg you, help save America before yours is a legacy of shame and horror. Don't destroy our children's future. We will support you. You must support us, your fellow Americans, and indeed, mankind.

Defend us from fundamentalism abroad but don't turn a blind eye to the fundamentalism of a diminished citizenry through loss of civil liberties, of dangerously heightened presidential autonomy through acts of Congress, and of this country's mistaken and pervasive belief that its "manifest destiny" is to police the world. We know that Americans are frightened and angry. However, sacrificing American soldiers or innocent civilians in an unprecedented preemptive attack on a separate sovereign nation, may well prove itself a most temporary medicine. On the other hand, should you mine and have faith in the best of this country to support your leadership in representing a strong, thoughtful, and educated United States, you may well triumph for the long haul. Lead us there, Mr. President, and we will stand.

with you.

Sincerely,

Sean Penn

San Francisco, California

In December 2002, Penn took a trip to Iraq to see the country, and try to gain a perspective that would inform his opposition to the war. This was not unlike what Jane Fonda had done during the Vietnam War. His visit was described in the *New York Times* like this (Burns 2002):

In Baghdad, Mr. Penn followed an itinerary set by the Iraqis, and was accompanied, like all foreigners here, by a government "minder," whose tasks include approving visits and monitoring conversations. The actor began with a visit to Al Mansour children's hospital on Friday, touring wards filled with children said by the Iraqis to have fallen sick with cancers, malnutrition and other afflictions caused by United Nations economic sanctions. Over the weekend, he visited a dilapidated school in the Baghdad suburbs, and a water treatment plant on the Tigris River bombed by American aircraft in 1991.

Then, three months after the invasion of Iraq had occurred in March 2003, Penn published a statement in June 2003, in response to his grave concerns over the policies of the Bush administration. He wrote the following (Walsh 2003):

> If military intervention in Iraq has been a grave misjudgment, it has been one resulting in thousands upon thousands of deaths, and done so without any credible evidence of imminent threat to the United States. Our flag has been waving, it seems, in servicing a regime change significantly benefiting US corporations.... We see Bechtel. We see Halliburton. We see Bush, Cheney, Rumsfeld, Wolfowitz, Powell, Rice, Perle, Ashcroft, Murdoch, many more. We see no WMDs. We see dead young Americans. We see no WMDs. We see dead Iraqi civilians. We see no WMDs. We see chaos in the Baghdad streets. But no WMDs....I am an American and I fear that I, and our people are on the verge of losing our flag...Yet, now here we are, just those five short years have passed, and that same flag that took me so long to love, respect, and protect, threatens to become a haunting banner of murder, greed, and treason against our principles, honored history, Constitution, and our own mothers and fathers. To become a vulgar billboard, advertising our disloyalty to ourselves and our allies.

The context of the post-9/11 period that scholars mark from September 12, 2001-May 1, 2011, when Osama bin Laden was killed (Rosenthal 2013), is frequently a factor in any film made in the early 2000s, as the country completely transformed. I discuss this context at great length in Chap. 4 with respect to the film, *Sicko* (2007). Although *21 Grams* makes no specific references to 9/11 or the post-9/11 wars that had begun at that time, its audience connected with the themes of grief, loss and revenge. The invasion of Iraq, and the subsequent consequences of the loss of American moral authority because of Abu Ghraib, enhanced interrogation and torture (Senate Intelligence Committee 2014), and an 11-year quagmire, was rooted in the desire to blame and get revenge for an attack on American soil in which lives were upended and obliterated in seconds. Similarly, this is Cristina's experience, and her singular focus on wanting to kill the driver who destroyed her family. The solid organ—one good heart—at first gives Paul his life back. But as Paul descends into the revenge plot of Cristina's, he begins to experience transplant rejection, and ultimately lives long enough to save the driver and his heart donor's widow from the grief and madness that is destroying her. The plot mirrors the descent into war and revenge that would swallow the United States for a generation (Wright 2006). Sean Penn's own journey simulates his "rejection" of transplanted American militarism in place of its heart and soul, while Iñárritu's journey simulates his deep experiences with his personal grief and anger over his son's death from medical complications, as he watched the United States go through similar stages of grief and loss.

The History of Medicine Context: Heart, Soul and Transplantation

When covering the history of medicine context for *21 Grams*, the major relevant milestones in solid organ transplantation are in the latter half of the twentieth century as successful transplants are linked to resolving immunity response and rejection of the organ: the development of the drug, cyclosporin, and other similar anti-rejection drugs, lead to the organ transplant system as we know the creation of United Network for Organ Sharing (UNOS), and the regulatory legislation that governs transplantation in the United States. This section covers that history, but starts with the curious 1907 experiment on the "weight of the soul". Transplant candidacy and recipient listing are discussed under Healthcare Ethics Issues.

Soul Subjects: The MacDougall Experiment

The title of the film refers to the now-debunked Duncan MacDougall experiment (Schwarcz 2019) in which he tried to measure the weight of the soul. In April 1907 the journal *American Medicine* published "Hypothesis Concerning Soul Substance Together with Experimental Evidence of The Existence of Such Substance." MacDougall, who practiced medicine in Massachusetts, sought to discover the "weight" of the soul, which would also prove that a human soul existed. In his experiment, he constructed a special scale and weighed six dying patients, and then weighed them again shortly after death, claiming that they weighed 21 grams less, which he calculated to be the weight of the soul. Macdougall apparently controlled for bodily fluid loss, air loss from the lungs, and so forth. This is why, when Paul dies, he states: "How many lives do we live? How many times do we die? They say we all lose 21 grams… at the exact moment of our death. Everyone. And how much fits into 21 grams? How much is lost? When do we lose 21 grams? How much goes with them? How much is gained? How much is gained? Twenty-one grams. The weight of a stack of five nickels. The weight of a hummingbird. A chocolate bar. How much did 21 grams weigh" (Kruszelnicki 2004)?

MacDougall repeated his experiment with 15 dogs, whose weight at death did not change. Before the academic paper was published, the *New York Times* ran the story on March 11, 1907 titled: "Soul Has Weight, Physician Thinks." (New York Times 1907). As a result, his work was treated as fact at the time, even though it has since been debunked as pseudoscience.

The MacDougall experiment is important to discuss in the context of a heart transplant, in particular, as the organ remains the universal symbol that embodies the "soul" or character. We refer to generous people as "big-hearted", and cruel people as "heartless". We love people with "all our hearts". We experience loss or betrayal as a "broken heart". For these reasons, almost all heart transplant recipients are intensely curious about their heart donors. They feel they are embodying the donor's soul. Conversely, the association of the heart and soul is also what creates some of the noted ethical dilemmas with organ donation after resource allocation decisions (see under Healthcare Ethics Issues).

Organ transplantation is defined as "a surgical operation where a failing or damaged organ in the human body is removed and replaced with a new one" (UMN Center for Bioethics 2004). *21 Grams* focuses on cadaveric, or deceased organ donation, but living organ donation is possible for a paired organ set (e.g. kidney) or a part of an organ that will still be able to function after the operation (e.g. liver, lung). For heart transplantation, organs can only come from a dead body, and patients needing cadaveric organ transplantation are put on a waiting list.

History of Transplantation

The history of transplantation that is relevant to the film occurred in the latter half of the twentieth century when the first kidney was transplanted from a donor who was an identical twin. However, there is a lengthy history that goes back to the nineteenth century in which several experiments surrounding homografts were done unsuccessfully for decades, particularly with human skin grafts. Finally, by World War II, a researcher "identified rejection as an immunological event, an original discovery and a novel insight" (Barker and Markmann 2013). To get around the immunological rejection problem, on December 23, 1954, Joseph Murray and John Merrill performed the first successful kidney transplant using the patient's identical twin as the living donor of a human kidney; the surgical procedure had been developed earlier (Barker and Markmann 2013). The recipient lived for 8 years after the transplant (Jonsen 2012).

The invention of hemodialysis was also a milestone in transplantation by prolonging the lives of potential recipients, but also led to rationing of dialysis because it was a scarce resource (see under Healthcare Ethics Issues). In 1963, renal transplantation was rare, and there were only three centers that performed it in carefully matched donors and recipients. In 1963, an immunosuppressant "cocktail therapy" had been invented with the drugs prednisone and azathioprine, which significantly reduced rejection. This led to a new era of transplant programs, and that protocol remained the standard of care until 1983 (Barker and Markmann 2013). With the cocktail therapy, liver and pancreas transplants were also attempted, and the first successful heart transplant took place in South Africa (1967) and the United States (1968) but the patients did not live long. The first procedure in South Africa took place December 3, 1967, and was performed by cardiac surgeon Christiaan Barnard; he transplanted a still-beating heart into Louis Washkansky, who lived for 18 days. A few weeks later, Barnard transplanted another heart into Philip Blaiberg, who lived 594 days (Jonsen 2012). Norman Shumway performed the first adult heart transplant in the United States on January 6, 1968, at Stanford University Hospital. Shumway and Barnard were colleagues, but prior to the Harvard Brain Death Criteria, it was illegal for Shumway to remove a beating heart from a donor; South Africa did not have such legal barriers, which permitted Barnard to do the first procedure (Richter 2008).

Organ rejection was still common, so transplant procedures were still limited. In 1968, the concept of brain death (see under Healthcare Ethics Issues) was introduced, which helped to establish criteria for organ retrieval. In 1978, the immunosuppressant drug, cyclosporin, was first tried in kidney transplants. Twelve years after the Harvard Brain Death criteria was published, in 1980, more guidance regarding brain death was issued through the *Uniform Determination of Death Act (UDDA)*, which defines death as either irreversible cessation of circulatory and respiratory functions or irreversible cessation of all functions of the brain, including the brain stem, and by 1981, the UDDA was turned into law in all states through the National Conference on Uniform State Laws in consultation with the American Medical Association (AMA), the American Bar Association (ABA), and the President's Commission on Medical Ethics. With greater clarity on definitions of death, in 1981 the first successful heart and lung transplant was performed. By 1983, cyclosporine was formally approved by the FDA to treat organ rejection (Barker and Markmann 2013) and transplant programs began to flourish. In fact, transplantation expanded during the time frame that HIV/AIDS had become an epidemic (see Chap. 2). For example, the excitement over advances in transplantation led UCLA to reject supporting faculty doing AIDS research so the institution could focus on expanding its transplant program. That same year, the Surgeon General C. Everett Koop convened the first workshop on solid organ procurement for transplant. To minimize the side effects of liver toxicity, azathioprine and corticosteroids were added to the cyclosporin regimen, which improved the lifespan of the organ transplant recipient.

In 1989, a strong drug, tacrolimus began to replace cyclosporine as the anti-rejection therapy, which had greater efficacy. However, drug-free immunosuppression would become a goal that was difficult to achieve in most recipients, and that includes the character of Paul Rivers, who begins to reject his donor heart.

Access to transplantation increased with the availability of dialysis and Medicare funding for end-stage renal disease (see under Healthcare Ethics Issues), so that patients requiring a kidney could live longer and be bridged to transplantation. Tissue typing, acceptance of brain death; improved organ preservation techniques, and established expertise in managing rejection led to much greater successes in organ transplantation, but many patients still would eventually reject their organs. In the early 2000s, expanding the donor pool led to Donation after Cardiac Death (DCD), which became an ethical problem for several practitioners (see under Healthcare Ethics Issues).

UNOS and Organ Procurement Agencies

In 1967, Paul Terasaki started the first organ sharing organization in Los Angeles, and then The Boston Interhospital Organ Bank followed in 1968 (Barker and Markmann 2013). The Southeastern Organ Procurement Foundation (SEOPF) next formed in 1969, comprising 12 hospitals, and in 1977 established the first database known as the "United Network for Organ Sharing". That service eventually led to UNOS (see further). In the early 1980s, as more transplant hospitals were opening, and more candidates were being put on the waiting list for transplantation, there were also concerns over oversight and organ stewardship. As a result, Congress passed the *National Transplant Act* in 1984, which called for an Organ Procurement and Transplantation Network (OPTN) to be created and run by a private, non-profit organization under federal contract, which was the birth of the United Network for Organ Sharing (UNOS), formed March 21, 1984, as a national entity formed to control organ allocation and placement, data collection, monitoring performance of transplant centers and organ procurement organizations. UNOS describes itself as an "independent, non-profit organization, committed to saving lives through uniting and supporting the efforts of donation and transplantation professionals. UNOS was first awarded the national contract in 1986 by the U.S. Department of Health and Human Services and functions as the established organ sharing system to "maximize the efficient use of deceased organs through equitable and timely allocation." It does this by creating a system to "collect, store, analyze and publish data pertaining to the patient waiting list, organ matching, and transplants" (UNOS 2020). In the pre-UNOS era, if an organ couldn't be used at hospitals that were local to the donor, there was no defined system to find matching candidates elsewhere, so this system greatly expanded access to transplantation.

Organ Procurement Agencies

UNOS next gave rise to individual Organ Procurement Organizations (OPOs). OPOs must be certified by the Centers for Medicare and Medicaid Services (CMS) and be members of the OPTN as well as the Association of Organ Procurement Organizations. OPOs are responsible for increasing the number of registered donors in a variety of ways, and coordinating the donation process, which includes approaching potential donor families for consent to donate on behalf of a patient who is appropriate for donation. When a potential donor becomes available due to medical criteria (such as brain death, for example), the responsible practitioners contact representatives from the local OPO, who will evaluate the potential donors, check the whether they are listed in the state donor registry, and will discuss donation with family members. This process is shown in *21 Grams*, in which Cristina is approached while she's still in shock, raising significant questions about consent to donation (see under Healthcare Ethics). Once a donor is secured, the OPO representative also contacts the OPTN computer system that matches donors and recipients, and looks for a recipient to match. The OPO then is involved with the recovery of the organs, and transport of the donated organs. OPOs also arrange for grief counseling and support to the donor families (Organdonor.gov 2020).

Healthcare Ethics Issues

Prior to the immunosuppressant era, the chief ethical dilemma regarding transplantation was housed under the Principle of Beneficence (maximizing benefits, minimizing harms) because of the dismal success rates of solid organ transplantation. However, once the problem of organ rejection was solved, the ethical dilemma regarding organ transplantation shifted to a Justice problem as demand exceeded supply, and organs became a scarce resource in which utilitarian frameworks for rationing a scarce resource (see Chap. 6) are ethically justified, which is an autonomy-limiting principle (see Chap. 6). In the transplant context, the healthcare provider is a steward of a scarce resource, which means that potential transplant recipients are put on the waiting list for a solid organ based on who would most benefit from the organ from a clinical standpoint, and who is most likely to adhere to the strict post-transplant regimen from a psychosocial standpoint. Transplant candidacy considers several behavioral and psychosocial factors, including comorbidities, smoking, alcohol abuse and other addictions. In the United States, access to post-transplant anti-rejection medications is also a critical factor; patients with no access to the medication post-transplant are not listed, which is not the case in countries with universal healthcare (see Chap. 4). Although the ethical framework for scarce resource allocation is discussed in Chap. 6, this section discusses transplant-specific ethical considerations, such as ethical strategies to increase the donor pool; organ preservation for potential donors who have progressed to brain death; donation after cardiac death; ethically challenging definitions of death and the "dead donor rule"; ethical dilemmas in assessing transplant candidacy; and the role of trauma medicine in identifying ideal donors who suffered catastrophic accidents. However, as *21 Grams* focuses on, the most basic ethical problem in heart transplant recipients—which can only come from a deceased organ donor—is *the nature of the heart transplant procedure itself*. The recipient is not only biologically connected to the donor, but spiritually connected to the donor, and yet has no social connection to the donor or the family.

The Donor Heart and the "Dead Donor Rule"

Bioethicist Albert Jonsen notes (Jonsen 2012):

> Heart transplantation not only startled the world, it raised the same ethical questions as kidney transplant, only in a louder register. Removal of a kidney from a living donor was partially justified by the fact that kidneys are paired organs; a person can live with only one. But removal of a viable heart definitely ends the life of its source. So the debate over the definition of death was revived: is it possible to assert that a person whose brain has ceased functioning is dead? [And] under what clinical conditions could a heart be removed from a person?

In 1966, the first major conference took place in London specifically to consider ethical issues in organ transplantation. Issues discussed ranged from living donors, obtaining consent on behalf of dead donors, donor compensation, criteria for evidence of death, and fair allocation of scarce human organs available for transplant (Jonsen 2012; Barker and Markmann 2013).

Uniform Death Criteria

The 1968 Harvard Brain Death criteria (see Chap. 3) stated that "obsolete criteria" for the definition of death was creating new problems for transplantation (Jonsen 2012). The new criteria included neurological signs of death: unresponsiveness, lack of movement or breathing, no reflexes, and evidence of irreversible coma. In 1982, the President's Commission on the Study of Ethics introduced a "uniform definition of death" that created two categories: cardiopulmonary death or "irreversible cessation of circulatory and respiratory function" and brain death or "irreversible cessation of all functions of the entire brain, including the brain stem" (Jonsen 2012).

This uniform definition was adopted in all U.S. states, which led to an increase in organs from a donor pool of individuals who had progressed to brain death, or death by neurological criteria. The terminology "brain death" would become highly problematic for donor families, as it gave the impression that death had not occurred in other parts of the body. The most common clinical ethics consultation involves flawed communication to families about "brain death", and poor understanding by families about what that means, leading to catastrophic cases such as the Jahi McMath case (Schmidt 2018). Some healthcare providers have asked for "consent" to remove life support from brain dead patients instead of just pronouncing the death and removing life support because of the obvious reason: the patient has died and does not require "treatment" or "life support".

Dead Donor Rule

Heart transplantation, unlike kidneys or livers, can only come from deceased donors, and are governed by what is known as the "dead donor rule"—what is generally used for all vital organs. Miller and Sade explain it this way (Miller and Sade 2014):

> Donors must be determined to be dead according to established legal criteria and medical standards prior to procurement of vital organs for transplantation. Most donors are determined to be dead on neurological criteria: the irreversible cessation of all functions of the entire brain…. In response to a shortage of "brain dead" donors, vital organs increasingly have been procured from donors declared dead according to circulatory criteria following withdrawal of life-sustaining treatment (LST). Protocols for donation after circulatory death (DCD) typically involve patients on mechanical ventilation with severe neurological damage short of "brain death,"…Hearts rarely have been procured under DCD protocols, although hearts of infants have been transplanted successfully in some controversial cases.

By 2003, when *21 Grams* was released, the shortage of organs led to new thinking about the "dead donor rule" as several potential donors who were "as good as dead"— meaning they had only minimal brain stem function, were declared to be futility cases, and had no chance of reasonable recovery, could be potential donors. Similarly, some patients whose hearts had stopped before brain activity ceased could also be potential donors but for the "dead donor rule". While the *Uniform Death Act* defined cardiac death as: "irreversible cessation of circulatory and respiratory functions," many patients could still be coded and brought back to life, and hence, practitioners were hesitant with the term "irreversible". Thus, "to satisfy the criterion of 'irreversibility' in its ordinary meaning, it must be impossible to restore circulation with available means of medical intervention" (Miller and Sade 2014). Patients who have a code status of Do Not Resuscitate (DNR) who die by cardiac criteria could be ethically justified to have their organs procured, but this remains controversial for practitioners who do not see such deaths as "irreversible" if cardiopulmonary resuscitation (CPR) were performed. Irreversibility is defined as the persistent cessation of function during a 5-minute period of observation meeting the definition of cardiac death.

In *21 Grams*, Cristina's husband dies by neurological criteria. Neurologist Eelco Wijdicks, in his book, Neurocinema (Wijdicks 2015) notes this:

> Two physicians approach Cristina and tell her, "Your husband suffered multiple skull fractures. We had to remove blood clots from around his brain….We are concerned that he's showing low brain activity." The next scene in the hospital shows her discussing organ donation with an organ donor coordinator, and the discussion is compassionate and real.

All hospitals have specific policies that outline the coordination of when to call the OPO, which is the agency empowered to get consent to donation from families (see further).

Donation After Cardiac Death (DCD)

DCD donor death occurs when respiration and circulation have ceased and cardiopulmonary function will not resume spontaneously. Patients or their surrogates who have elected to withdraw life support to be able to donate organs can begin the process after the patient is declared dead by cardio-pulmonary criteria. In this case, a deceased donor has been declared dead on the basis of traditional cardio-pulmonary criteria (permanent cessation of circulatory and respiratory function), rather than on neurological "brain death" criteria (permanent cessation of whole brain function). DCD candidates include those with a non-recoverable and irreversible neurological injury resulting in ventilator dependency but not fulfilling brain death criteria; end stage musculoskeletal disease, pulmonary disease, or those with severe spinal cord injury. However, in many cases withdrawal of life support may not lead to death in a short enough time frame that permits organ retrieval.

Typically, when potential DCD patients are being evaluated for withdrawal of life support for any reason, the organ procurement agency is contacted, which then begins to determine suitability for organ donation before addressing donation with the patient's surrogate decision-maker.

Consenting Traumatized Adults: Talking to Donor Families

21 Grams depicts a realistic consent to donation discussion (Wijdicks 2015) after a potential donor has progressed to brain death:

> Transplant Coordinator: As you know, the doctors did everything they could to save your husband's life, but he has shown no brain activity. We're here to help you with some of the final decisions that need to be made. We have a patient who is gravely ill. I am here to give you some information on organ donation. Are you willing for your husband to donate his heart?
>
> Cristina's Sister: Can we discuss this another time?

Transplant Coordinator: I'm afraid not. I can give you time to discuss it, but this is a decision that needs to be made soon.

Let's face it: For patients with no advanced directives surrounding organ donation, there is just no good time to get consent; it almost always involves consenting a surrogate at the worst time, who is likely in shock (if the death is sudden and through trauma), and/or grieving, even if the death and dying process has taken much longer. Capacity to consent to organ donation is a real concern, but there are few workarounds, given the level of trauma associated with donor circumstances.

Elements of consent to donation must include an overview of the process with ample opportunity for the family to ask questions and to demonstrate understanding and appreciation. The discussion may also include the need for additional testing to determine suitability for donation; consent to procedures or drug administration for the purposes of organ preservation or the retrieval procedure. In the case of DCD (see above), the consent discussion also must convey possibility that donation will not take place if cardiopulmonary death does not occur within one hour following removal of life support, and there must be consent to DNR.

The donor's surrogate next signs a consent form for Authorization for Removal of Anatomical Gifts for organ and tissue donation, and a copy of the signed consent form is added to the donor's medical record. Typically, the donor family is offered emotional and spiritual support through the hospital, which may include pastoral care, social work, and sometimes a clinical ethicist.

Organ retrieval and transplantation is a choreographed process in which two teams are essentially "blinded" from one another. In many cases, retrieval is done at a different institution than the transplant procedure, and the organ is transported by plane or helicopter. But if the surgical teams are in the same institution, one surgical team is devoted to the donor and family members, who perform the retrieval surgery in a separate operating room, and work with the family for seamless post-death processes. Meanwhile, another surgical team is in another operating room and is devoted to the recipient family. Essentially, the donor family is in hell, while the recipient family is in heaven—feeling as though a miracle was granted. For these reasons, the transplant ecosystem ensures that donor families do not have contact with the recipients or their families as such contact can lead to unintended social outcomes.

If donation involves withdrawal of life support, it will usually take place in the operating room only. The family will be given the option of saying goodbye to their loved one in the intensive care unit (ICU) or sometimes can accompany the donor to the operating room, but they have to leave once the death is pronounced. In the United States, donor families are not responsible for any costs associated with the donation.

Altruistic Organ Donation

By 1984, when the *National Organ Transplant Act* passed (see previous section), roughly 200,000 persons had been declared "brain dead" using the uniform definition of death, greatly expanding available hearts. The legislation also established that only compensation for medical or out-of-pocket costs would be permitted so that "organ selling" would not become attractive to vulnerable populations. In theory, organ selling is banned in most democratic countries, but in practice, organ selling has become an occult practice in many countries amongst desperate people. Theoretically, altruistic organ donation is what OPOs encourage with their "gift of life" rhetoric. The idea that organs are, in fact, donated, is a principle concept in transplantation: "organs are donated in a spirit of altruism and volunteerism and constitute a national resource to be used for the common good" (Jonsen 2012). Altruistic donation is designed to "prevent commercialization of organs and exploitation of the healthy poor and to promote equality in organ distribution." (Jonsen 2012). But altruism also limits the supply of organs available; in the United States, organ donation is thus an "opt in" arrangement. Potential donors can make preferences about donation known through advance directives or signing their consent to donation on their driver's license, but most organs are procured on-site by OPO representatives who seek consent in real time from the families, which is what we see in *21 Grams* At these junctures, informed consent is often coercive because it is being sought from an emotionally distraught family member whose loved one may have suffered a sudden death from trauma. There have even been cases where families refuse to consent only to discover that the donor has, in fact, opted in through his/her driver's license, and OPOs have, at times, sued refusing surrogates and won. Research demonstrates that opt-in policies may not be ideal for increasing the donor pool. However, according to Calne (2006) in a retrospective on transplantation in the *Lancet*:

> Whenever something is wanted but in short supply, there will be pressure to obtain the commodity by payment…There has been much discussion about the payment of donors for organs, whether the donor or his or her family should be paid directly or through a government agency, or whether payment should be forbidden, in which case there is a danger of illicit payment or other means of coercion. The actual practice of organ transplantation is viewed differently according to whether the transplant team are paid individually for each operation, as is the case for both donor and recipient teams in some centres in the USA…. The donor teams have to be available at all times, often for long journeys, frequently at night, to a centre where their presence is not exactly welcomed.

"Opt-out" policies, which several democratic countries have adopted, is the only ethical solution for increasing donation. These are systems in which everyone is presumed a donor unless s/he specifically "opts out". In Austria, for example, the law ensures that organ donation is the default option at the time of death, but patients can explicitly "opt out". In opt-out countries, more than 90% of people register to donate their organs, but in countries with explicit opt-in policies, only 15% of citizens are registered organ donors (Davidai et al. 2012). To sweeten the deal to donors, opt-in countries, such as Israel, give priority for being listed as recipient to registered donors (Zaltzman 2018). Ultimately, countries with an opt-out policy help to instill that everyone who stands to benefit should also be willing to donate in the spirit of maximum benefit and equal access.

On the Receiving End: Ethical Issues with Transplant Candidacy

The decision to donate an organ is governed by the Principle of Autonomy: it is solely based on a patient's or surrogate's preference to opt-in, or to expressly opt-out. On the recipient side, because organs for transplant are a scarce resource, transplant candidacy is based on the utilitarian principle of "providing the most benefits to the most people". Several factors are taken into consideration other than solely clinical appropriateness for candidacy. A patient's likelihood of long-term benefit is based, in the United States, on access to the medication regimen that prevents organ rejection, as well as a patient's willingness to change lifestyle factors, such as smoking or substance and alcohol abuse, which are contraindicated. In *21 Grams*, we see the transplant candidate, Paul Rivers, smoking in his bathroom, which arouses ire. Naturally, he lies about his smoking, but continues to do so after the transplant prior to suffering from organ rejection. He is clearly not compliant or adhering to the recommended lifestyle regimen that would optimize the transplant's success.

Adherence to medication is difficult, but some organs are more successful than others without immunosuppression therapy. According to Calne (2006).

> Besides their expense, conventional drug regimens can cause great hardship to patients and non-compliance is common. Some patients with liver transplants stopped taking their drugs and performed a clinical experiment demonstrating immunological tolerance, surviving many years with good liver function despite the absence of any maintenance immunosuppression. Other patients were not so lucky, and this weaning of maintenance immunosuppression is far more likely to be successful with liver than with kidney transplants, which is consistent with experimental demonstration of liver tolerance without any drugs after orthotopic liver transplantation in pigs and rodents.

For heart transplantation, there are certain medical requirements to being listed as a recipient. First, a patient's heart disease must be in its final stage, which means that without a heart, certain death is imminent. Next, a patient's life expectancy after the transplantation must be several months to several years so that if s/he suffers from other severe illnesses that reduce life expectancy, s/he does not meet the criteria for being listed as a recipient. Mental health is also a factor, as patients who have poor coping strategies or underlying mental health problems that are difficult to manage or treat are not suitable candidates. A "maximum benefit" rule, in keeping with utilitarian principles is the framework for transplant candidacy. This takes both medical need and probability of success of the transplant into dual consideration. Some bioethicists argue that refusing transplants to patients with addictions, or who are obese is not just, and in fact, implies a character judgement and bias, as we do not ration other expensive medical procedures based on such criteria (Ho 2008). For example, patients who need orthopedic surgery for a broken leg are not denied surgery based on risky behaviors, such as skiing (Ho 2008); nor do we deny insulin to patients with type 2 diabetes who are obese due to their diet and sedentariness. However, in the transplant context, rationing is based on ethically sound stewardship of a very scarce resource; orthopedic surgery is not a scarce resource, but financial rationing occurs due to the U.S. healthcare system anyway (see further).

In *21 Grams*, Paul Rivers meets the medical need category, and because he is young, he should have several years of life as a result of the transplant. But because he smokes, the success of the transplant is clearly threatened. In fact, the bathroom smoking scene was inspired by the real-life experiences of Charlotte Gainsbourg's father (see earlier), who when hospitalized for cardiovascular problems, hid his cigarettes in a pill bottle as the character, Paul Rivers does. Ordinarily, people with drug addiction, smokers, alcoholics or obese patients are typically not good candidates for donor hearts because these behaviors shorten the efficacy of the donor heart. But the United States imposes another rationing category that countries with universal healthcare (see Chap. 4) don't: insurance coverage for immunosuppressant therapies post-transplant. Patients who have no financial access to the post-transplant medications, and no access to the necessary medical follow up, and no insurance or payor to cover their transplant surgeries, are not listed as potential recipients. Additionally, multiple listing is legal at different transplant centers, which favors the wealthy who can travel around to different transplant centers, while each transplant evaluation can cost tens of thousands of dollars. Paul Rivers, a math professor, who is concomitantly seeking out fertility treatment with his wife for an in vitro fertilization procedure can afford his transplant. His continued smoking, however, and lack of empathy for his estranged wife, leaves the viewer wondering if he is "deserving" of the donor heart. We wonder, whether someone with fewer financial resources but with no addictions who has equal medical need, may have been a better candidate.

Thus, Americans are subjected to financial rationing in the transplant context, which harkens back to the early days of dialysis, when the first dialysis committee rationed based on "social worth" criteria, reserving access to dialysis to essentially tax-paying white males with families, calculating how many dependents, for example, would be left if the patient were to die (Jonsen 2007). The first dialysis committee was in Seattle, and was exposed in a *Life* magazine article November 9, 1962, (Alexander 1962) with the headline: "They Decide Who Lives, and Who Dies?" Alexander was describing The Admissions and Policy Committee of the Seattle Artificial Kidney Center, which comprised seven members of the community: minister, lawyer, businessman, homemaker, labor leader and two physicians. Each month they reviewed about 12 charts of patients who had end-stage renal disease and selected roughly two out of a dozen for the scarce resource of dialysis (Jonsen 2007). The history of dialysis and transplantation was the subject of the early 1980s book, Courage to Fail, by sociologist Renee Fox (Fox and Swazey 1978). In fact, the birth of healthcare ethics can, in part, be traced to the allocation of dialysis, as such committees sprung up in all hospitals that had limited numbers of dialysis machines in the early days. Similarly, allocation of solid organs by transplant committees make similar types of decisions, and because of the for-profit U.S. healthcare system (see Chap. 4), access to organs still takes "social worth criteria" such as financial means into consideration. Notes bioethicist Albert Jonsen (Jonsen 2007):

> The era of replacing human organs and their functions began with chronic dialysis and renal transplantation in the 1960s. These significant medical advances brought unprecedented problems. Among these, the selection of patients for a scarce resource was most troubling.

In Seattle, where dialysis originated, a "God Committee" selected which patients would live and die. The debates over such a committee stimulated the origins of bioethics.

Renal exceptionalism: Access to dialysis

When the American public learned about the "God Committee" in Seattle through the *Life* expose, the public found it hard to accept that rationing of a life-saving resource was going on in the richest country in the world. Hemodialysis was invented by Belding Scribner in 1960, who "devised Teflon arteriovenous conduits for long-term vascular access" (Kaporian and Sherman 1997; Barker and Markmann 2013). In 1966, the technology was improved, and by the late 1960s, only a few hospitals offered chronic dialysis to a limited number of patients. In 1972, directly consequent to the *Life* expose, Congress approved Medicare funding for chronic dialysis, and the technology became widely available for patients with end-stage renal disease. Thus, the history of transplantation and living kidney donation is tied to the exception of federal funding for dialysis, which is a bridge to kidney transplantation. No such exclusive federal funding exists for other types of solid organ transplantion. However, transplant candidates for solid organs from living donors are also instructed to solicit organs through social media, looking for matches, as finding their own donors permits earlier transplants. Patients in need of transplants typically advertise their need in the hopes of securing a willing, altruistic donor. What is concerning is the "mission creep" into organ selling. Organ selling is not discussed here, but could be discussed with the film, *Dirty Pretty Things*, mentioned at the start of this chapter.

The Donor-Recipient Relationship

The most important transplant ethics issue in *21 Grams* is the violation of confidentiality between the donor and recipient. The following scene between the doctor and Paul reveals the most common question recipients ask:

> Paul: I have a question for you: Whose heart do I have?
> Doctor: I can't tell you. It's hospital regulations, just like the donor's family doesn't know your name either.
> Paul: But you know?
> Doctor: Yes. This is your heart now, that's all that should matter to you.
> Paul: I want to know who saved my life.

But Paul won't accept this "not knowing" because any recipient of a donor heart naturally wants to know about the donor as "the emotional and physical bond that organ transplantation brings about establishes a kinship identification based on a common imaginary of blood ties…Yet, it is Paul's desire to know whose heart is inside him now that drives the narrative and the chaotic form" (McCormack 2016).

Notes Wijdicks (2015):

> The physician appropriately says he cannot tell, but one of the nurses suggests [Paul] write an anonymous letter through the [OPO]. This is correctly portrayed, as transplant recipients can express gratitude through anonymous correspondence, but confidentiality must be maintained. It is a commonly asked question, and [OPOs] have clear policies. However, Paul is not so happy with this policy and hires a private investigator, who provides him with medical details [about the donor] and the actual address of the donor's wife, Cristina.

For any healthcare provider who works in transplantation, the relationship between Paul and Cristina in *21 Grams* is the transplant ethics nightmare scenario, and the raison d'etre of why confidentiality is strict. McCormack explains (McCormack 2016):

> While initial transplants were conducted under the eyes of an avid press with surgeons keen to speak of their successes, often with recipients at their sides, the contemporary practices for transplantation demand and enforce a strict policy of anonymity. Transplant teams actively discourage face-to-face contact between donor family and organ recipient, stressing that such exchanges generally lead to unnecessary complications. Thus, very little information is given to either the recipient or the donor family regarding the identity of the other…Today, recipients are encouraged to send letters to the donor kin (where these letters are always vetted by the transplant coordinator to ensure [no personal identifiers]…Transplant teams labour to keep these two groups apart to the extent that such relationships are assessed and understood as pathological…. The fear that relationships will become intrusive, obsessive and thus inappropriate emerges from the idea that people who are grieving are largely irrational in their emotional responses and needs, and that those who have undergone the traumatic experiences of transplantation may feel an overwhelming sense of guilt that manifests itself as a so-called unhealthy psychological identification with the organ donor and the family.

And this is exactly what happens in *21 Grams*. Paul is driven to the donor family like a moth to a flame, and can't live a separate life with the heart as he did in his pre-transplant existence; the donor heart is experienced as a disruption to the self, rather than a continuation of his life (McCormack 2016).

Social scientists have noted that for many recipients, the "gift of life" is difficult to accept because the recipient does not know how to reciprocate it (McCormack 2016). As Paul becomes sexually involved with Cristina, she coerces him into attempting a revenge killing of the man who caused the death of her husband, and even says: "You owe it to Michael [the donor]. You've got his heart. You're in his house fucking his wife and sitting in his chair. We have to kill him."

Other Ethics Subplots

When teaching *21 Grams*, it may also be worth exploring the many ethics subplots in the film, which I will only briefly summarize here.

First, there is a strong fertility subplot that knits together the themes of life and death. Pre-transplant, Paul, who is very ill, has agreed to participate in fertility treatment and IVF to help his estranged wife get pregnant so she can have his child if he dies, and presumably, he can live on through his progeny. But during this process, Paul discovers that his estranged wife is having trouble getting pregnant because of complications from an abortion she had from a pregnancy she withheld from him when they had separated prior to his illness. We learn that after they separated, his estranged wife returned to be his caregiver. Paul thus abandons the IVF process—creating life outside the body—with his estranged wife. In his post-transplant life, Paul's sexual relationship with Cristina results in her becoming pregnant, which helps to replace her lost family, but Paul dies from transplant rejection, yet still procreating and living through progeny.

Other ethics subplots revolve around Jack, the driver responsible for the accident that yields the donor heart. As a "Born Again Christian", he is trying to redeem for past lives and past sins, and suffers profound spiritual and existential suffering. The viewer is forced to examine the profound moral injury of being a perpetrator of a deadly crash. Jack takes life and gives life at the same time.

Finally, Cristina's own journey includes her admission that she had once attempted suicide, and a presumption that she was a recovered addict when she was married. In her grief, the viewer confronts the grittiness of poor coping skills, and relapse into self-harm and addiction.

Conclusions

21 Grams is a messy transplant ethics film that forces the viewer to confront the underbelly of transplant ethics rather than the "gift of life" rhetoric used by OPOs. The film is about the intimate ties between a heart donor's family and the heart recipient's family. The film examines what transplant programs go to great lengths to prevent: the "pathologized relationship between organ recipients and the families of cadaveric donors" (McCormack 2016) and "brings to the fore both the urgent desire of the organ recipient to be close to the donor family and the purported pathological ramifications of such encounters" (McCormack 2016).

Iñárritu dramatizes the heart donor-recipient relationship to make clear that while two "unrelated lives and histories are brought together in one body" (McCormack 2016), the heart of the ethical issues in transplantation have very little to do with the medical procedure of organ retrieval and transplantation. Transplant ethics remains the Frankenstein story in which the carefully constructed ecosystem of transplant centers and OPOs try to control these out of body experiences in which someone's death allows another to live.

Theatrical Poster

21 Grams (2003)

Directed by: Alejandro González Iñárritu
Produced by: Alejandro González Iñárritu and Robert Salerno
Written by: Guillermo Arriaga
Story by: Alejandro González Iñárritu and Guillermo Arriaga
Starring: Sean Penn, Naomi Watts, Benicio Del Toro, Charlotte Gainsbourg
Music by: Gustavo Santaolalla
Production Company: That Is That Productions
Distributed by: Focus Features
Release Date: November 21, 2003
Running time: 124 min

References

5 Ways '21 Grams' Amazes 15 Years Later. (2018). Alejandro González Iñárritu's masterpiece is still a cinematic heavyweight. *Focus Features*, November 21, 2018. https://www.focusfeatures. com/article/fifteenth-anniversary_-irritu_21-grams.

Alexander, S. (1962). They decide who lives, who dies. *Life*, November 9, 1962.

Barker, C., & Markmann, F. (2013). Historical overview of transplantation. *Cold Spring Harbor Perspectives Medicine, 3*(4), a014977. https://doi.org/10.1101/cshperspect.a014977.

Burns, J. F. (2002). Actor Follows His Own Script on Iraq and War. *New York Times*, December 16, 2002. https://www.nytimes.com/2002/12/16/world/threats-and-responses-hollywood-actor-follows-his-own-script-on-iraq-and-war.html.

Calne, R. (2006). Essay: History of transplantation. *The Lancet, 368*, s51–s52. https://doi.org/10.1016/S0140-6736(06)69928-5. https://www.thelancet.com/journals/lancet/article/PIIS0140-6736(06)69928-5/fulltext, December 01, 2006.

Curiel, J. (2003). Director of '21 Grams' says his film was shaped by 9/11. *San Francisco Chronicle*.

Davidai, S., Gilovich, T., & Ross, L. (2012). The meaning of default options for potential organ donors. In *Proceedings of the National Academy of Sciences* (pp. 15201–15205). https://stanford. app.box.com/s/yohfziywajw3nmwxo7d3ammndihibe7g.

Ethics of Organ Transplantation. (2004). University of Minnesota Center for Bioethics, March 27, 2007. http://www.ahc.umn.edu/img/assets/26104/Organ_Transplantation.pdf.

Fox, R., & Swazey, J. (1978). *The courage to fail: The social view of organ transplants and dialysis*. University of Chicago Press.

Ho, D. (2008). When good organs go to bad people. *Bioethics, 22*(2), 77–83. https://doi.org/10.1111/j.1467-8519.2007.00606.x. https://www.ncbi.nlm.nih.gov/pubmed/18251767. https://www.ncbi.nlm.nih.gov/pmc/articles/PMC4100619/.

Hunter, S. (2003). '21 Grams': The chaotic order of the Universe. *Washington Post*, November 26, 2003. https://www.washingtonpost.com/wp-dyn/content/article/2003/11/26/AR2005033116464.html.

IMDB. (2020). *21 Grams*. https://www.imdb.com/title/tt0315733/.

Jonsen, A. (2007). The God squad and the origins of transplantation Ethics and policy. *Journal of Law, Medicine and Ethics, 35*(2), 238–240. https://journals.sagepub.com/doi/10.1111/j.1748-720X.2007.00131.x.

Jonsen, A. (2012). The ethics of organ transplantation: A brief history. *AMA Journal of Ethics, Virtual Mentor, 14*(3), 264–268. https://doi.org/10.1001. https://journalofethics.ama-assn.org/article/ethics-organ-transplantation-brief-history/2012-03.

Kaporian, T., & Sherman, R. A. (1997). A brief history of vascular access for hemodialysis: An unfinished story. *Seminars in Nephrology, 17,* 239–245.

Kruszelnicki, K. (2004). 21 Grams. *Australian Broadcasting Company (ABC).* https://www.abc.net.au/science/articles/2004/05/13/1105956.htm.

McCormack, D. (2016). Transplant temporalities and deadly reproductive futurity in Alehandro Gonzalez Inarritu's 21 Grams. *European Journal of Cultural Studies, 19*(1), 51–68.

McGowan, T. (2011). *Out of time: Desire in atemporal cinema.* Minneapolis, London: University of Minnesota Press. www.jstor.org/stable/10.5749/j.ctttprf.

Miller, F., & Sade, R. (2014). Consequences of the dead donor rule. *Annals of Thoracic Surgery.* https://doi.org/10.1016/j.athoracsur.2014.01.003.

Mitchell, E. (2003). Review: hearts incapacitated, souls wasting away. *New York Times,* October 18, 2003. https://www.nytimes.com/2003/10/18/movies/film-festival-review-hearts-incapacitated-souls-wasting-away.html.

Monty Python website. (2020). *The Meaning of Life Sketch.* http://montypython.50webs.com/scripts/Meaning_of_Life/9.htm.

https://www.sfgate.com/entertainment/article/Director-of-21-Grams-says-his-film-was-shaped-2548160.php, November 21, 2003.

Organ Procurement and Transplantation Network. (2020). *Tranplantation history.* https://optn.transplant.hrsa.gov/learn/about-transplantation/history/.

Organdonor.gov. (2020). U.S. Government Information on organ donation and transplantation.

Penn mightier than the sword? (2002). *Washington Post,* October 19 2002. https://www.washingtonpost.com/archive/lifestyle/2002/10/19/names-38/8dee1ca2-f553-44ce-836e-b959bb6124ab/.

Rediff. (2016). Why you need to know this man. Rediff.com, February 22, 2016. https://www.rediff.com/movies/report/why-you-need-to-know-this-man/20160222.htm.

Richter, R. (2008). What have we done? Forty years of heart tranplants. *Stanford Medicine.* http://sm.stanford.edu/archive/stanmed/2008fall/article10.html.

Rosenthal, M. S. (2013). The end of life experiences of 9/11 civilians: Death and dying in the world trade center. *Omega Journal of Death and Dying, 67*(4), 323–361.

Schmidt, S. (2018). Jahi McMath, the Calif. girl in life-support controversy, is now dead. *Washington Post,* June 29, 2018. https://www.washingtonpost.com/news/morning-mix/wp/2018/06/29/jahi-mcmath-the-calif-girl-declared-brain-dead-4-years-ago-is-taken-off-life-support/.

Schwarcz, J. (2019). The real story behind '21 Grams'. *McGill Office for Science and Society.* https://www.mcgill.ca/oss/article/did-you-know-general-science/story423behind-21-grams.

Scottish Council on Human Bioethics (SCHB). (2020). http://www.schb.org.uk/films/details.php?films_id=197&prev_cat_id=23.

Soul has weight, Physician thinks. *New York Times,* March 11, 1907. https://www.nytimes.com/1907/03/11/archives/soul-has-weight-physician-thinks-dr-macdougall-of-haverhill-tells.html.

UNESCO. (2018). *Teaching Bioethics.* http://www.teachingbioethics.org/-Unit-2-.

United Network for Organ Sharing (UNOS). (2020). *Transplant History.* https://unos.org/transplant/history/.

United States Senate. Senate Intelligence Committee Study on CIA Detention and Interrogation. (2014). https://www.intelligence.senate.gov/sites/default/files/publications/CRPT-113srpt288.pdf.

Walsh, D. (2003). Sean Penn's Times statement: patriotism and the struggle against US militarism. *World Socialist,* June 13, 2003. https://www.wsws.org/en/articles/2003/06/penn-j13.html.

Wikipedia. (2020). 11/09/01, September 11, 2020. https://en.wikipedia.org/wiki/11%2709%2201_September_11.

Wijdicks, E. F. M. (2015). *Neurocinema: When film meets neurology.* New York: Taylor and Francis.

Wright, L. (2006). *The looming tower.* New York: Alfred A. Knopf.

Zaltzman, J. (2018). Ten years of Israel's organ transplant law: Is it on the right track? *Israel Journal of Health Policy Research, 7,* 45. https://doi.org/10.1186/s13584-018-0232-1.

Part III
Professionalism and Humanism

Chapter 8
A Classic for a Reason: *The Doctor* (1991)

Medical educators have been teaching with *The Doctor* (1991) since it was released in the George H. W. Bush (Bush 41) era. Based on the autobiography, A Taste of My Own Medicine, by Rosenbaum (1988), this classic healthcare ethics film is about a doctor's experience with illness, his new approach to patient care based on what he learns from the other side of the bed, and his realization that professionalism is not at all the same as humanism. *The Doctor* is about teaching the "hidden curriculum" to healthcare trainees—how to teach them to have compassion for patients in a learning environment that not only breeds contempt and burnout, but teaches them to "care less". An essential part of healthcare ethics is about organizational ethics and culture and teaching ethical behavior, professionalism and humanism to healthcare trainees. *The Doctor* is a professional ethics "delivery system" that models an array of medical mentor archetypes that will teach your students well, and allow you to dive into the hidden curriculum with ease. This chapter discusses the origins of the film, its social location, the history of medicine context, and the professional ethics issues raised in the film.

Origins of *The Doctor*

The film, *The Doctor* is an original screenplay based on A Taste of My Own Medicine: When the Doctor is the Patient, written by Ed Rosenbaum (1915–2009). Rosenbaum was 73 when his book was published, and opens it with this (Rosenbaum 1988):

> I have heard it said that to be a doctor, you must first be a patient. In my own case, I practiced medicine for fifty years before I became a patient. It wasn't until then that I learned that the physician and patient are not on the same track. The view is entirely different when you are standing at the side of the bed from when you are lying in it….
>
> On my seventieth birthday [in 1985], I reported to the hospital to have a biopsy. I had practiced medicine at this hospital in Portland, Oregon, for more than forty years. I had been the chief of medicine and president of the staff…On fifteen thousand previous visits

© Springer Nature Switzerland AG 2020
M. S. Rosenthal, *Healthcare Ethics on Film*,
https://doi.org/10.1007/978-3-030-48818-5_8

I had entered through a private door like a king…But today was different. I was one of the common herd….

Rosenbaum was diagnosed with laryngeal cancer (throat cancer) in 1985, and underwent radiation therapy and surgery to treat it. After months of unexplained hoarseness, his tumor was initially missed because it was situated in a spot that could only be seen with a fiberoptic pharyngoscope (see under History of Medicine), which had only been in use for four years at the time of his diagnosis (Rosenbaum 1988). Rosenbaum noted that he had initially gone to a doctor friend around the same age who wasn't familiar with the tool before going to a younger doctor who was up to date.

Rosenbaum was born in 1915 and accepted to medical school in 1934 when there was a quota system on Jewish medical students (see Chap. 9). He graduated medical school in 1938, interned at Jewish Hospital in St. Louis (1938–9), did his residency at Michael Reese Hospital in Chicago (1939–40), and his Fellowship at Mayo Clinic (1940–41; 1946–48). His training was interrupted by his service as a doctor in World War II. In 1948 Rosenbaum moved to Oregon and began teaching internal medicine and cardiology at the University of Oregon Medical School before becoming a rheumatologist. He practiced in Oregon until he retired in 1986. He initially began diary entries about his throat cancer experience, which was expanded into a book. Rosenbaum's book was part of a burgeoning genre known as "narrative medicine" also seen in Norman Cousins' Anatomy of Illness (1979). Cousins and Rosenbaum were the same age; Cousins died in 1990, a year before *The Doctor* was released. Similarly, the book, Heartsounds (Lear 1980), is based on the experiences of Manhattan urologist, Harold Alexander Lear, who deals with callous and egregious treatment from a complex medical system from the point-of-view of the patient. It was made into a film in 1984.

Rosenbaum's book tracks along the same timeline as the AIDS crisis (see Chap. 2), in a time frame where men did not easily discuss their experiences with illness. According to a 1991 interview (Rarick 1991):

An amateur author for years, Rosenbaum had long thought of writing a book on the silly things patients do and say. But he ended up writing one defending the patient's point of view… The book and movie have made Rosenbaum something of a spokesman for patient advocacy…Shielded by a comfortable and sometimes insular profession, doctors have lost sight of many everyday concerns, Rosenbaum believes. And because they are often treated by a colleague for free and without appointment, they have little experience receiving, rather than giving, medical care.

We lose sight of what the average person feels and thinks,' he says. 'We as doctors really don't know how patients feel until we are patients ourselves.

Rosenbaum says modern doctors must spend so much of their time learning about high-tech wizardry that they sometimes lose track of what he calls 'the art of medicine.'

In another interview in 1991, Rosenbaum added (Baum 1991):

The very first day of medical school we go into an anatomy lab and stick a knife into that cadaver…That's not a normal human act. And once we do that we're different. We develop a shell around ourselves. The isolation intensifies when a doctor starts his practice…We socialize among ourselves. We're in a better economic group. Most of us are political y conservative. We really don't realize what the average person is thinking.

A Taste of My Own Medicine was published March 1988, and the options were sold to the Disney corporation upon a galley proof (Rarick 1991). The book was then re-released as a paperback with the title The Doctor (Rosenbaum 1991), to coincide with the film. The film *The Doctor* premiered July 24, 1991, opened in theaters August 2, 1991, and became an instant hit, making $7.2 million in its first 10 days (Rarick 1991).

Production and Cast

The Doctor was based on an original screenplay by Robert Caswell, inspired by Rosenbaum's book, but its plot and characters were original, with the exception of a doctor undergoing treatment for throat cancer. Dr. MacKee (played by Will am Hurt) is nothing like Rosenbaum in any other way. One review noted (Baum 1991):

> Rosenbaum's 1988 book has none of the romantic details of the movie. There were nc problems with his wife and no emotional entanglement with a woman dying of a brair tumor. And Rosenbaum, who retired in 1988, was not a heart surgeon, as Hurt's character is.

> And a colleague, Dr. Joseph Matarazzo, said Rosenbaum needed no transformation from arrogance to understanding.

> "He didn't have a patient in Portland who won't tell you he was the most caring doctor they'd ever had," said Matarazzo, chairman of the university's medical psychology department.

Caswell kept a few names the same, but needed to invent a story as Rosenbaum's book was essentially his personal musings, memories of his patients and family members, and a recounting of his specific treatment experience. The book is essentially a stream of consciousness work that functions as a diary, and can be somewhat disjointed. In the film, some doctor names are the same, but they are written as entirely different characters. Meanwhile, the patient "June" was only briefly mentioned in the book, and was developed into a very different relationship in the film. While Rosenbaum's book had the "germ" of a screenplay, the screenplay for *The Doctor* was not really adapted, but created from whole cloth.

The Doctor has a unique perspective: it was directed by a woman, which was unusual for 1991 Hollywood. Randa Haines had previously directed the film version of *Children of a Lesser God* (1986) starring deaf actress Marlee Matlin and William Hurt, which was a critically acclaimed Reagan era film that came out a year before *Dirty Dancing* (1987—see Chap. 5), and which won Matlin the Academy Award for Best Actress. (Matlin and Hurt began an actual romance after the film.) Haines paired with William Hurt again for *The Doctor*, which also won considerable critical praise, but which eventually became a "cult classic" for medical educators and mentors. Ironically, Hurt moved to Oregon—the same as Rosenbaum—years after starring in *The Doctor*.

By 1991, Hurt had a significant career, having won critical praise for several major films in the 1980s, including his debut as a scientist in the research ethics film, *Altered States* (1980); *Body Heat* (1981); *The Big Chill* (1983); *Kiss of the Spider Woman* (1985) for which he won the Academy Award for Best Actor; *Broadcast News* (1987); and *The Accidental Tourist* (1988).

Rosenbaum had a cameo walk on part in *The Doctor* in which he played a passerby physician in the corridor, saying "Hi" to Dr. MacKee (William Hurt). However, Hurt needed to coach Rosenbaum, who was unable to do the task in several takes (Rarick 1991).

Hurt's role as the initially entitled and pompous Dr. Mackee, clearly understood that his character was being transformed by a mortality moment. Hurt stated (IMDB 2019):

> The simple fact of existence, of being aware that you are aware; this to me is the most astounding fact. And I think that it has something to do with dying. When you are a kid you are beset by fears and you think, 'I'll solve the fear by living forever and becoming a movie star.' But I am not going to live forever. And the more I know it, the more amazed I am by being here at all. I am so thrilled by the privilege of life, and yet at the same time I know that I have to let it go.

The film's release in the Summer of 1991 was timely, and was part of a genre of "white male illness" films reflecting cultural anxieties over AIDS, in which white males were dying in large numbers (see Chap. 2). The *New York Times* summarized it this way (Maslin 1991):

> "The Doctor," which greatly resembles the current "Regarding Henry" in its tale of a rich, cavalier professional who is made to face his own frailty, has a more realistic outlook and a more riveting figure in its central role. Mr. Hurt, making the most of his sleek good looks and stately bearing in the film's early scenes, presents Jack MacKee as a charming, cocksure doctor who specializes not only in difficult surgical procedures but also in careful, deceptively breezy intimidation….
>
> Colleagues and patients alike are thrown off-balance by Jack's mixture of false casualness, crisp professionalism and cutting wit. The film makes it clear that Jack, having spent his entire career perfecting this bedside manner, has lost track of his inner thoughts, and learned to concentrate solely on getting the job done. "Caring's all about time," he tells someone, when discussing whether work like this ought to engage the emotions. "When you've got 30 seconds before some guy bleeds out, I'd rather cut more and care less."…
>
> [The film] opens with a quick succession of scenes that encapsulate Jack's professional life. He teases, jokes and plays rock-and-roll oldies in the operating room, setting a mood that is quickly broken when Ms. Haines offers a shot of the patient's battered face. He tells a patient who is dismayed by a new scar that she looks like a magazine centerfold, staples and all…And he also finds after a party that he is coughing blood onto the shirt of his tuxedo.

The Other Doctors in *The Doctor*

Aside from the main character of Dr. Jack MacKee, there are five other archetypal doctor characters whose professionalism and behaviors you can dissect with your students. Jack's closest colleague, Dr. Murray Kaplan, is a cardiologist played by

Mandy Patinkin, (who later went on to play "Saul" in the Showtime series, *Home-land*). Dr. Kaplan is funny, charming and trying to get Jack to lie for him in a lawsuit where Dr. Kaplan clearly made a medical error. At the same time, Dr. Kaplan mocks Dr. Eli Blumfield (played by Adam Arkin), who is truly compassionate and humanistic, and continues to talk to patients under anesthesia in the event they can hear something. MacKee's ear, nose and throat (ENT) surgeon is a female physician who is extremely professional and competent, but cold: Dr. Leslie Abbott, played by Wendy Crewson. Abbott represents a physician who came of age, and did her training at the peak of the second wave feminism movement (see Chap. 5), and has perfected a professional shield that ultimately is impenetrable. At the top of her field in 1991, Dr. Abbott would have been in medical school in the early 1970s; completed internships, residencies and fellowships by the mid-1980s, and we meet her when she is clearly at least 5 years into her subspecialty practice as a senior ENT expert. Dr. Abbott is a familiar archetype in academic medicine, and we've all met her. (See further under Women in Medicine under Social Location.) Dr. Charles Reed, the radiation oncologist, was played by Zakes Mokae, a black South African actor who rose to fame amidst the Apartheid system. Mokae also starred in *A Dry White Season* (1989), about the brutal system of South African racism. He presented the type of radiation oncologist who was able to care for a physician-colleague, but also helped to fill in the diversity of academic medical centers. The *New York Times* describes the cast of "doctors" and professional diversity this way (Maslin 1991):

> The film's best scenes, those in the hospital, deftly capture the give-and-take of doctors striving to balance seriousness and dark humor. Mandy Patinkin is especially good as the friend who is Jack's partner…Adam Arkin is memorable as a younger colleague who shows his mettle when Jack is in distress…Zakes Mokae, as a radiologist, presides knowingly over the episodes that most powerfully alter Jack's character. And Ms. Crewson is especially interesting for the ambiguity of a medical manner that initially suggests cool competence, and later reveals to Jack everything he himself has done wrong.

Synopsis

The Doctor has a macro story and a micro story. At the macro level, *The Doctor* is about how organizational ethics and organizational culture produce the types of professional behaviors of healthcare providers in large academic institutions, which can range from cold and cut off to caring and compassionate. We clearly see students modelling mentors, and if the mentors do not demonstrate compassion and humanism, neither will the students. The micro story is about mid-career epiphanies of a cardiac surgeon at the top of his game, whose own professional behaviors are mirrored back to him, and he doesn't like what he sees at all. He also realizes his personal relationships have suffered because he has become so cut-off. He befriends June, a brain tumor patient who teaches him about the realities of being a patient in the lousy U.S. healthcare system (see Chap. 4), and who ultimately dies prematurely because the insurance company would not cover the correct diagnostic test that would have caught her tumor early, when it was treatable. Dr. Jack MacKee needs to learn how to be a humanistic doctor from other patients, and from his patienthood experience. As a result, Dr. MacKee changes his teaching methods and mentorship by requiring his residents to experience a mock inpatient stay, and experience the same tests and vulnerability that patients do.

Healthcare trainees love this film because the ending actually teaches them as well. Rosenbaum initially didn't like the ending, but came around when he saw how it resonated with audiences. The macro and micro stories of *The Doctor* show examples of unprofessional behavior in the clinical setting with patients; show examples of unprofessional behavior with colleagues; and provide examples of role models demonstrating ideal professionalism and humanism. Healthcare providers who view this film recognize themselves, their mentors, and the organizational ethics systems problems inherent in Big Medicine.

The Doctor's Social Location: The Reagan-Bush Transition

The historical time frame of Rosenbaum's book and the film *The Doctor* spans 1988–1991, which is the 1988 election year and the full transition from the Reagan presidency into the George H. W. Bush presidency—a generational change—also comprising the Gulf War, which began August 2, 1990 and ended February 28, 1991. Rosenbaum's book comes out March 1988, Reagan's last year in office. Rosenbaum, born in 1915, hints in interviews that he likely voted for Reagan (as a Conservative), and Reagan, born in 1911, is only four years older than Rosenbaum. In fact, when Rosenbaum's laryngeal tumor is found in 1985, President Reagan was having polyps removed from his colon, and had invoked the 25th Amendment while in surgery, handing the keys over to his Vice President, George H. W. Bush. In Chap. 2, I discuss many aspects of the transitional period between the Reagan and Bush presidencies with respect to health policies in the context of HIV/AIDS, while Chap. 4 provides a picture of the discriminatory healthcare system that prevailed during this period in which "healthism" and "wealthcare" are practiced. Rosenbaum's book, however, focuses on what these systems were doing to the healthcare providers themselves, and how healthcare delivery had devolved into a transactional relationship devoid of healthcare provider compassion and virtues. But it was also a reflection of sharp generational changes. Rosenbaum's experience is located in the Reagan era, when he is in his 70s, close to retirement and fully aware as a practitioner that many young men were sick at the same time, dying at the prime of their lives because of AIDS. He is also a beneficiary of the Reagan era, however, as the 1980s class wars (see Chap. 5) have produced a greater wealth gap, demonstrated in a number of films dubbed "Reaganite" (see Chap. 5), such as *Wall Street* (released December 1987), which preaches the "Greed is Good" mantra. When Rosenbaum's book is released, the Iran-Contra scandal is going on (see Chap. 5), Margaret Thatcher becomes the longest serving British Prime Minister; South African Apartheid is in full bloom;

and the 1988 Democratic Convention (July 18–21) in Atlanta features some coming attractions: a riveting speech from then 27-year-old John F. Kennedy Jr. (July 19, 1988) emulating aspirations of his generation—tail end Baby Boomers (1960–4); a terrible speech by early Baby Boomer Governor Bill Clinton, (July 18, 1988), and ultimately, the nomination of Governor Michael Dukakis (then 58) as the Democratic Nominee, along with the very senior Senator Lloyd Bentsen then 70), as the Vice Presidential nominee, who was polling well ahead of presumptive Republican nominee, the sitting Vice President. George H. W. Bush was painted as being disconnected, and "born with a silver spoon in his mouth" (Richards 1988). Gary Hart, who had been the frontrunner, was forced to drop out over publicized womanizing (see Chap. 5). The Republican Convention, held August 15–18 that year officially nominated the Vice President as the Republican nominee, along with Dan Quayle as the Vice Presidential running mate. It was in 1988 when the famous exchange between Bentsen and Quayle occurred in which Bentsen destroyed Quayle's credibility. Quayle defended his limited experience: "I have as much experience in the Congress as Jack Kennedy did when he sought the presidency" (Quayle 1988); and Bentsen responded: "Senator, I served with Jack Kennedy. I knew Jack Kennedy. Jack Kennedy was a friend of mine. Senator, you're no Jack Kennedy" (Bentsen 1988). It was followed by prolonged applause. The 1988 Bush campaign was advised by Roger Ailes, who would eventually run Fox News; Ailes, working with campaign manager Lee Atwater, was responsible for the infamous "Willy Horton" ad, noted for its blatant racism. The ad suggested that under Dukakis, African American criminals would be let out of prison on the weekends and rape white women (Time 2019; Baker 2018). But what ultimately ended Dukakis' chances was a debate misstep regarding a question about the death penalty, and a disastrous press photo, in which he posed for cameras in military gear on a tank, and looked completely out of place and "unpresidential" by 1988 standards (King 2013).

The 1988 election was essentially a third term for Reagan, and represented the first time the incumbent party won a third term in the White House. Bush won with a decisive victory of 426 electoral college votes, and the first Bush era would begin in 1989, and the Cold War would end with the fall of the Berlin Wall that November. By 1990, when Rosenbaum's book was being developed into a film, the Gulf War began, which was the first major U.S. war since the Vietnam War had ended, and which would cast a long shadow into the next century. Themes from *The Doctor* were mirrored through the Gulf War: Americans saw a hugely victorious military response to Saddam Hussein's invasion of Kuwait, in which 697, 000 troops were deployed to the Persian Gulf. American soldiers appeared well trained, invincible and full of bravado. At the same time, Gulf War Syndrome symptoms began to afflict roughly 250,000 Gulf War veterans. The Gulf War veteran experiences were also layered onto the AIDS crisis (see Chap. 2). By the time *The Doctor* was released in 1991, the film resonated on a number of levels—machismo and bravado could be touched by vulnerability and illness. At the same time, more Americans were living longer and were experiencing illness in a now complicated and unfriendly healthcare system. Rosenbaum's experiences with cancer at age 70—the same age Reagan was when he became President in 1981—was also tracking with Ronald

Reagan's covert dementia, which had become apparent in his second term, spanning 1984–88 (Pilkington 2011; Ellis 2016). By the time the Bush 41 era had begun, Reagan began to retire from public life, only disclosing his Alzheimer's disease by 1994, when it was already quite severe.

The Doctor hits theaters as Americans are thinking very carefully about their own health experiences and healthcare, also tracking with the popular non-fiction bestseller, Final Exit (Humphry 1991), about do-it-yourself suicide and which tracks with the media interest in Dr. Jack Kevorkian, whose first patient in 1990 was in the very early stages of Alzheimer's disease (see Chap. 3). Despite Bush's Gulf War victories, growing anxieties over health and the healthcare system, which *The Doctor* clearly presents, paves the way for the 1992 Presidential run of Bill Clinton, who promises to focus on healthcare reform (see Chap. 4).

Rosenbaum's 1988 book was about generational changes in attitudes about healthcare delivery, and a dying paternalism and patriarchy that he clearly demonstrates in his memoir. The film *The Doctor*, on the other hand, presents a cast of healthcare providers that looked like a 1991 hospital with Baby Boomer physicians running the show, who had been handed the torch of Rosenbaum's and Reagan's generation. By 1992, Bill Clinton would be the first Baby Boomer elected President of the United States, while his wife, not unlike Dr. Abbott, would not be received well. Together, the Clintons would attempt the first major overhaul in healthcare reform in a generation, which would not go well (see Chap. 4).

Women in Medicine

The Doctor also hits theaters as women entering the medical profession peaks, upending decades of sexist training in medicine in which women in the field were discriminated against and harassed. In 1950, a mere 6% of physicians were women; by 1990, the number increased to 17%; and by 2000, it increased to 22.8%. This provides a picture of the demographics in 1991, in which Leslie Abbott was clearly among very few women in surgical residencies at the time of her training. The first scene of Dr. Abbott is her being ogled for her looks by male doctors. This was not hyperbolic or unrealistic but very representative of the female experiences in medicine. Women in medicine were also ridiculed for working fewer hours when balancing the "double duties" of children and families.

In 1991, mentors in charge of training the next generation of physicians were trained in an era where palliative care and clinical ethics education was a rare offering in hospitals (Morrison and Meier 2015). Females who trained in the 1980s and 1990s had very different, often negative trainee experiences due to their gender (Kirk 1994; Zimmerman 1987).

Many women in medicine trained in an era where the book House of God by Samuel Shem (1978), was seen as reality rather than fiction, and where diversity was absent in the workforce. Yet, by the time many were in mid-career themselves, they face increased incidents of burn-out and increased emotional exhaustion (Durbye et al. 2005), characteristics that have been associated with unresolved moral distress issues (Rushton et al. 2013; Rosenthal and Clay 2013).

When teaching *The Doctor*, it's important to point out that these women—the Leslie Abbotts—then were responsible for teaching the next generation of women in medicine, now in mid-career. In 2003, the ratio of female to male students was 50.8%, but has been hovering around 47% since (Rosenthal and Clay 2017). There has even been the suggestion that there are "too many" women in medicine (Rarick 1991) because they tend to choose more family-friendly subspecialties (e.g. family medicine; internal medicine), leaving vacuums in more demanding specialties, such as surgery. Current female medical students viewing *The Doctor* may not be seeing appropriate ratios of females in medical leadership due to a lag in organizational ethics policies surrounding promotion of females to such positions as Chairs or Deans (Rosenthal and Clay 2017). Ultimately, understanding how Leslie Abbott developed her professional veneer is a key teachable moment.

History of Medicine Context

The main history of medicine topic to discuss when teaching *The Doctor* is the overall history of professionalism and humanism in medicine, which was around as long as medicine, but which resurged in the 1990s when physicians began to refuse to treat AIDS patients (see Chap. 2), culminating in specific courses that started to pop up on "Medical Professionalism".

The earliest known "rules" of practice was "Hammurabi's Code of Laws" (1795-1750 BC): "If a surgeon performs a major operation on a [nobleman], with a lancet and caused the death of this man, they shall cut off his hands." This is the very first set of rules for physicians, discovered in Iran in 1901, on a tablet. This is considered the first documented legal code for the medical profession. The code also describes a scaled fee schedule for surgical services and makes clear that there must be documentation of diseases and therapies, including a description of therapeutic benefits. There are various penalties for poor medical outcomes, using an "an eye for an eye" system of justice. The code fully explained patients' rights, and malpractice was recognized and punishable by law. The Code of Hammurabi can be considered the genesis of the current concepts of healthcare regulation (King 2019).

Hippocrates, of course, is credited with establishing medicine as a distinct profession. The Hippocratic work On the Physician recommends that physicians always be well-kempt, honest, calm, understanding, and serious. The most famous "professional oath" is known as the "Hippocratic Oath", generally taken by doctors swearing to ethically practice medicine. It is widely believed to have been written by Hippocrates, or by one of his students. The Oath reads:

> I will respect the hard-won scientific gains of those physicians in whose steps I walk, and gladly share such knowledge as is mine with those who are to follow.

I will apply, for the benefit of the sick, all measures [that] are required, avoiding those twin traps of overtreatment and therapeutic nihilism.

I will remember that there is art to medicine as well as science, and that warmth, sympathy, and understanding may outweigh the surgeon's knife or the chemist's drug.

I will not be ashamed to say "I know not," nor will I fail to call in my colleagues when the skills of another are needed for a patient's recovery.

I will respect the privacy of my patients, for their problems are not disclosed to me that the world may know…Above all, I must not play at God.

There are various versions of the Hippocratic Oath, but above are excerpts from the Modern Version, which does not actually contain the phrase "First, Do No Harm". Instead, more central to the Hippocratic Oath are virtues such as intellectual honesty; collegiality; compassion and humbleness.

There are other physicians that caution doctors about ego and greed. Sun Simiao, credited with authoring the earliest Chinese Encyclopedia for Clinical Practice, was born in the 6th Century. Sun Simiao completed two 30-volume works on medical practice, and was especially concerned about physicians being influenced by a desire for rewards, including financial rewards, fame, or favors granted them. Sun Simiao is considered among the first to depict the characteristics of a great physician, and cautioned physicians about behavior that was inappropriate to their profession: Sun Simiao (581-682 AD) states (Rosenthal 2010):

Whenever a great physician treats diseases, he has to be mentally calm and his disposition firm. He should give way to wishes and desires, but has to develop first a marked attitude of compassion…

A great physician should not pay attention to status, wealth or age…he should not desire anything and should ignore all consequences [of reward].

It is inappropriate to emphasize one's reputation to belittle the rest of the physicians, and to praise one's own virtue.

Another famous physician who worried about professionalism was Maimonides, a Jewish physician (1138–1204). Considered one of the most influential contributors to philosophy and Jewish law, Moses Maimonides was a prominent physician to both Jews and Muslims. He moved to Cairo, Egypt in 1166, where he began to practice and teach medicine. Maimonides composed 10 works in medicine that are considered authentic, including a Physician Oath, written in 12th century (Rosenthal 2010):

May neither avarice nor miserliness, nor thirst for glory or for a great reputation engage my mind… Grant me the strength, time and opportunity always to correct what I have acquired…, extend its domain…Today he can discover his errors of yesterday and tomorrow he can obtain a new light on what he thinks himself sure of today.

By the late nineteenth century, Sir William Osler (1849–1919) changed medical school teaching to demonstrate bedside manners to students. Osler created medical school training as we know it today, which involves bedside teaching and rounds with medical students. This was not based as much in teaching science, but in teaching professional behaviors at the bedside, or bedside manner. This process also enabled mentoring physicians to teach the "unintended" or "hidden" curriculum. Osler's introduction of bedside teaching and the establishment of medical residency coincided with the Flexner Report on Medical Education in the United States and Canada, released in 1910. This report was based on a careful survey of 155 medical schools of that time, and revolutionized medical education in the United States and Canada, creating national standards for medical education that included admission standards, and stricter curriculum standards. The report called for adopting a research medical university model. However, this led to an erosion in teaching "humanism" as evidenced-based medicine was established (Duffy 2011).

By the 1980s, the physician who was responsible for making professionalism and humanism in medicine a major topic was Edmund Pellegrino, who transitioned from a nephrologist and Chair of Medicine (University of Kentucky) to the field of health-care ethics. Pellegrino's famous essay, "The Good Physician" (Pellegrino 2002), as well as earlier works (Pellegrino and Relman 1999), discussed character traits of the "Good Physician". Such traits included fidelity to trust; benevolent behaviors; intellectual honesty; compassion and truthfulness. By 1991, *The Doctor* explodes into medical school teaching as an entire subfield surrounding "professionalism in medicine" emerges, which becomes a critical core area for medical curriculum guidelines. *The Doctor*, in fact, contributes to the renewed interest in virtue medicine. Pellegrino describes virtue as a necessary "complement" to the "moral life." How we carry out our duties and obligations are all shaped by the character of the "moral agent". But how, Pellegrino asks, does virtue survive in a culture where "self-interest, not altruism, is the rule of success." He states: "Medicine should be a moral enterprise…we need leadership that eschews self-interest and truly advances the welfare of patients" (Pellegrino 2002).

A year before Rosenbaum's book was published, a PBS NOVA documentary crew followed seven medical students at Harvard who had started as first year medical students 1987, and followed them through their training, re-interviewing them 20 years later. *The Doctor Diaries* (PBS 2009) is a cinema verite (reality) documentary that tracked the subjects through four years of training and shows us the results when they are mid-career in 2007. All of these students would have been in training when *The Doctor* was released. In one particular scene, we see a female medical student learning about death at the bedside—not from a textbook, but in the operating room, where a complication has just claimed the patient's life. She is told that she cannot approach the family in tears, and needs to pull herself together. It is part of the unintended curriculum, which *The Doctor* is focused on.

The History of Laryngeal Cancer

Depending on your trainee audience, you may want to cover some material on laryngeal cancer by supplementing it with one or two articles that review etiology, diagnosis and treatment. Indeed, Jack McKee, as a male over 40, is representative of the patient population with this cancer, which is diagnosed primarily in males. Any review article will help to validate that *The Doctor's* handling of Dr. McKee's diagnosis and treatment is accurate, and the treatment options provided to him remain similar today, as Koroulakis and Agarwal note in this description (Koroulakis and Agarwal 2019):

> Laryngeal cancers represent one-third of all head and neck cancers and maybe a significant source of morbidity and mortality. They are most often diagnosed in patients with significant smoking history, who are also at risk for cancers in the remainder of the aerodigestive tract. They can involve different subsites of the larynx, with different implications in symptomatic presentation, patterns of spread, and treatment paradigm. Early-stage disease is highly curable with either surgical or radiation monotherapy, often larynx-preserving.

Evaluation remains "biopsy during direct laryngoscopy" (Koroulakis and Agarwal 2019), which is exactly how Dr. Abbott handles it, while Dr. Abbott recommends radiation first to help preserve the voice.

Healthcare Ethics Issues: Professional Ethics

The Doctor is really about professional ethics—a critical part of healthcare delivery. Professional ethics can be further broken down into distinguishing between "professionalism"—*behaving* ethically—and "humanism"—*being* an ethical person, which is often expressed as virtue ethics in medical curricula. Students need to understand that healthcare providers can behave ethically and still lack the character traits of a compassionate and ethical person. It is the difference between Dr. Abbott and Dr. Blumfield and Dr. MacKee pre-illness and Dr. MacKee post-illness. This section covers professionalism, humanism, bedside behaviors, mentorship, and a range of other factors that can erode healthcare provider behaviors, impacting healthcare delivery overall.

Professionalism Versus Humanism

There is no one definition of professionalism, but there is consensus that professionalism means acting in the patient's best interests and not the practitioner's best interests, even when it means the practitioner is inconvenienced, risks expressing unpopular views, or risks damage to his/her career (Rosenthal 2010). It would be easy for Dr. MacKee, for example, to back up Murray Kaplan's version of the truth in a lawsuit because it is more convenient. But once he considers the patient's perspective, he doesn't agree.

Meanwhile, humanism can be defined as genuine concern over the interests, values and dignity of the patient. *The Doctor* helps teach the intersection between professionalism and humanism. A humanistic healthcare provider has moral integrity, respects boundaries, and keeps the patient's best interests at heart without compromising his/her own values or moral compass.

Professionalism and humanism may also be defined as the dividing line between healthcare providers "acting" like they care about patients and *actually* caring about patients. In *The Doctor*, the physician who actually embodies humanism and virtuous character traits, Dr. Blumfield, is ridiculed by his peers, who are more interested in the business of medicine. But how can virtue can be taught? The answer is mentorship and modelling good behaviors, but also addressing triggers of moral distress, which can erode professionalism (see further). Additionally, the vast majority of medical students will be admitted because of their academic performances in the sciences, when professionalism and humanism are actually enforced through humanities-based majors, not the typical applicant (Dienstag 2008; Chen 2013; Witzburg and Sondheimer 2013; Burkhardt et al. 2016; O'Neill et al. 2013).

Bedside Manner and Boundaries

There are several examples of bedside behaviors in *The Doctor*, ranging from Dr. Blumfield talking to patients under anesthesia to Dr. Abbott's cold yet professional style. Abbott may be overcompensating for gender stereotypes by exhibiting a cold affect toward her male patient and colleague; however, she also reinforces gender stereotypes by checking her nails when telling him he has cancer. Meanwhile, MacKee's egregious handling of his patient's concerns with surgical staples (telling her she's like a centerfold) actually passed in 1991 as bedside manner. But MacKee has a renewed spirit of humanism when he deals with his Hispanic patient completely differently in a culturally competent manner.

Understanding boundaries is another important aspect of professionalism that is not always straightforward. In *Regarding Henry (1991)*, a film that opened at the same time as *The Doctor*, featuring a man recovering from a brain injury, a physical therapist offers friendship and a hug. Healthcare providers are people, and can find it difficult to draw boundaries with patients they genuinely care about, and even regard as friends. On the flip side, boundaries may be crossed by *patients*, and healthcare providers may be caught off guard by a patient's anger, abusive treatment or simply, inappropriate treatment, as demonstrated by Dr. MacKee's tantrum in the radiation oncology waiting room; Dr. Reed demonstrates both professional and collegial behavior to deal with Dr. MacKee's outburst.

Boundary discussions are especially critical for mental health professionals in training. Films demonstrating mental health boundaries range from One *Flew Over the Cuckoo's Nest* (1975) and *Awakenings* (1990), both discussed in my book Clinical Ethics on Film (2018), as well as *Good Will Hunting* (1997), in which an ethical psychiatrist struggles with a very difficult patient. Finally, boundary discussions in the "MeToo" era are particularly significant (see further).

The Hateful Patient

There has always been what is known as the "hateful patient" or "difficult" patient, in which the practitioner is continuously goaded and challenged by the patient. A famous 1978 paper titled "Taking Care of the Hateful Patient" (Groves 1978) first identified that such patients frequently have borderline personality traits and other sociopathic traits. The article offers helpful advice in establishing contracts and boundaries with patients. At times, risk managers may need to be involved to help establish behavior contracts with patients. However, too often, organizational ethics and systems issues lead to patients feeling abandoned or ignored, while institutionalized "healthism", "wealthcare", sexism, and racism may lead to patients being mislabeled as difficult.

Sexual Harrassment/MeToo

Overt unprofessional behavior with the opposite sex should be obvious to most, and needs no definition. However, *The Doctor* displays what used to "pass" for okay that may seem shocking today. For example, in the opening scene, Dr. MacKee in the operating room asks his nurse to sing along to a Jimmy Buffett song: "Why Don't We Get Drunk ". She is clearly uncomfortable with this and remains silent. In the scene introducing ear nose and throat specialist Dr. Leslie Abbott, her male colleague says to MacKee "she gives good throat"—a nod to the pornographic film, *Deep Throat* (1972), while all of the male colleagues are openly admiring her body. To a 1991 male medical audience, none of these comments seemed all that inappropriate. At that time, medical educators were still recommending *The Hospital* (1970), which was for years the definitive "organizational ethics" film used in healthcare ethics. In *The Hospital*, overt sexual assault is glorified, making the film unusable (see this book's Introduction). However, with respect to consensual sex, *The Hospital* is instructive, and warns of dire unintended consequences. The film opens when a young intern meets accidental death after he has sexual relations with a nurse in an empty hospital bed. After falling asleep, he's mistaken for a diabetic patient, and is killed by an intravenous insulin drip that is hooked up to him in error. This satiric commentary on disorganized hospital systems infers that unprofessional behavior can have unintended consequences.

When discussing *The Doctor* in the context of a broader discussion of professional ethics, it's important to discuss consensual relationships for trainees in the workplace; in fact, many healthcare providers meet their life partners through their training or professional environment, since so much time is spent together. But even married couples need to maintain a professional relationship if they are in the same workplace. Hospital romances have long been a theme in popular television shows and films for good reason—they occur commonly. But the consequences of indiscretion or romantic drama can adversely affect patient care.

Covering sexual harassment is also critical, if it is not being covered in other mandated training (it is not discussed here). But in the era of the Larry Nasser scandal (Houser and Zraick 2018), covering non-consensual relationships with patients is also critical, and remains one of the most common forms of sexual misconduct.

In the film *Critical Care* (1997), a young doctor is seduced not by a patient, but by the daughter of an ICU patient, who is unwittingly filmed during the act. The patient's daughter then uses their encounter in an extortion scheme. The 1997 film was a satire on hospital organizations, frivolous medical malpractice suits, and dysfunction. It is perhaps more prescient in light of the digital age, where healthcare providers are being filmed or recorded without their knowledge. Prior to the mass use of email, texting, and social media, what we said about our patients to our colleagues; what we said about colleagues to other colleagues; workplace gossip; rumor mongering, and the like—were typically not on display for the world to see. Use of social media in healthcare environments is an organizational ethics issue that needs to become part of mainstream medical training.

Dynamics are also shifting as women are beginning to outnumber men in certain specialities, such as Internal Medicine. An interesting pairing with *The Doctor* may be *Smart People* (2008), in which a male patient (played by Dennis Quaid), asks his female emergency room doctor (played by Sarah Jessica Parker) out on a date. We would like to presume that she has terminated the therapeutic relationship before sleeping with him, which would be the only way she could ethically accept the date.

Moral Distress, Burnout and Moral Injury

One critical problem facing medical education is moral distress, moral injury and healthcare providers burning out, which erodes professionalism and humanism. Moral distress, initially defined by Andrew Jameton (Jameton 1984; Epstein and Hamric 2009), refers to a situation where the healthcare provider recognizes a moral problem but is constrained from acting on it, or resolving it. There may be external constraints (legal or patient rights-based) or internal constraints, which have more to do with organizational hierarchies, and trainees having no moral agency to make professional decisions (Rosenthal and Clay 2013). Moral residue is a term initially defined by George Webster and Francoise Bayliss (Webster and Bayliss 2000) as: "that which each of us carries with us from those times in our lives when in the face of moral distress we have seriously compromised ourselves or allowed ourselves to be compromised." Moral residue thus refers to the "lingering feelings" after the morally distressing event has passed (Epstein and Hamric 2009). Ultimately, it leads to a crisis in professionalism and humanism. Moral distress is not the same as feeling "sad" about a case, or being vicariously traumatized by another's tragedy; rather, it is when the right action is identified but cannot be carried out. Reducing moral distress for students involves re-tooling and properly equipping the mentors so that students have the appropriate pilots on-board to help them with a smooth landing into new territory; in some cases, it means that mentors with too much baggage themselves—who were trained by a pre-illness Dr. MacKee, for example, are probably not the right pilots. It also requires understanding of what skillsets healthcare trainees are bringing on-board, and what tools they will need when they land. When teaching *The Doctor*, medical educators should be mindful of discussing mentorship experiences with students? I argue that failure to reduce rates of moral distress in medical training can lead to a variety of consequences, and may even trigger or exacerbate depression and anxiety, now a recognized problem in medical school (Drake 2014; Puthran et al. 2016; Rosenthal and Clay 2013; Rotenstein et al. 2016).

Moral Injury

While conversations about moral distress and moral residue are part of any professional ethics curriculum, healthcare providers are also prone to moral injury, a type of injury that has tended to be associated more with military activities (Moral Injury Project 2020), and is defined like this: "Moral injury is the damage done to one's conscience or moral compass when that person perpetrates, witnesses, or fails to prevent acts that transgress one's own moral beliefs, values, or ethical codes of conduct" (Moral Injury Project 2020). This definition is similar to that of moral distress, but frequently involves being the perpetrator of an immoral act, which many physicians believe they are when participating in the U.S. healthcare system (see Chap. 4). In healthcare, the moral injury occurs when healthcare providers cannot provide high-quality care and healing in the context of health care. Talbot and Dean (2018) note:

> Navigating an ethical path among such intensely competing drivers is emotionally and morally exhausting. Continually being caught between the Hippocratic oath, a decade of training, and the realities of making a profit from people at their sickest and most vulnerable is an untenable and unreasonable demand. Routinely experiencing the suffering, anguish, and loss of being unable to deliver the care that patients need is deeply painful. These routine, incessant betrayals of patient care and trust are examples of "death by a thousand cuts." Any one of them, delivered alone, might heal. But repeated on a daily basis, they coalesce into the moral injury of health care.

> Physicians are smart, tough, durable, resourceful people. If there was a way to MacGyver themselves out of this situation by working harder, smarter, or differently, they would have done it already. Many physicians contemplate leaving heath care altogether, but most do not for a variety of reasons: little cross-training for alternative careers, debt, and a commitment to their calling. And so they stay — wounded, disengaged, and increasingly hopeless.

In *The Doctor*, the constant "sorry—busy day" that Jack McKee eventually calls out as wholly disingenuous, becomes a euphemism for the inability to provide the time and quality of care to patients in an increasingly complicated for-profit business model.

Moral Courage and Whistleblowing

Speaking up or standing up for what's right—even when it is politically unpop-ular—is known as moral courage, and if it's a moral action of last resort, it's known as "whistleblowing" in severe situations. There are times when professionalism demands moral courage, and taking an ethical stand, particularly when patient or public safety is at risk. It's important to make clear that moral courage is the antidote to moral distress but may come with consequences. However, it may also be neces-sary to preserve one's moral integrity. There are several examples of moral heroism that arose after *The Doctor* had become a cult classic in medical schools.

Jeffrey Wigand, for example, blew the whistle on Big Tobacco (CBS 1996). Wigand lost his job and family over death threats and corporate bullying to do what was right. His information, however, was critical to public health and safety, and to the outing of unethical practices in the tobacco industry. He stated he was compelled to speak out and would do it again.

Sanford Klein, MD, at Robert Wood Johnson University Hospital complained about patient safety in radiology and lost his clinical privileges (Klein v. University of Medicine 2005). He stated he would do it again because it was an ethical issue. David Lemonick, MD, at Western Pennsylvania Hospital, complained in 2004 about serious issues affecting patient safety in the emergency department, and was termi-nated (AMed News 2005). John Ulrich, Jr., MD, at Laguna Honda Hospital, also complained about practices that put patients at risk, and was then informed that an investigation into his competency had been launched. He resigned, and was awarded a $4.3 million dollar settlement (Urlich v. Laguna Honda Hospital 2002).

One of the ugliest whistleblowing incidents occurred at the University of Toronto in 1997. Dr. Nancy Olivieri was testing deferiprone for treatment of thalassemia. Olivieri had concerns the drug was toxic, and the probable cause of progression of liver fibrosis in some patients. She reported her concerns to Apotex, the generic drug company and sponsor of her research. She told them she had a duty to disclose the risks, and also reported the risks to her university's institutional review board, which amended consent forms. Apotex terminated the trial and forbid her to publish or risk legal action. Ultimately, the University of Toronto declined to support Olivieri; it was revealed that Apotex was about to donate $30 million to University of Toronto. But it was the bioethicists at the University of Toronto who remained silent at the time, except for one nurse-ethicist who was not supported in speaking out (Baylis 2004). This ugly incident revealed deep problems with conflicts of interest.

Conflicts of Interest

By the 1990s, conflicts of interest in medicine had taken center stage, coinciding with *The Doctor*'s popularity. In general, a conflict of interest in the clinical setting exists when a personal relationship, or any inducement for professional or financial reward, interferes with a healthcare provider's ability to be objective in a patient care situation. Obvious examples are when doctors are influenced by pharmaceutical companies, or have stock in products they are recommending. By the early 2000s, many academic medical centers began passing strict rules surrounding industry relationships which could be construed as conflicts of interest. However, several insider narrative medicine books began to emerge, authored by physicians who were at the top of their careers in 1991, when *The Doctor* came out, who exposed the entangled business of medicine, such as Jerome Kassirer's On The Take: How Medicine's Complicity with Business can Endanger Your Health (Kassirer 2005). Eventually, the entire rot of the U.S. healthcare system would be exposed in the 2007 film *Sicko* (see Chap. 4).

Teaching Virtue and Humanism

Ultimately, the most important lessons from *The Doctor* revolve around what lessons we are really teaching in medical school, and what is "virtue". For example, Pellegrino states "A virtue-based physician would recognize pro bono work as important" but to be virtuous also requires a "community of virtue" (Pellegrino 2002). Teaching by example only works when mentors can emulate the qualities of virtue ethics. To make teaching virtue medicine effective, medical educators may need to go beyond their institutions to find examples.

Paul Farmer, who is the subject of the book, Mountains beyond Mountains (Kidder 2003), meets Pellegrino's criteria for a "good physician"; Farmer is not known for charm, but for medical humanism. Paul Farmer established a global health entity, Partners In Health, in the late 1990s, long before it was popular to do medical charity work in Haiti.

In the documentary, *The English Surgeon* (2007), brain surgeon Henry Marsh has been going to the Ukraine for over 15 years to help improve upon the medieval brain surgery he witnessed there during his first visit in 1992. His Ukrainian colleague Igor Kurilets sees him as a guru and a benefactor. But for all the direct satisfaction he gets from going, Henry also sees grossly misdiagnosed patients, children who he can't save, and a lack of equipment and trained supporting staff. But he gives his time and skills to help benefit patients in the global community. In *Chernobyl Heart*, (2003) an American heart surgeon holds back tears because he feels he is performing too simple an operation for the level of gratitude and accolades that are bestowed upon him when he does charity corrective heart surgeries in the Ukraine and Belarus, which were affected by the Chernobyl disaster, giving rise to a heart defect known as "Chernobyl heart".

When shown real examples of medical humanism and virtue medicine, healthcare providers are typically giving back in a global healthcare setting to patients in less developed countries. But medical humanism in the United States can also be demonstrated by providing care in communities devastated by health disparities.

In 1991, just after *The Doctor* had opened in theaters, UPI published an interview with Rosenbaum in which he stated: "We began to be so good we began to be suspect of the doctor who practiced the art of medicine… We'd gone too far off the path and now we're just beginning to come back…".

The article continued to summarize Rosenbaum's interview like this (Rarick 1991):

> Doctors should give more information to patients; be more willing to ask questions and listen; be less fearful of expressing their emotions; be frank and honest with patients.
>
> Although [Rosenbaum] admits he was the 'worst offender in the world about this,' he now says patients must be kept waiting less, if at all. 'Every minute they wait is torture,' he says. 'To us it's routine.' And then there are small things, such as the hospital gowns. Rosenbaum says he recently visited a clinic in which the gowns had been redesigned to be longer and more modest. 'There are little things that mean a lot to the patient,' he says.
>
> [Rosenbaum] even learned some things from the movie version of his own experiences. Originally, he didn't like a scene (added for the film) in which the post-illness Hurt makes his residents take off their clothes and walk around a hospital in gowns so they see how patients feel.
>
> 'I thought that was corny,' Rosenbaum says. 'I can't believe how the audience reacted to it. They loved it.'
>
> Rosenbaum does not blame the doctors who treated him. He says to some degree the very nature of the profession forces physicians to keep a distance from their patients.
>
> 'You have to develop that shell in order to exist,' he says. 'You're dealing with death and dying and bad news every day of your life.'
>
> The doctors who treated him were 'no better or worse that I was myself,' he says. 'They were doing exactly what I did as a doctor.'

Since Rosenbaum's prescriptions above, teaching medical humanism has revolved around transforming organizational cultures so that a community of virtue medicine can thrive. Based on a consensus from experts (Rosenthal and Clay 2013), the following organizational programs have shown to improve the moral community of medicine for healthcare providers: clinical ethics consultation services and ethics rounding; moral distress debriefings and other forums; "Schwartz Center Rounds", a specific type of panel-based rounds open to all trainees that focuses on health care provider emotions and experiences (www.theschwartzcenter.org).

Conclusions

The Doctor functions as both a healthcare ethics film and clinical ethics film. One could focus on the patient experiences of Dr. MacKee and present the film within more classic clinical ethics themes, such as autonomy, consent and capacity, and so forth. However, the doctor-as-patient story is really about professional ethics epiphanies and changing diehard practice habits by having a change in perspective. At times, I have taught this film by inviting a physician-colleague battling illness, which has been transformative for students. For example, in a private note to me dated November

3, 1987, another doctor, my grandfather, Jacob (Jack) Lander (1910–1989) who was the same generation as Rosenbaum, wrote this while being treated for esophageal cancer (Lander 1987):

> …[O]ne has to contend with facing the label stuck on one – namely – CANCER! It's certainly different, when treating hundreds of patients over the past 50 years – objectively – and having to face it subjectively! Tomorrow I see the two doctors who will tell me [and my daughters] what I have to face in the future. For the present I am taking two potent pain killers regularly. Perhaps that is why there is some deficient lucidity in this letter…

What Rosenbaum wrote, cited at the beginning of this chapter, almost mirrors my grandfather, Dr. Lander's 1987 note to me; they are both communicating a universal experience for aging physicians facing their own mortalities. Rosenbaum admits, however, that doctors are even more avoidant of illness than the average patient (Rosenbaum 1988):

> [W]henever I got sick, I treated myself. Doctors were not for me. I realize now that without verbalizing it, all my life I had avoided consulting them for my own medical problems because I was afraid of what they might tell me – and I knew their limitations. I had just been lucky. Eventually my day came, and I was trapped by what I must have known, almost from the start, was a grave illness.
>
> When I became ill, like my patients, I wanted my doctors to be gods – and they couldn't be. But I also wanted them to understand my illness and my feelings and what I needed from my physicians.

When teaching *The Doctor* to healthcare trainees, communicating to them that they, too, will someday be sick and vulnerable is a potent way to transform healthcare delivery and health policy.

Theatrical Poster

The Doctor (1991)

> Director: Randa Haines
> Producer: Laura Ziskin, Edward S. Feldman, Michael S. Glick
> Screenplay: Robert Caswell
> Based on: *A Taste of My Own Medicine: When the Doctor Is the Patient* by Edward Rosenbaum
> Starring: William Hurt, Christine Lahti, Mandy Patinkin, Elizabeth Perkins, Adam Arkin, Charlie Korsmo, Wendy Crewson
> Music: Michael Convertino
> Production Company: Touchstone Pictures, Silver Screen Partners IV
> Distributor: Buena Vista Pictures
> Release Date: July 24, 1991
> Run time: 122 min

References

Baker, P. (2018). Bush made Willie Horton an issue in 1988. *The New York Times* , December 3, 2018. https://www.nytimes.com/2018/12/03/us/politics/bush-willie-horton.html.

Baum, B. (1991). Oregon doctor inspires new film. *Deseret News*, August 20, 1991. https://www.deseretnews.com/article/178925/OREGON-DOCTOR-INSPIRES-NEW-FILM.html.

Baylis, F. (2004). The Olivieri debacle: Where were the heroes of bioethics? *Journal of Medical Ethics, 30*, 44–49.

Bentsen, L. (1988). *Vice presidential debate*, October 5, 1988. https://www.c-span.org/video/?4127-1/1988-vice-presidential-candidates-debate.

Burkhardt, J. C., DesJardins, S. L., Teener, C. A., Gay, S. E., & Santen, S. A. (2016). Enrollment management in medical school admissions: a novel evidence-based approach at one institution. *Academic Medicine, 91*(11), 1561–1567.

CBS. (1996). Interview with Jeffrey Wigand. *60 Minutes*, February 4, 1996.

Chen, P. W. (2013). The changing face of medical school admissions. *New York Times*. May 2, 2013. http://well.blogs.nytimes.com/2013/05/02/the-changing-face-of-medical-school-admissions/?_r=0. Accessed January 4 2017.

Cousins, N. (1979). *Anatomy of an Illness*. New York: WW, Norton.

Dienstag, J. L. (2008). Relevance and rigor in premedical education. *New England Journal of Medicine, 359*(3), 221–224.

Doctor says hospital fired him in retaliation. *American Medical Association News (AMed News)*, June 27, 2005. https://amednews.com/article/20050627/profession/306279960/7/.

Drake, D. (2014). How being a doctor became the most miserable profession. *Daily Beast*. April 14, 2014. http://www.thedailybeast.com/articles/2014/04/14/how-being-a-doctor-became-the-most-miserable-profession.html. Accessed January 4, 2017.

Duffy, T. (2011). The Flexner report—100 years later. *The Yale Journal of Biology and Medicine, 84*(3), 269–276. https://www.ncbi.nlm.nih.gov/pmc/articles/PMC3178858/.

Dyrbye, L. N., Thomas, M. R., & Shanafelt, T. D. (2005). Medical student distress: Causes, consequences, and proposed solutions. *Mayo Clinic Proceedings, 80*(12), 1613–1622.

Ellis, J. (2016). How Ronald Reagan dealt with his Alzheimer's diagnosis. *Newsweek*. https://www.newsweek.com/ronald-reagan-alzheimers-disease-442711.

Epstein, E. G., & Hamric, A. B. (2009). Moral distress, moral residue, and the crescendo effect. *Journal of Clinical Ethics, 20*(4), 330–342.

Groves, J. (1978). Taking care of the hateful patient. *The New England Journal of Medicine, 298*, 883–887. https://www.nejm.org/doi/full/10.1056/NEJM197804202981605.

Houser, C., & Zraick, K. (2018). Larry Nassar sexual abuse scandal. *The New York Times*, October 22, 2018. https://www.nytimes.com/2018/10/22/sports/larry-nassar-case-scandal.html.

Humphry, D. (1991). *Final exit*. The Hemlock Society.

IMDB. (2019). *William Hurt biography*. https://www.imdb.com/name/nm0000458/bio.

Jameton, A. (1984). *Nursing practice: The ethical issues*. Englewood Cliffs, N.J.: Prentice-Hall.

John R. Ulrich, Jr., M.d., Plaintiff-appellant, v. City and County of San Francisco; Laguna Honda Hospital; Maria v. Rivero, M.d.; Theresa Berta, M.d.; Melissa Welch, M.d., Defendants-appellees, 308 F.3d 968 (9th Cir. 2002).

Kassirer (2005). On the take: How medicine's complicity with business can endanger your health. New York: Oxford University Press.

Kidder, T. (2003). *Mountains Beyond Mountains*. Penguin Random House, New York.

King, J. (2013). Dukakis and the tank. *Politico*, November 17, 2013. https://www.politico.com/magazine/story/2013/11/dukakis-and-the-tank-099119.

King, L. W. (2019). The code of Hammurabi. *The Avalon Project, Yale Law School*. https://avalon.law.yale.edu/ancient/hamframe.asp.

Kirk, J. (1994). Gender inequality in medical education. In B. Bolaria & R. Bolaria (Eds.), *Women, medicine, and health*. Halifax, Nova Scotia: Fernwood.

Klein v. (2005). University of Medicine and Dentistry of New Jersey, 878 A.2d 856, 185 N.J. 35.

Koroulakis, A., & Agarwal, M. Cancer, Laryngeal. [Updated 2019 Nov 23]. In: *StatPearls* [Internet]. Treasure Island (FL): StatPearls Publishing; 2020 January. Available from: https://www.ncbi.nlm.nih.gov/books/NBK526076/.

Lander, J. J. (1987). Personal letter, November 3, 1987.

Lear, M. (1980). *Heartsounds*. New York: Simon and Schuster.

Maslin, J. (1991). William Hurt as doctor whose spirit heals when he falls ill. *The New York Times*, July 24, 1991. https://www.nytimes.com/1991/07/24/movies/review-film-william-hurt-as-doctor-whose-spirit-heals-when-he-falls-ill.html.

Morrison, R. S., & Meier, D. E. (2015). America's care of serious illness: 2015 state by state report card on access to palliative care in our Nation's hospitals. In M. Appellof & L. Morgan L (Eds.), New York, NY: Center to Advance Palliative Care; 2015. https://reportcard.capc.org/wp-content/uploads/2015/08/CAPC-Report-Card-2015.pdf. Accessed January 5, 2017.

O'Neill, L., Vonsild, M. C., Wallstedt, B., & Dornan, T. (2013). Admission criteria and diversity in medical school. *Medical Education, 47*(6), 557–561.

PBS Nova. (2009). *The doctor diaries*. Air date: April 7, 2009. https://www.pbs.org/wgbh/nova/video/doctors-diaries.

Pellegrino, E. D., & Relman, A. (1999). Professional medical associations: Ethical and practical guidelines. *JAMA, 282*(10), 984–986.

Pellegrino, E. D. (2002). Professionalism and the good physician. *The Mount Sinai Journal of Medicine, 69*(6), 378–384.

Pilkington, (Ed.). (2011). Ronald Reagan had Alzheimer's while president, says son. *The Guardian*. January 17, 2011. https://www.theguardian.com/world/2011/jan/17/ronald-reagan-alzheimers-president-son.

Puthran, R., Zhang, M. W., Tam, W. W., & Ho, R. C. (2016). Prevalence of depression amongst medical students: A meta-analysis. *Medical Education, 50*(4), 456–468.

Quayle, D. (1988). *Vice presidential debate*, October 5, 1988. https://www.c-span.org/video/?4127-1/1988-vice-presidential-candidates-debate.

Rarick, E. (1991). Dr. Edward Rosenbaum spent half a century practicing. *United press international*, August 24, 1991. https://www.upi.com/Archives/1991/08/24/Dr-Edward-Rosenbaum-spent-almost-half-a-century-practicing/1081683006400/.

Richards, A. (1988). Keynote speech. *Democratic convention*, July 18–21, Atlanta, GA.

Rosenbaum, E. (1988). *A taste of my own medicine: When the doctor is the patient*. New York: Random House.

Rosenthal, M. S. (2010). *Introduction to clinical ethics, Part 5: Professionalism and humanism* (Training Module). University of Kentucky College of Medicine.

Rosenthal, M. S., & Clay, M. (Eds.). (2013). University of Kentucky program for bioethics. *The Moral Distress Education Project*. www.moraldistressproject.org.

Rosenthal, M. S., & Clay, M. (2017). Initiatives for responding to medical trainees' moral distress about end-of-life cases. *AMA Journal of Ethics, 19*(6), 585–594.

Rotenstein, L. S., Ramos, M. A., Torre, M., et al. (2016). Prevalence of depression, depressive symptoms, and suicidal ideation among medical students: A systematic review and meta-analysis. *JAMA, 316*(21), 2214–2236.

Rushton, C. H., Kaszniak, A. W., & Halifax, J. S. (2013). A framework for understanding moral distress among palliative care clinicians. *Journal of Palliative Medicine, 16*(9), 1074–1079.

Shem, S. (1978). *House of god*. New York: Richard Marek Publishers.

Talbot, S., & Dean, W. (2018). Physicians aren't 'burning out.' They're suffering from moral injury. *Stat Reports*, July 26, 2018. https://www.statnews.com/2018/07/26/physicians-not-burning-out-they-are-suffering-moral-injury/.

The Moral Injury Project. (2020). Syracuse University. http://moralinjuryproject.syr.edu/about-moral-injury/.

Time. (2019). *Top 10 Campaign Ads*. http://content.time.com/time/specials/packages/article/0,28804,1842516_1842514_1842557,00.html.

Webster, G., & Bayliss, F. (2000). Moral residue. In S. Rubin & L. Zoloth (Eds.), *Margin of error: The ethics of mistakes in the practice of medicine*. Hagerstown, MD: University Publishing Group Inc.

Witzburg, R. A., & Sondheimer, H. M. (2013). Holistic review—shaping the medical profession one applicant at a time. *New England Journal of Medicine, 368*(17), 1565–1567.

Zimmerman, M. K. (1987). The women's health movement. In M. Farle & B. Hess (Eds.), *Analyzing gender: A handbook of social science research*. Newbury Park, CA: Sage Publications.

Chapter 9
Collegiality and Racism in *Something the Lord Made* (2004)

Institutional diversity is an important aspect of healthcare delivery, while institutionalized racism is a potent healthcare ethics issue that has led to health disparities in medical education, healthcare delivery, and patient care. *Something the Lord Made* (2004) is, on one hand, a perfect history of medicine film about the first cardiac surgery performed in 1944 to correct Tetrology of Fallot, a congenital heart problem also known as "blue baby syndrome" or congenital cyanotic heart disease. The surgery was performed by Dr. Alfred Blalock (1899–1964); he perfected the surgical technique with his research partner and lab assistant, Vivien Thomas (1910–1985), who was by his side when Blalock operated on the first cardiac surgical patient. Dr. Helen Taussig (1898–1986), one of the few female physicians of her generation who founded the field of pediatric cardiology, brought the "blue baby" problem to Blalock. The surgical innovation was developed within a translational research framework (see under History of Medicine). Today, what is known as the Blalock-Taussig Shunt is the procedure featured in this film, and Blalock and Taussig made medical history together to solve "blue baby syndrome". Taussig is one of the few women in science and medicine at the time who was not denied her place in history (see further).

But that's not what the film is about. This film is about who was erased from history: Vivien Thomas, an African American lab technician who partnered with Blalock for years. Thomas was a significant contributor to the Blalock-Taussig Shunt but he never received any credit as part of that team until years later. *Something the Lord Made*—the name Blalock gave to Thomas' surgical skills—focuses on the Blalock-Thomas research relationship. This remarkable partnership spanned three decades (1930s–60s), and speaks to the issue of organizational and structural racism in medical institutions and medical education, whitewashing of medical history, as well as the subject of reparations that began in the 1970s. When teaching professional ethics and collegiality, this story deeply resonates with a wide range of diverse students who have struggled with overt or covert racism, admission quotas in higher education and medical schools, and enduring health disparities that result from the

© Springer Nature Switzerland AG 2020
M. S. Rosenthal, *Healthcare Ethics on Film*,
https://doi.org/10.1007/978-3-030-48818-5_9

absence of diversity in healthcare providers. This chapter discusses the origins of the film *Something the Lord Made*, the social location of the Blalock-Thomas relationship, the History of Medicine context and the healthcare ethics issues raised in the film.

Discovering the Blalock-Thomas Story: Origins of the Film

The Blalock-Thomas story was very well known to the institution of Johns Hopkins School of Medicine and anyone who had trained there. The 34-year partnership produced hundreds of surgical descendants who went on to prestigious careers of their own, and who owed their skills to Thomas, in particular. After Blalock died in 1964 (see further), Peter Olch, a medical historian, interviewed Vivien Thomas about the historic blue baby surgery (McCabe 1989), and Thomas clarified for medical history that the translational research question was posed by Taussig, "who came to Blalock and Thomas looking for help for the cyanotic babies she was seeing. At birth these babies became weak and 'blue,' and sooner or later all died. Surely there had to be a way to 'change the pipes around' to bring more blood to their lungs, Taussig said" (McCabe 1989). The Johns Hopkins medical history archive on Vivien Thomas reveals that by 1970, the institution was invested in trying to recognize Thomas' achievements (Johns Hopkins Medical Archives 2020). as the entire country was beginning to deal with its shameful past regarding segregation and the Jim Crow era, recognizing that professionals such as Thomas were simply barred from the same access to education than far less talented white males of his generation. Correspondence began about commissioning his portrait for the main lobby, which was an institutional tradition in which revered medical giants of Johns Hopkins were captured through portraiture, including Sir William Osler (see Chap. 8), and by the 1950s, Alfred Blalock. By February 27, 1971, the Thomas portrait was unveiled. According to McCabe (McCabe 1989):

> It was the admiration and affection of the men he trained that Thomas valued most. Year after year, the Old Hands came back to visit, one at a time, and on February 27, 1971, all at once. From across the country they arrived, packing the Hopkins auditorium to present the portrait they had commissioned of "our colleague, Vivien Thomas."

Historical archives reveal that an honorary doctorate was in the works around the same time as the portrait, but presumably there was more administrative paperwork and hurdles to overcome, which delayed the honorary doctorate, and an appointment to the medical school faculty until May 21, 1976–35 years after his arrival at the same institution that had separate "colored" entrances and bathrooms. The 1976 honorary doctorate for Thomas was not well-publicized in the medical press. Unless you were an attendee at that event, or happened to catch small stories in the five small news publications that carried the story, you missed it.

When Thomas retired in 1979, he began to work on his autobiography with a Johns Hopkins colleague, Mark Ravitch as editor. It was published posthumously—two days after Thomas died from pancreatic cancer. The original title of his book was

Pioneering Research in Surgical Shock and Cardiovascular Surgery: Vivien Thomas and His Work with Alfred Blalock (Thomas 1985). The book was not a bestseller, but received great reviews in the *Annals of Surgery* in 1986 (McGoon 1986); and the *Annals of Thoracic Surgery* (Bahnson 1987). The *Journal of Medical History* in 1987 also reviewed the book and notably referred to Thomas as an "American Negro" (Swain 1987). A few other obscure publications reviewed his autobiography, too. But by that time, several African Americans in medicine and science had begun to discover Thomas on their own; the first Ph.D. thesis on Thomas was done in 1973 (Johns Hopkins medical archive), and several other students first learned of him by seeing his portrait.

The day after Thomas died, which coincided with his book's release, journalist Katie McCabe learned about him for the first time—probably from his *Washington Post* obituary which reported (Washington Post 1985):

> Dr. Vincent Gott, a Hopkins cardiologist, said that Mr. Thomas "was almost legendary among cardiac surgeons. It is safe to say there is not a cardiac surgeon over 40 who doesn't know of him and his tremendous contributions to the specialty."
>
> Another Hopkins heart surgeon, Dr. J. Alex Haller, said, "This is a great loss. Dr. Blalock once said that Vivien Thomas' hands were more important to him in the development of the blue-baby operation than his own—and he meant it."

McCabe was intrigued, and began work on an award-winning article in the *Washingtonian*, published in 1989. The article was gripping from start to finish; here is a sample of McCabe's beautiful journalistic prose (McCabe 1989):

> Say his name, and the busiest heart surgeons in the world will stop and talk for an hour. Of course they have time, they say, these men who count time in seconds, who race against the clock. This is about Vivien Thomas. For Vivien they'll make time.... No, Vivien Thomas wasn't a doctor, says Cooley. He wasn't even a college graduate. He was just so smart, and so skilled, and so much his own man, that it didn't matter. And could he operate. Even if you'd never seen surgery before, Cooley says, you could do it because Vivien made it look so simple.... "You see," explains Cooley, "it was Vivien who had worked it all out in the lab, in the canine heart, long before Dr. Blalock did Eileen, the first Blue Baby. There were no 'cardiac experts' then. That was the beginning." But in the medical world of the 1940s that chose and trained men like Denton Cooley, there wasn't supposed to be a place for a black man, with or without a degree. Still, Vivien Thomas made a place for himself....
>
> Blalock and Thomas knew the social codes and traditions of the Old South. They understood the line between life inside the lab, where they could drink together in 1930, and life outside, where they could not. Neither one was to cross that line. Thomas attended Blalock's parties as a bartender, moonlighting for extra income. In 1960 when Blalock celebrated his 60th birthday at Baltimore's Southern Hotel, Thomas was not present..."Dr. Blalock let us know in no uncertain terms, 'When Vivien speaks, he's speaking for me,'" remembers Dr. David Sabiston, who left Hopkins in 1964 to chair Duke University's department of surgery. "We revered him as we did our professor."
>
> To Blalock's "boys," Thomas became the model of a surgeon. "Dr. Blalock was a great scientist, a great thinker, a leader," explains Denton Cooley, "but by no stretch of the imagination could he be considered a great cutting surgeon. Vivien was."

A Production with Teeth

McCabe's article won the 1990 National Magazine Award for Feature Writing, and had the teeth—literally and figuratively—to inspire the feature film named for the article: a retired dentist, Irving Sorkin, whose hobby was history, read the McCabe article and took it to connections he had in Hollywood to get the film made, which he co-produced (Washington Post 2007; McCabe 1989, 2007). While Sorkin was successful in getting HBO to make the film, the McCabe article also inspired a PBS *American Experience* documentary, *Partners of the Heart* (2003), narrated by Morgan Freeman, which also won an award for the Best History Documentary of that year (OAH 2020). The documentary aired as the feature film was being made.

Something the Lord Made aired May 30, 2004, on HBO, right in the middle of the 2004 election campaign (see further). By then, HBO had tackled several medical history/bioethics films, including *And the Band Played On* (1993), discussed in Chap. 2; *Miss Evers' Boys* (1997), discussed in Chap. 1, and *Wit* (2001), discussed in Clinical Ethics on Film (Rosenthal 2018). A few years later, HBO would also produce *You Don't Know Jack* (2010), discussed in Chap. 3. The history of HBO films is discussed more in Chap. 1, but *Something the Lord Made* was directed by Joseph Sargent, who had also directed *Miss Evers' Boys*. Sargent's career is discussed in detail in Chap. 1, but films about race were familiar territory for him. The screenplay was written by Peter Silverman and Robert Caswell; Caswell had also written *The Doctor* (1991), discussed in Chap. 8.

The Cast

HBO paired Alan Rickman (1946–2016), a British actor, as the Southern Blalock with rapper artist, Mos Def (who later changed his name to Yasiin Bey). Rickman had criss-crossed from British to American films continuously throughout the 1980s and 1990s. He was most known for his role as a German terrorist in *Die Hard* (1988), who takes over a high rise tower, and his romantic lead in the British film, *Truly, Madly Deeply* (1990). When cast as Blalock, his last film had been part of the enduring ensemble cast for another British classic, *Love Actually* (2003); Rickman was cast as Emma Thompson's husband Harry, and played straight man to the iconic "giftwrapping" scene with Rowan Atkinson. Rickman's flawless performance as Blalock earned critical acclaim (see further), and ironically, Rickman died prematurely of pancreatic cancer, just as Vivien Thomas did (IMDB 2019).

Def/Bey (born in 1973 as Dante Terrell Smith) was essentially a "triple threat" performer. He had started out as a child actor, embarked on a successful hip-hop/rap music career in the 1990s as both a performer and producer; and also appeared in several films as an actor by the early twenty-first century. He is also well-known as an activist, particularly on the topic of police brutality. In fact, the same year that Vivien Thomas' portrait was unveiled (1971), Marvin Gaye's "What's Going On" was topping the charts, one of the first songs about police brutality targeting the

African American community—decades before the Rodney King beating and the Black Lives Matter movement. Def/Bey's father was a Nation of Islam follower, and a practicing Muslim. Def grew up in Brooklyn during the crack epidemic of the 1980s and escaped many of the scourges of drug and gang violence through studying the Arts and acting. Def/Bey completely embodied Vivien Thomas and received high praise for his performance (IMDB 2012; Birchmeier 2019). Ironically, just five days after Rickman died suddenly from pancreatic cancer on January 14, 2016, Def/Bey announced his retirement from entertainment completely. He announced his retirement on Kanye West's website as follows: "I'm retiring from the music recording industry as it is currently assembled today, and also Hollywood effective immediately" (Schwartz 2015). However, he became active again in recent years under the name Yasiin Bey.

Dr. Helen Taussig, the first female cardiologist who was deaf in one ear (see further), was played by Mary Stuart Masterson. Masterson was best known for two iconic Tomboy roles: "Watts" in the 1980s "Reaganite" film (see Chap. 5), *Some Kind of Wonderful* (1987) opposite Eric Stoltz, and "Idgie" in *Fried Green Tomatoes* (1991).

Synopsis

Something the Lord Made follows the life and career of Vivien Thomas, born and raised in Nashville, and his relationship with Alfred Blalock. We meet Thomas around 1929 when he is working as a carpenter, and the stock market crashes, taking all of his life savings, interfering with his plans for pursuing a medical career. He lands a job at Vanderbilt University as a janitor to clean and maintain Dr. Alfred Blalock's laboratory, which includes animal care duties as Blalock works with dogs in his research (see further). Soon Blalock realizes that Thomas is really a talented research scientist in janitor's clothing, and transforms his "janitor" into an integral research collaborator. Early in their relationship, a dramatic misstep where Blalock yells at Thomas is quickly resolved when Thomas abruptly walks out, resulting in a humble apology by Blalock, in which he makes clear that he will always treat Thomas with respect from that point on. Together they work on treating hypovolemic shock (a.k.a. traumatic shock), a critical discovery prior to World War II. Blalock then secures Thomas a job as a lab technician at Johns Hopkins University Hospital when he accepts an offer to be chief of surgery. There, they embark on the translational research problem of corrective cardiac surgery for "blue baby syndrome"—a problem that Dr. Helen Taussig presents to them. The film brings Thomas' and Blalock's segregated world to life; Thomas continues to struggle with racism and poverty and unequal treatment outside the lab, but he also realizes that Blalock has limitations when it comes to intellectual honesty and collegiality, as he refuses to acknowledge Thomas when sharing credit for his work. The film follows both of them through Blalock's expression of regret to Thomas, Blalock's death, and Thomas receiving his portrait and honorary doctorate in 1976 at a Johns Hopkins ceremony. The film's last shot is of portraits of Alan Rickman and the actor previously known as Mos Def, which then fades to the actual portraits.

The Social Location of *Something the Lord Made*

There are three time frames to discuss in the Blalock and Thomas story. First, their own social locations during their lifespans, during segregation and after the *Brown v. Board of Education* (1954) decision. Second, it's important to examine the time frame of McCabe's *Washingtonian* article in 1989 (see earlier) from the standpoint of race in the United States in the George H. W. Bush era. Finally, examining racial issues in the time frame that the film was made in 2003–4 is also critical.

The Plessy Era: 1899–1954

Alfred Blalock's lifespan maps exactly onto the lifespan of segregation in the United States. He was born in 1899, three years after *Plessy v. Ferguson* (1896) was decided (see Chap. 1), and he died in 1964, the same year that the *Civil Rights Act* was passed. He essentially knew nothing other than segregation and institutionalized racism in the United States. In fact, Alfred Blalock lived roughly the same time span as Eugene Dibble (1893–1968), the African American physician involved in the Tuskegee study (see Chap. 1). Dibble lived a little longer than Blalock and died at age 75 on June 1, 1968,—two months after Martin Luther King was assassinated. Blalock was thus from the "Plessy Generation" (see Chap. 1), and even though he lived another decade after *Brown v. Board of Education* (1954), its impact would not be felt until after he died (see further).

Vivien Thomas was a decade younger than Blalock, born in 1910 and was in mid-life when things began to change: he was 44 when *Brown* was decided, 54 when the *Civil Rights Act* was passed, 55 when the *Voting Rights Act* was passed (1965), and received an honorary doctorate from Johns Hopkins in 1976—only three years after the Tuskegee Study (see Chap. 1) was formally closed. Thomas retired in 1979 and died in 1985. But his lifespan mapped onto Eunice Rivers (1899–1986), the African American nurse involved in the Tuskegee study (see Chap. 1), who was older than Thomas but died a year later. Thomas' generation was a bridge between *Plessy* and *Brown*, but Thomas lived almost half his life under segregation and racist policies. Because *Something the Lord Made* focuses on his struggles to attend medical school, a review of access to medical school for African Americans is critical.

Medical Education for the Plessy Generation

The first black colleges were established before the Civil War as private institutions in the North. Before the *Plessy* decision, the *First Morrill Act*, a.k.a. *National Land-Grant Colleges Act* of 1862 made postsecondary education more accessible to white American citizens, but the act was revised in 1890 (aka *Second Morrill Act*) to expand establishment of nineteen black colleges. After the passage of the Act, public land-grant institutions specifically for African Americans were established in each of the southern and border states. As a result, some new public black institutions were founded, and a number of formerly private black schools came under public control; eventually 16 black institutions were designated as land-grant colleges. These institutions offered courses in agricultural, mechanical, and industrial subjects, but few offered college-level courses and degrees. In 1896, after the *Plessy v. Ferguson* decision, it was clear that African Americans would need to expand black-only colleges, and so the expansion of black colleges went from one in 1837 to over 100 by 1973, and most of were founded post-*Plessy* (Duke University 2020; State University 2019).

Between 1868 and 1904, several black medical schools opened, but many needed to close around 1910 as a result of The Flexner Report, (see Chap. 8). Nonetheless, through the decades, graduate and professional programs at black colleges grew out of other cases that continued to uphold and expand upon *Plessy*: In *Sinuel v. Board of Regents of University of Oklahoma* (1948), the courts stipulated a state must offer schooling for African Americans as soon as it provided it for whites; in *MacLaurin v. Oklahoma State Regents* (1950) the courts stipulated that black students must receive the same treatment as white students; and in *Sweatt v. Painter* (1950), the court stipulated that the state must provide facilities of comparable quality for black and white students. This resulted in African American students being admitted to traditionally white graduate and professional schools if their program of study was unavailable at a comparable black college. But part of the proliferation of black colleges was due to a desire by the states to avoid admitting black students into their traditional white colleges (State University 2019).

By 1932, most African Americans seeking medical degrees and other professional degrees were educated through the black college and university system, which resulted in a thriving and growing population of black professionals who were otherwise barred from educational, economic and social opportunities. A year before the landmark *Brown v. Board of Education* decision, more-than 43,000 African American college students were enrolled in black colleges and universities. About 32,000 were going to private black institutions, such as Fisk University, Hampton Institute, Howard University, Meharry Medical College, Morehouse College, Spelman College, and the Tuskegee Institute. About 11,000 were going to smaller black colleges located in southern and border states. These students became teachers, ministers, lawyers, and doctors serving their community in a racially segregated society (State University 2019).

By the early 1950s, black colleges and universities provided about 90 percent of the higher education training to African Americans: 75% of all doctoral degrees; the majority of law degrees granted, educating 80% of future African American federal judges; and finally, 85% of physicians, (State University 2019), which included Eugene Dibble Jr., who would become the medical director of the Tuskegee Institute (see Chap. 1).

It would be another notable Tuskegee figure, its second President, Robert Russa Moton (see Chap. 1)—who would also be linked to *Brown v. Board of Education* (Encyclopedia Virginia 2020):

> The all-black Robert Russa Moton High School in Farmville, near Moton's birthplace, was also named for him. In 1951, students walked out of classes to protest the school's inadequate facilities, prompting a lawsuit that eventually led to the landmark U.S. Supreme Court decision *Brown v. Board of Education of Topeka, Kansas* (1954). The ruling mandated the desegregation of public schools. The high school now houses the Robert Russa Moton Museum [See: http://www.motonmuseum.org/about/].

At around the same time as the Moton school walkout, Reverend Oliver Brown wanted his 7-year-old daughter, Linda, to attend their neighborhood all-white school, in Kansas, which was following the state's segregation laws. Thurgood Marshall, the attorney representing the National Association for the Advancement of Colored People (NAACP), represented Reverend Brown, and argued the case before the U.S. Supreme Court. On May 17, 1954, segregated education was upended with the Supreme Court ruling of *Brown v. Board of Education of Topeka, Kansas*. This landmark decision hinged on the sudden death of Supreme Court Justice Frederick Vinson (Totenberg 2003) who would never have challenged the legal doctrine of segregation, but President Eisenhower saw his death as an opportunity for a moral correction. He appointed a much more progressive Chief Justice—Earl Warren—who was willing to overturn *Plessy*'s "separate but equal" ruling. The *Brown* decision made clear that separate education for African Americans in public schools was unconstitutional because separate facilities were inherently unequal. The decision held that racially segregated public schools deprived African American children of equal protection guaranteed by the Fourteenth Amendment. Marshall's arguments included a famous social science experiment known as the "Clark Doll Test", a study conducted by psychologist, Kenneth Clark, who was also noted for being the first African American to earn his doctorate at Columbia University. Clark introduced the psychological effects of segregation on African American children. The *Brown* decision ordered that desegregation be done with "all deliberate speed". But according to the Legal Defense Fund (LDF 2020):

> Unfortunately, desegregation was neither deliberate nor speedy. In the face of fierce and often violent "massive resistance", [the NAACP's Legal Defense Fund] sued hundreds of school districts across the country to vindicate the promise of Brown. It was not until LDF's subsequent victories in *Green v. County School Board* (1968) and *Swann v. Charlotte-Mecklenburg* (1971) that the Supreme Court issued mandates that segregation be dismantled "root and branch," outlined specific factors to be considered to eliminate effects of segregation, and ensured that federal district courts had the authority to do so.

Many states exploited the *Brown* decision by closing its black schools, and firing all its African American teachers (Eckhert et al. 2018; Gladwell 2017) because the *Brown* decision had nothing to say about integration of African American teachers into white schools. Thus, many argue that African American students ultimately wound up being unsuspecting "human subjects" in a social experiment that exposed them to many unintended risks and harms. Instead of being educated in safe environments with outstanding African American mentors and champions, the students were thrown into hostile environments facing daily harassment, racial bias, and often racist white teachers. Notes Wheaton College faculty on the occasion of the death of Linda Brown Thompson—the named plaintiff (Eckhert et al. 2018).

> In the *Brown v. Board* decision, teachers are only mentioned once in the main text. We are still paying the price for that glaring omission as schools lost many excellent African-American teachers. There were approximately 82,000 African-American teachers across the South at the time of the Brown decision. As schools were integrated, those serving African-Americans were closed, and their teachers were fired.

The impact of *Brown* actually led to far fewer African Americans applying to graduate schools because of the painful racism they endured in their primary and secondary education years, which also affected motivation and grades. For segregated black colleges and universities at the time like The Tuskegee Institute, the *Brown* decision actually made things worse for African American teachers and administrators (Gladwell 2017) who would lose their jobs and struggle to find work in traditional white institutions.

In 2008, then-Presidential Candidate Barack Obama noted in a Philadelphia speech: "Segregated schools were, and are, inferior schools 50 years after *Brown v. Board of Education*—and the inferior education they provided, then and now, helps explain the pervasive achievement gap between today's black and white students" (LDF 2020). As for patient care, fewer African American physicians led to significant health disparities in healthcare delivery, too as there were fewer black university hospitals. There were a few white-run hospitals that would see African American patients; one good example is Henrietta Lacks, who was seen at Johns Hopkins in the 1950s— at the same time Blalock and Thomas were there—because that hospital had a ward for African American patients. (Biopsying Lacks' tumor and using it to make a cell line, called HeLa, actually had little to do with her race.)

With respect to medical training for African Americans, some physicians had trained at white institutions in the North, which accepted the occasional African American applicant, but most trained at a number of black colleges and universities that grew as a result of segregation, reinforced by the *Plessy* decision. As discussed in Chap. 1, African American professionals born in the Plessy Generation, who were raised in the segregated South, believed in the philosophy of "racial betterment" (see Chap. 1). Although the 1954 *Brown* decision legally ended segregation, as discussed in Chap. 1, it would be well into the 1970s before anything really changed in the lives of African Americans living in South. Meanwhile, professional organizations would remain segregated for years; the American Medical Association (AMA) only began admitting African American professionals as members in 1968.

The Harold Thomas Case

One of the most notable cases surrounding Plessy era discrimination was a lawsuit filed by Harold Thomas in 1941 because white teachers were being paid more than African American teachers for the same job. Harold Thomas was Vivien Thomas' brother. He sued the Nashville Board of Education and was represented by Thurgood Marshall, who would later argue the *Brown* case to the Supreme Court (Wood 2008). McCabe noted (1989):

> Vivien's older brother, Harold, had been a school teacher in Nashville. He had sued the Nashville Board of Education, alleging salary discrimination based on race. With the help of an NAACP lawyer named Thurgood Marshall, Harold Thomas had won his suit. But he lost his job. So Vivien had learned the art of avoiding trouble.

Medical School Quotas

An important topic to cover is the medical school quota system that routinely barred women, Jews, and African Americans from enrolling in medical schools. The film depicts a camaraderie between Taussig and Thomas based on a shared experience of being the outsider in a white male culture. Taussig's experience was typical, and Thomas was trapped in the segregated system of *Plessy v. Ferguson* (see Chap. 1). Several medical students are still finding more hidden enrollment quotas, which more recently are using race as a way to be inclusive in enrollment (Blake 2012). Notes Sokoloff (1992):

> Quota systems had several expressions. Some were geographic and derived from the charters of state schools dedicated to residents. Sectarian medical schools had their own constituencies. Prestigious private institutions viewed themselves as producers of the nation's future leaders and wanted their student bodies to represent a diversified, broad population. Ivy League schools led the way. At Columbia University, President Nicholas Murray Butler devised a policy of "selective admissions".

Quotas for Women in Medicine

The history of women in medicine tracks somewhat with the history of reproductive rights (see Chap. 5). A surge of women entering medicine began in the nineteenth-century, when women made up more than 10 percent of enrollments in most U.S. medical schools. By 1900, 18% of medical school enrollments were women in some cities (Morantz 1978). As women were increasingly struggling for autonomy and self-determination over their bodies, the attraction to medicine and an independent profession was a way to escape the oppressive existences of marriage and compulsory motherhood (see Chap. 5). The male medical professionals did not begin to bar women from medicine until the twentieth century, which some scholars attributed to a "male backlash". This, again, was similar to male opposition to reproductive control.

As women's enrollment increased, the quotas on women's enrollments increased, too, which was what Taussig experienced (see earlier). Notes Morantz (1978):

> The failure of women doctors to institutionalize nineteenth-century gains made resistance to male attacks increasingly difficult. Lacking proper support, all but one of the women's medical schools closed down. Although co-education remained an alternative, equal opportunities failed to materialize. Medical schools established female quotas, which rarely exceeded a token 5 percent… [T]he conflict between feminism and professionalism forced [many] to choose between supporting inferior medical education for women or limiting the number of women physicians.

By 1960, women's enrollment in medical school was down to 9%, and reached levels on par with 1900 by the late 1970s; these trends historically tracked with Helen Taussig's career.

Quotas for African Americans

As discussed in Chap. 1, there were several black colleges and universities that had fine medical schools, which included the Tuskegee Institute. In 1934, Carter G. Woodson, noted in his book The Negro Professional Man in the Community (Woodson 1934), that being a physician was the highest achievement in the "Negro race" and it was still true by 1972 (Sorenson 1972). In the *Journal of Negro Education*, Andrew Sorenson (1972) discussed that despite the existence of black medical schools, the routine barring of African Americans from white medical schools through the decades led to a public health problem in the black community as there were a disproportionately lower number of physicians compared to the growth of the African American population. A 1953 article about the issue of quotas opened like this (Bloomgarden 1953):

> America's single most important welfare problem—most authorities have long agreed—is the endangering of the nation's health by a shortage of medical care and particularly of doctors; and this supposedly organizationally ingenious nation has not yet found a way of improving this situation. Worse, we use wastefully the scarce physicians' training facilities we possess: some of the best candidates we have for the all too few places in the medical schools are barred in favor of less fitted individuals. The racist discrimination in admissions policy that every pre-med schoolboy knows exists is, thus, not primarily the problem of those discriminated against: it is a blow against the nation's health.

By the late 1960s, after the *Civil Rights Act* was passed, established white medical schools began to expand admissions for African Americans, and these quotas began to disappear, which also tracks historically with Thomas' eventual recognition in the 1970s (see further).

Quotas for Jewish Students

In addition to firm quotas for women and African Americans, quotas limiting Jewish applicants to medical school were widespread. Notes Barron Lerner (2009):

A central reason that colleges and medical schools established quotas in the early twentieth century was the immigration of millions of Eastern European Jews to New York and other cities. When children from these families pursued higher education, the percentage of Jewish applicants increased. Quotas for Jewish medical students and physicians disappeared fairly rapidly after World War II, partly in response to Nazi atrocities against the Jews.

Within the immigrant population coming to the United States, Jewish immigrants were particularly interested in medicine. "In 10 major cities surveyed over five-year periods, the number of Jewish physicians graduated had increased from seven in 1875–1880 to 2,313 in 1931–1935…in 1934…more than 60% of the 33,000 [medical school] applications on file were from Jews (Sokoloff 1992). In 1931, Heywood and Britt (Sokoloff 1992; Heywood Broun and Britt 1931) looked at the Jewish quota system, and wrote:

> Yet the question of discrimination against Jews in medicine is the most delicate and difficult chapter in the whole history of prejudice in America. There is less frankness here, more cross currents and division of opinion, greater danger that an intrusion of comment may bring down the wrath alike of those who discriminate and those discriminated against…. It would be going too far to ascribe belligerent hatred to the medical colleges. They are honestly disturbed by what they consider menacing conditions.

Ultimately, in addition to the Holocaust and returning Jewish veterans who applied to medical school, the Jewish quota system also ended when New York and Philadelphia investigated the problem, and New York State established four publicly supported nondiscriminatory medical schools, which accepted many New York Jewish applicants (Halperin 2019).

Blalock and Thomas 1930–1964

Blalock and Thomas met in 1930 and worked together until Blalock's death. They first met at Vanderbilt University, where Thomas was hired as a lab technician by Blalock. Due to the Depression and failed banks (see Chap. 1), Thomas lost all of his money which was in a Nashville bank account. At that juncture, his plan to go to college and medical school was untenable, and he was grateful to just have a job in Blalock's lab. He was officially hired as a janitor at first (Thomas 1985). Blalock and Thomas did a number of experiments together and developed several surgical innovations (see History of Medicine). In 1941, Johns Hopkins offered Blalock the position of Professor and Chief of Surgery, as well as directing the surgery curriculum for the medical school. Blalock accepted the position on the condition he could bring Thomas with him, who was the only African American at Hopkins who was not a janitor. It was at Hopkins where Blalock and Thomas developed their Blue Baby shunt surgery (in 1944).

Baltimore was extremely segregated in 1941, when Thomas moved there; he dealt with housing issues and more intense racism than he experienced in Nashville (Thomas 1985). When he was at work, even his wearing a lab coat created stares, and was scandalous. What was not depicted in the film is that to make extra money because of his low pay, he actually worked as a bartender for Blalock at his parties (Thomas 1985), and Blalock eventually got him a raise by 1946, when he became the highest paid technician at that time at Johns Hopkins. Thomas often dreamed of getting a terminal degree throughout the 1940s and 1950s and considered the historically black university, Morgan State University but they would not give him credit for his work to date, and it was not feasible (McCabe 1989).

As *Something the Lord Made* beautifully demonstrates, Blalock and Thomas shared a close collegial relationship, however, institutionalized and government-organized racism interfered with academic equality and in the sharing of academic credit (see under Healthcare Ethics Issues). To understand the historical location of Blalock's and Thomas' collaboration, it essentially spanned almost the entire time frame of the Tuskegee study (1932–72), which I discuss exhaustively in Chap. 1, including the social and historical context of racism at that time, racism in medicine, which included false premises about the African American body, and the experiences of African American medical professionals in this time frame. Blalock lived to see the 1963 March on Washington, and the Kennedy Assassination; but the last medical article on the Tuskegee study was published in 1964 (see Chap. 1), the year Blalock died—just two months after he had retired. Like the racist policies endemic in medicine at that time, which spanned Blalock's entire career, the Tuskegee study would continue into the next decade as a symptom of the blatant racism coursing through the veins of the medical world in which Blalock and Thomas still managed to collaborate.

Thomas Going Solo: 1965–1979

Vivien Thomas' career continued at Johns Hopkins until he retired in 1979. Despite having had no secondary school education or a college degree, he managed in the Plessy era to become a world class surgeon who mentored hundreds of white surgical residents who would go on to prominence themselves (Thomas 1985). Thomas' solo career at Hopkins essentially began when the Civil Rights period reached a fever pitch in the late 1960s. Thomas was at Hopkins when Martin Luther King was assassinated, followed by terrible rioting in every major U.S. city, including Baltimore, and then two months later, he witnessed Robert F. Kennedy's assassination. But Thomas also witnessed the moon landing, which, like his own improbable career, was a story of perseverance. By 1970, when discussions about formally recognizing and honoring Thomas ensued (see earlier), the "Sixties" was now a euphemism for dramatic sociological change; a new reconstruction period had occurred by 1970 in which the Civil Rights period would be digested more fully, while Women's Rights was just getting started (see Chap. 5). Culturally, the ground had completely shifted with respect to how African Americans were perceived. This was reflected in a genre of films known as "blaxsploitation," starting with *Shaft* (1971) featuring Richard Roundtree, and the amazing theme song by Isaac Hayes. Now, African Americans were starring as law enforcers, private detectives in a "super hero" role. On television, *Mod Squad* (1968–1973) had a cult following, and featured Clarence Lincoln III as "Link"—demonstrating mixed race working relationships.

In 1972, Thomas would have read the Jean Heller story that broke in July 1972 surrounding the Tuskegee study (see Chap. 1), just a month after the first story on the Watergate burglary in the *Washington Post* by Bob Woodward and Carl Bernstein (June 22 1972). By the time Gerald Ford was inaugurated in 1974, Thomas was living in a country that was unrecognizable from the one in which he first met Blalock in 1930.

In 1976, the book Roots by Alex Haley was published, and spent 22 weeks as the number one the *New York Times* bestselling book, and forty-six weeks on the bestseller list. (Haley had also co-authored the Autobiography of Malcom X (1965)). The same year millions of Americans were reading Roots, Johns Hopkins made an organizational ethics decision to finally recognize Vivien Thomas as a doctor, and awarded him an honorary doctorate, so that Dr. Thomas could finally take his place as an institutional treasure, along with his 1971 portrait. By 1977, *Roots* was turned into a miniseries, and was a watershed cultural event for American audiences; many were only first learning about the slave trade and the consequences of slavery in the United States. By 1979, amidst a flurry of affirmative action policies throughout the government and private sectors, Thomas retired. It was the same year The Belmont Report was published (National Commission 1979), which established the first ethical principles for human subject research. "Vulnerable populations", such as economically disadvantaged African Americans, would be formally recognized as a protected class of patients and potential medical research subjects, but it would be decades before anyone connected the barriers to formal medical education that Thomas overcame to the social production of health disparities, augmented by the dearth of African American physicians post-*Brown*. Thomas died in 1985, but would live to see a widening of health and economic disparities in the African American community during the Reagan years (see Chap. 2).

The White Savior: 1989

Although Thomas wrote an autobiography in 1985 (Thomas 1985), his life story would not be well known or come alive until Katie McCabe's *Washingtonian* article in 1989 (see earlier). The Thomas story functions as a "white savior" genre in which the personal and emotional stories of African Americans who lived during segregation and the Jim Crow period were often not known until white writers brought their stories to various media, raising questions about who owns these narratives. The classic example is Rebecca Skloot's The Immortal Life of Henrietta Lacks (2011) which has made millions for the white author. But this genre was just beginning in 1989, when the *Washingtonian* article hit. By then, Steven Spielberg's *The Color Purple* (1985)—discussed more in Chap. 1—and *Mississippi Burning* (1988) were integral films for white audiences about race. Spike Lee was a groundbreaking African American filmmaker who exploded onto the scene with his film, *Do The Right Thing* (1989) about boiling racial tensions in Brooklyn. In 1989, just as the first Bush era was beginning (see Chap. 8), racial tensions had become more subterfuge, and in many ways, the Blalock-Thomas relationship resonated more in 1989 than during their lifetimes. But as medical schools began to realize they had serious diversity problems, and doctor shortages led to thousands of foreign-trained healthcare providers working in the United States with J-1 visas, Thomas' experiences as the only African American lab technician at Johns Hopkins began to tell universal stories of non-whites of every shade working in labs and experiencing racism—especially after 9/11.

2004: Post 9/11 and Non-whites

When *Something the Lord Made* first aired May 30, 2004, it resonated on levels to new audiences no one anticipated. The socio-political landscape of post-9/11 is discussed at great length in Chap. 4, but with respect to racism, the focus began to shift to Islamophobia and anyone with "brown skin". Regardless of profession, character or background, anyone "brown" became a terrorist suspect or someone to be feared or despised. Americans or U.S. residents from India, Pakistan, and a wide range of majority Muslim countries or Middle-Eastern countries were suddenly the targets of overt discrimination. But by 2004, American medicine had become dependent on foreign-born scientists; longstanding problems that Thomas experienced with access to higher education, uneven educational opportunities and widening economic disparities prevailed. There were simply not enough American-born scientists and doctors to keep the United States competitive in the STEM fields, which required importing from other countries. When the Blalock-Thomas story was revisited on film in 2004, thousands of immigrant scientists born in other countries were working in U.S. laboratories and academic medical centers on J-1 visas, or had become permanent residents and citizens. Historically, science and immigration had always been an "American love story" (Rosenthal 2017), and almost every major American scientific achievement—from the atomic bomb (involving Jewish refugee scientists who sought asylum) to the moon landing (Germany's Werner Von Braun) or even the iPhone (Steve Jobs' biological father was a Syrian immigrant granted asylum in the United States)—had involved an immigrant scientist or one whose parents had emigrated (Rosenthal 2017).

The world of science and medicine by 2004 was filled with foreign-born scientists and physicians, many of whom were practicing Muslims. Anyone with brown skin could be called a "terrorist", told to "go back to their country" and in many hospitals, would be openly rejected by patients. The best way to explain what the United States felt like in 2004 is to view the disturbing film, *Crash* (2004), which won the Academy Award for Best Picture. *Crash* was an updated "toxic L.A." genre film (see Clinical Ethics on Film) that painted the reality of the post-9/11 cultural anxieties in Los Angeles. This was an era in which Muslim hate crimes escalated, and "Muslim-looking" patrons were routinely kicked out of retail establishments and told to "Go plan Jihad somewhere else"; Hispanics were distrusted, seen as a lazy, burdensom, and treated as second class citizens. Asians—especially Koreans in L.A.

who owned many of the small businesses—were both the target, and sometimes the perpetrators of discrimination. *Crash* set all of these contexts against the chronic and enduring racism towards African Americans in a city that never properly dealt with festering wounds from the 1990s, which produced the Rodney King beatings and the chaotic O.J. Simpson trial, which made legal analyst Greta Van Sustren a star, and who by 2004 had become a mainstay on Fox News, which fomented hatred toward immigrants and Muslims (Bloom 2018).

A month before *Something the Lord Made* debuted on HBO, the American atrocities at the Abu Ghraib prison in Iraq were disclosed by the press (CBS 2004), and the iconic pictures of young American soldiers torturing Iraqi citizens held without due process were broadcast: Muslim males naked, wearing hoods, and piled on top of other naked men while young American soldiers took "selfies" to commemorate their deeds. This was the America in which the Vivien Thomas story would come alive on film. One week later, on June 5, 2004, news that former President Ronald Reagan had died dominated the headlines. Reagan's last speech was about the strengths of American immigration and that it was a beacon of freedom to refugees seeking asylum (Reagan 1989). His death would mark the end of the Republican party as it was once known in the 1980s and 1990s—a party that was actually pro-immigration because it was pro-business. It was also a party that had once denounced white supremacy. Although Ronald Reagan was alive during 9/11, no one would ever know what he thought about the country in its aftermath because he did not remember that he was ever a president by then; he had totally succumbed to Alzheimer's disease (Altman 1997). During the summer of 2004, as the film was being reviewed and nominated for awards (see further), the 2004 election cycle was in full swing, and at the Democratic Convention on July 27, 2004, the next Democratic President of the United States, Barack Obama, would give a remarkable speech on the convention floor in which he described his improbable American journey that allowed him to pursue greatness (Obama 2004). In some ways, it was the political version of Vivien Thomas' story: despite everything, Thomas was allowed to pursue greatness with his 1944 colleagues.

History of Medicine

When teaching the history of medicine context of *Something the Lord Made*, the obvious topics to cover are a review of Blalock's and Thomas' careers and contributions to the fields of surgery, and the founding of cardiac surgery. However, it's equally important to review the career of Helen Taussig, as both a "Women in Medicine" topic, and a pediatrics topic, as Taussig indeed founded the field of pediatric cardiology, and brought the problem of "blue babies" to Blalock and Thomas in the first place. Finally, a discussion of the consequences of "revisionist" medical history is critical, as new ethical obligations fall to medical educators to correct the medical record for historical and moral posterity, and put back the women, African Americans and other minorities into the medical history books.

Helen Taussig

With respect to Helen Taussig, she is an outsized figure from the perspective of Women in Medicine. Taussig had dyslexia, and had also contracted tuberculosis at age 11, when her mother died from it (McLaren 2020). She was determined to get an education just around the time that first wave feminism and the fight for the vote was heating up. She studied at Radcliffe College in 1917, then University of California Berkeley where she earned a Bachelors of Arts in 1921. She wanted to study medicine at Harvard but due to quotas (see earlier), she was denied admission, and went to Boston University instead (1922–24). Taussig's mentor was another woman in medicine—Canadian pathologist Maude Abbott, who had an interest in congenital heart disease that inspired Taussig (McLaren 2020). When Taussig began practicing at Johns Hopkins, at age 32, she was made the director of John Hopkins' Harriet Lane Clinic, a pediatric health center, which made Taussig one of the most successful women in medicine at that time. But shortly after, Taussig became deaf and used her hands to guage cardiac rhythms, which is how she discovered blue baby syndrome (McLaren 2020).

Ultimately, Taussig was the founder of pediatric cardiology and the surgery, as it was she who brought the problem to Blalock and Thomas, and helped them to innovate the surgery to correct the congenital heart defect that causes the syndrome. In the early 1960s, it was Dr. Taussig who helped to prevent the drug thalidomide from being used in the United States when she testified to the Food and Drug Administration on its terrible effects based on the data from Europe (National Library of Medicine 2015). In 1965—the year Blalock died—Taussig became the first woman president of the American Heart Association (McLaren 2020).

Correcting the Historical Records

Revisionist history of medicine is everywhere in the medical or STEM curricula. Vivien Thomas was essentially erased from the annals of medical history archives and books until the 1980s. In the film, when *Life* magazine gathers the blue baby team for the historic picture and article, Vivien Thomas isn't in the picture or the article, nor do his colleagues yell "Wait" to the photographer to make sure he is included. When he looks at the article about his work that has not mentioned him, his wife, Clara states: "Just cuz you weren't in the news, doesn't mean you weren't there." Thomas was not in the 1949 *Time* magazine article, either, or in any of the subsequent media or medical journal articles that were written about the blue baby surgeries when Blalock was alive, and could have corrected the record. The Vivien Thomas story is an example of "whitewashing" medical history, which means to slap clean white paint over the history to conceal perceived "defects"—remove individuals who are inconvenient or politically incorrect; cover up unethical or illegal details, or sanitize the record. Classic examples of whitewashing are ignoring the professional contributions of African Americans, other minorities (e.g. Jews, Asians, Native Americans) and women (e.g. Rosalind Franklin) to the history of medicine. Other forms of whitewashing are to "sanitize" medical history to remove anything that would pass for "unethical research" today, including the contribution of non-consenting vulnerable populations as human subjects. A classic example of "sanitized" medical history is James Marion Simms, who tortured female African American slaves and used them without consent to perfect gynecological surgeries and procedures (Washington 2007). Although the classic term "whitewashing" was derived from the actual use of white paint to cover something up, in recent decades, it began to be used in Hollywood specifically to refer to casting white individuals into roles that could be, or should be, played by African Americans instead, and by extension, began to refer to "whites only" in history. Ultimately, whether the traditional or more recent use of the term whitewashing is applied, *Something the Lord Made* is one of the best examples you can find.

The role of medical educators is to scrutinize the history of medicine record and provide a full accounting, as well as the historical context that led to the whitewashing in the first place. In some cases, as in the history of the insulin discovery (Bliss 1982) or the discovery of HIV/AIDS (see Chap. 2), whitewashing occurs because of greed and competition for glory that has nothing to do with race or gender, but in ego—taking credit and ownership that should be shared with others. Credit sharing is a classic publication ethics issue discussed further under Healthcare Ethics Issues. One could make the case that Blalock's reluctance to include Vivien may have been more similar to Robert Gallo's behaviors in *And the Band Played* On (see Chap. 2) by his refusing to acknowledge his colleagues as co-discoverers of HIV. Blalock, however, does give credit in public to all of his white colleagues, including his white female colleague. One could also muse whether Blalock's reluctance was really due to the fact that Vivien didn't have the proper perceived credentials to participate, worried about "what would the neighbors think". It is more realistic in retrospect to view his concerns as multifactorial. It is documented by McCabe (1989) and Thomas (1985) that Blalock, as he states in the film had regrets. He knew what he did, but he didn't know how to quite correct it. Correcting the historical record fell to his surgical descendants as a moral correction, and Johns Hopkins in the 1970s should be credited for "doing the right thing".

Healthcare Ethics Issues

In my teaching, *Something the Lord Made* has become a "sleeper hit" with my students—especially those with diverse backgrounds. When teaching *Something the Lord Made*, there are many healthcare ethics dimensions. The macro healthcare ethics issues revolve around disparities in medical education, such as quotas, which then create health disparities for several minority communities. One could also point out research ethics issues in *Something the Lord Made*, such as the ethical frameworks for animal models and human subjects; some students may even be shocked to see how dogs were used as the animal models for Blalock. However, I use this film to teach about the micro-ethics issues, which comprise the underbelly of academic medical centers: institutional racism, microaggressions, collegiality, credit-sharing, and organizational ethics.

Racism and Microaggressions

When teaching this film to current students, overt racism still occurs, but most experience racism these days as microaggressions—patronizing comments, questions about ethnicity or heritage, silent eye rolls or distinct body language indicating disdain, disapproval, or harsh judgement. Microaggressions are sneaky, and few of us escape them. The *Journal of the American Medical Association* published a study of microaggressions in 2018 (Osseo-Asare 2018), which was widely reported (Carroll 2018). The study author noted (Osseo-Asare 2018):

> Minority residents appear to face extra challenges during a time already marked by considerable stress. Residency program leaders and accreditation bodies should work to address these challenges not only as an important wellness issue, but also to minimize potential damage to the minority pipeline…Although blacks, Hispanics and Native Americans together make up one third of the nation's population, these three minorities constitute just nine percent of physicians… And part of that disparity may be due to biases minorities encounter in the pipeline.

The Vivien Thomas experience of the 1940s was being replicated in 2018. According to the study authors (Osseo-Asare 2018):

> A common issue for these residents was being mistaken for support staff, even for janitors, despite wearing a white coat, stethoscope, and identification badge showing they were physicians…One young physician described such an encounter. A patient's aunt had been visiting and when the mother came in, the aunt told her about the "janitor" who had been taking care of the patient. The mom told the resident about her conversation with the aunt: "She told me, 'Oh the janitor was so smart. He was telling everybody else what to do. He really knew his stuff.'"

> Other residents talked about how people somehow couldn't manage to tell them apart: "Six of us are black women. They're constantly interchanging our names, constantly interchanging people that don't even look alike."

> Sometimes the aggression was more overt. One Hispanic resident talked about an encounter with a xenophobic patient, who said, "someone like you should go back to where you came from. You're taking advantage of our resources, and there's all these students that would like to get into medical school that are here and from the United States and don't get in. And then you people come, and you take our places, and you take our jobs. And you don't even have citizenship, and you don't even speak English.'"

One professor recalled the experiences of a minority surgeon who entered the operating room and was told what needed to be cleaned (Carroll 2018; Osseo-Asare

2018). In the film, Thomas quits briefly because of the microaggressions, but eventually returns to work with Blalock; similarly, many minorities in healthcare quit because they find the experience so difficult (Carroll 2018; Osseo-Asare 2018).

Microaggressions are prevalent in patient care, too, in which minority patients are treated unequally and differently, but need more healthcare because of barriers to healthcare in the first place (see Chap. 4). One physician noted (Tello 2017):

> [M]ost physicians are not explicitly racist and are committed to treating all patients equally. However, they operate in an inherently racist system…We now recognize that racism and discrimination are deeply ingrained in the social, political, and economic structures of our society…For minorities, these differences result in unequal access to quality education, healthy food, livable wages, and affordable housing. In the wake of multiple highly publicized events, the Black Lives Matter movement has gained momentum, and with it have come more strident calls to address this ingrained, or structural, racism, as well as implicit bias….

> A [Muslim] colleague of mine recently wrote about her experiences treating patients at our own hospital. She has been questioned, insulted, and even attacked by patients, because she is a Muslim woman who wears a headscarf. She is not alone. Recent published reports include overt bigotry expressed towards doctors of black, Indian and Jewish heritage. Several medical journals have just published guidelines for doctors with titles like "Dealing with Racist Patients" and "The Discriminatory Patient and Family: Strategies to Address Discrimination Towards Trainees." It's sad that we need these guides.

Implicit bias and microaggression is also expressed in admissions practices, a well-known phenomenon. At my institution, where African American representation in healthcare training is still low, after viewing *Something the Lord Made*, I ask my students to walk around the institution and note who is a healthcare trainee, student or provider, and to then observe who makes up the custodial staff so they are aware of problems with diversity in higher education and medical education. The exercise for my students is sobering because the Vivien Thomas story is alive and well, and resonating deeply.

Collegiality

Early in the film when Blalock and Thomas had a successful breakthrough with shock surgery, Thomas had not set the "smoked drums". This referred to a "kymograph, consisting of a revolving drum, bearing a record sheet (usually of smoked paper) on which a stylus moved up and down displaying the effects of the drugs on contractile tissues, and was commonly used in experimental physiology and pharmacology laboratories in teaching institution" (Khilnani and Thaddanee 2013). Blalock went into a fit of rage and, asked Thomas if he had "shit for brains". Thomas walked out. Blalock then realized Thomas had, instead, recorded the experiment in detailed notes, and immediately ran after Thomas and apologized insisting "It won't happen again." That scene demonstrates, oddly, the right thing to do with a colleague. If you lose your temper or insult someone, take responsibility and apologize. Apology is actually rare, and in itself, can repair a relationship. McCabe (1989) reported the incident this way:

> Something went wrong," Thomas later wrote in his autobiography. "I no longer recall what, but I made some error. Dr. Blalock sounded off like a child throwing a temper tantrum. The profanity he used would have made the proverbial sailor proud of him. ... I told him he could just pay me off... that I had not been brought up to take or use that kind of language. ... He apologized, saying he had lost his temper, that he would watch his language, and he asked me to go back to work.

> From that day on, said Thomas, "neither one of us ever hesitated to tell the other, in a straightforward, man-to-man manner, what he thought or how he felt. ... In retrospect, I think that incident set the stage for what I consider our mutual respect throughout the years."

What Blalock apologized for was crossing professional boundaries with a colleague. Generally, crossing boundaries with those in subordinate positions occurs when authority figures behave disrespectfully toward a fellow co-worker and human being, by shouting, using obscenities in their presence, or even making comments about their personal appearance. When teaching around these issues, it's important to point out that anger can erupt in high stress situations, but one can correct it through apologizing, which is demonstrated in *Something the Lord Made*.

Lab Culture

When reviewing the Blalock-Thomas collaboration, it should also be viewed from the lens of a laboratory culture versus a hospital culture, which I discuss more in Chap. 8. Buchanan (2018) examined collegial interactions under laboratory conditions. Today, most labs are comprised of a team, rather than a single research assistant, as Thomas was. In addition to the principal investigators, there are faculty, post-docs, graduate students, lab technicians (what Thomas was initially designated as at Johns Hopkins), and even lab managers. However, few Principal Investigators are trained in leadership or management skills or even organizational ethics (Buchanan 2018), which can create all kinds of personality conflicts. Typically, there is a large focus in research ethics on ensuring ethical practices, but there are no mechanisms to "improve protections for its workers against work-related mental health issues, harassment, or disenfranchisement" (Buchanan 2018). In fact, it is even outside the role of most Institutional Review Boards (IRBs) to deal with workplace relationships in the lab. Thus, diversity issues, microaggressions and even overt racism would become a Human Resources issue and not part of the purview of the Office of Research Integrity. Similar to physician burnout (see Chap. 8), lab culture can foster horizontal violence and burnout as well. Buchanan (2018) notes that: Distress in the lab can occur from "unclear expectations, unrealistic goals, overwork, lingering disagreement regarding experiment design, publication readiness, hypothesis validity, lack of trust among team to uphold shared ethical norms; lack of communication about status of experiments, mishaps, failures; and unclear plan in place for when experiment results do not match hypothesis. Many institutions now have research ethics consultation services, which can address some of these issues. When teaching *Something the Lord Made* to healthcare trainees who spend time in labs, some focus on collegiality in the lab environment may be useful.

Credit Sharing

The issue of credit sharing is also discussed at length in Chap. 2. With respect to this film, Thomas confronted Blalock regarding the credit situation: "I'm invisible to the world; I thought it was different in here." Blalock responded by blaming the segregated world they inhabit. Thomas says: "I'm not talking about them; I'm talking about you!" A review of the historical archive in both the lay and medical press documents that Blalock missed opportunities time and again to correct the record and give Thomas credit for their partnership; he never did. Institutionally, their partnership was transparent—not even an "open secret". McCabe (1989) reported on Blalock's regrets, but it seemed to revolve more around not helping Vivien with getting a terminal degree.

> During his final illness Blalock said to a colleague: "I should have found a way to send Vivien to medical school." It was the last time he would voice that sense of unfulfilled obligation.
>
> Time and again, to one or another of his residents, Blalock had faulted himself for not helping Thomas to get a medical degree. Each time, remembers Dr. Henry Bahnson, "he'd comfort himself by saying that Vivien was doing famously what he did well, and that he had come a long way with Blalock's help."…Blalock's guilt was in no way diminished by his knowing that even with a medical degree, Thomas stood little chance of achieving the prominence of an Old Hand. His prospects in the medical establishment of the 1940s were spelled out by the only woman among Blalock's "boys," Dr. Rowena Spencer, a pediatric surgeon who as a medical student worked closely with Thomas.
>
> In her commentary on Thomas's career, published this year in A Century of Black Surgeons, Spencer puts to rest the question that Blalock wrestled with decades earlier. "It must have been said many times," Spencer writes, "that 'if only' Vivien had had a proper medical education he might have accomplished a great deal more, but the truth of the matter is that as a black physician in that era, he would probably have had to spend all his time and energy making a living among an economically deprived black population."

When I teach this story, historical context is everything, and the Plessy era did not end in the South until the late 1960s, beyond the *Voting Rights Act* (1965). Blalock's world view was shaped by racial segregation, racism in medicine, and overt racism everywhere in the Jim Crow era. Blalock could not go beyond a certain point when it came to crediting Thomas outside the institution. But after his death, acknowledging this historical wrong became an *institutional decision*.

It's important to note it was not Blalock's fault that Vivien did not have the opportunity to get a medical degree, nor did he have a moral obligation to help him. They both understood that there were unfair societal arrangements that kept Vivien Thomas from obtaining his credentials. It was the failure of Dr. Blalock to acknowledge Thomas' efforts as a colleague, which was the moral transgression, yet his regrets continued to revolve around his failure to assist him to complete his credentials. This demonstrates a cognitive dissonance in which Blalock must have rationalized the connection between sharing credit with Thomas and Thomas' lack of credentials, instead of what it really was: institutionalized racism. Clearly, Blalock must have felt that sharing credit with an African American lab assistant would only be accepted by his peers had Vivien demonstrated the right credentials.

Honorary Practices, Reparations and Organizational Ethics

The study of apology and reparations is a subset of organizational ethics. On a large scale, many scholars in this area point to Germany as one of the first examples of how to make amends. After the Holocaust, Germany was unable to escape its crimes against humanity because the allies ensured there was ample documentation. The country embarked on enshrining reparations, apology, ownership of its past and preventative ethics policies in its constitution to ensure that racism and right winged ideology could not flourish again, or at least dominate. Turkey, on the other hand, never acknowledged the Armenian Holocaust, which had been Dr. Jack Kevorkian's legacy (see Chap. 3). In the United States during the 1970s, policies to counteract the legacy of racism became organizational ethics issues, too. A more recent effort to address structural racism and white supremacy began in several institutions in the wake of Black Lives Matter protests during the spring of 2020 (see Afterword).

At Johns Hopkins, the decision to honor Thomas started as a lobbying effort by his colleagues in surgery, and eventually led to the portrait and honorary degree discussed earlier. The portrait was commissioned in 1969 (Thomas 1985), painted by Bob Gee, and the unveiling of the portrait as a formal institutional ceremony occurred in February 1971. For many institutions, reparation may have stopped there, but Johns Hopkins completed the task with an honorary degree that recognized Thomas by giving him a doctorate of laws, (not an M.D., as the film presents). This enabled him to be appointed in the Department of Surgery as an Instructor, and to be called "Dr. Thomas". In this respect, Hopkins was probably ahead of its time in making reparations, but its organizational ethics decision to do this at all helped to morally correct the historical wrong of Thomas' exclusion in history. By 2005, Thomas' name was chosen as one of four named Colleges of Medicine within the institution.

Conclusions

Medical educators are continuously confronted by issues of race and diversity in healthcare delivery, which have become more pronounced in the Trump era. Healthcare trainees comprise students from every culture and race, and new challenges have emerged over the issue of healthcare trainees who came out of the shadows in the Obama era because of DACA (Deferred Action on Childhood Arrivals), who lost their protected status in the Trump era, but are trained healthcare providers now.

Something the Lord Made allows educators a point of entry: a heartfelt story about race and collegiality for the past that is more relevant today than when it was first released.

Theatrical Poster

Something the Lord Made (2004)

Written by: Peter Silverman and Robert Caswell
Directed by; Joseph Sargent
Starring: Mos Def, Alan Rickman, Kyra Sedgwick, Gabrielle Union, Mary Stuart Masterson
Producers: Robert W. Cort, David Madden, Eric Hetzel, Juian Krainin, Mike Drake
Production company: HBO Films, Nina Saxxon Film Design
Original Network: HBO
Original Release: May 30, 2004

References

Altman, L. (1997). Reagan's twilight. *The New York Times*, October 5, 1997. https://www.nytimes.com/1997/10/05/us/reagan-s-twilight-a-special-report-a-president-fades-into-a-world-apart.html.

Bahnson, H. T. (1987). Pioneering research in surgical shock and cardiovascular surgery: Vivien Thomas and his work with Alfred Blalock. *Annals of Thoracic Surgery, 44*, July 1987. https://doi.org/10.1016/S0003-4975(10)62364-8.

Birchmeier, J. (2019). *Biography of mos def*. https://www.allmusic.com/artist/mos-def-mn0000927416/biography.

Blake. (2012). Affirmative action and medical school admissions. *AMA Journal of Ethics. 14*(12), 1003–1007. (https://journalofethics.ama-assn.org/article/affirmative-action-and-medical-school-admissions/2012-12.

Bliss, M. (1982). *The Discovery of Insulin.* University of Toronto Press, Toronto.

Bloom, A. (2018). *Divide and conquer: The story of roger ailes.* A&E Indie Films.

Bloomgarden, L. (1953). Medical school quotas. *Commentary Magazine*, January 1953. https://www.commentarymagazine.com/articles/lawrence-bloomgarden/medical-school-quotas-and-national-healthdiscrimination-that-hurts-us-all/.

Broun, H., & Brit, G. (1931). *Christians only: A study in prejudice* (pp. 137–145). New York: Vanguard.

Brown v. (1954). *Board of Education of Topeka,* 347 U.S. 483

Buchanan, C. (2018). *Lab values: Professional ethics in the laboratory.* University of Kentucky Research Ethics Lecture.

Carroll, L. (2018). Minority doctors in US residency programs routinely face racism. *Reuters health,* September 28, 2018. https://www.reuters.com/article/us-health-minorities-medical-trainees/minority-doctors-in-us-residency-programs-routinely-face-racism-idUSKCN1M822I.

Duke University. (2020). *Black history month: A medical perspective—Exhibited February–March 1999 and February–March 2006.* https://guides.mclibrary.duke.edu/blackhistorymonth/education.

Eckhert, J., Johnson, K., & Pykkonen, B. (2018). Wheaton faculty experts on the impact of Brown v Board of education. *Wheaton News*, April 3, 2018. https://www.wheaton.edu/news/recent-news/2018/april/the-impact-of-brown-v-board-of-education/.

Encyclopedia Virginia. (2020). *Robert Russa Moton*. 2020. https://www.encyclopediavirginia.org/Moton_Robert_Russa_1867-1940#start_entry.

Gladwell, M. (2017). Miss Buchanan's period of adjustment. *Revisionist History (podcast)*. June 28, 2017. http://revisionisthistory.com/episodes/13-miss-buchanans-period-of-adjustment.

Halperin. (2019). Why did the United States medical school admissions quota for Jews end? *American Journal of Medical Sciences*, *358*(5), 317–325. https://doi.org/10.1016/j.amjms.2019.08.005.

IMDB. (2019). *Alan Rickman biography*. https://www.imdb.com/name/nm0000614/.

IMDB. (2012). *Mos Def biography*. https://www.imdb.com/list/ls050496143/.

Johns Hopkins Medical Archives. (2020). https://medicalarchives.jhmi.edu:8443/finding_aids/vivien_thomas/vivien_thomasd.html.

Khilnani, G., & Thaddanee, R. (2013). The smoked drum. *Indian Journal of Pharmacology, 45*(6), 643–645. https://doi.org/10.4103/0253-7613.121394.

Legal Defense Fund. (2020). *Landmark: Brown v. Board of education*. http://www.naacpldf.org/case/brown-v-board-education.

Lerner, B. (2009). In a time of quotas, a quiet pose in defiance. *The New York Times*, May 26, 2009. https://www.nytimes.com/2009/05/26/health/26quot.html.

McGoon, DC. (Nov 1986). Pioneering research in surgical shock and cardiovascular surgery: Vivien Thomas and his work with Alfred Blalock. *Annals of Surgery*, *204*(5), 607. https://www.ncbi.nlm.nih.gov/pmc/articles/PMC1251350/?page=1.

McCabe, K. (1989). Something the lord made. *The Washingtonian*, August 1989. http://reprints.longform.org/something-the-lord-made-mccabe.

McCabe, K. (2007). Like something the lord Made: The Vivien Thomas story. *The Washingtonian*, October 29, 2007. https://www.washingtonian.com/2007/10/29/like-something-the-lord-made-the-vivien-thomas-story/.

McLaren, K. (2020). Helen Brooke Taussig. *Encyclopedia Britannica*, May 20, 2020. https://www.britannica.com/biography/Helen-Brooke-Taussig.

Morantz, RM. (Jun 1978). Review: Women in the medical profession: Why were there so few? Review of "Doctors wanted-no women need apply": Sexual barriers in the medical profession, 1835–1975. by Mary Roth Walsh. *Reviews in American History*, 6(2), 163–170. https://www.jstor.org/stable/2701292.

National Commission for the Protection of Human Subjects. (1979). *The Belmont Report*. http://www.hhs.gov/ohrp/humansubjects/guidance/belmont.html.

National Library of Medicine. (2015). *Changing the face of medicine. Dr. Helen Brooke Taussig*. https://cfmedicine.nlm.nih.gov/physicians/biography_316.html.

Obama, B. (2004). Remarks at the democrat national convention. July 27, 2004. https://www.nytimes.com/2004/07/27/politics/campaign/barack-obamas-remarks-to-the-democratic-national.html.

Organization of American Historians. (OAH). (2020). *OHA Awards and Prizes*. 2020. https://www.oah.org/awards/.

Osseo-Asare, A., Balasuriya, L., Huot, S. J., et al. (2018). Minority resident physicians' views on the role of race/ethnicity in their training experiences in the workplace. *JAMA Netw Open., 1*(5), e182723. https://doi.org/10.1001/jamanetworkopen.2018.2723.

Plessy v. (1896). *Ferguson*, 163 U.S. 537.

Reagan, R. (1989). Farewell address. *The New York Times*, January 12, 1989. https://www.nytimes.com/1989/01/12/news/transcript-of-reagan-s-farewell-address-to-american-people.html.

Rosenthal, M. S. (2017). *Research and immigration: An American love story*. February 14, 2017, University of Kentucky Research Ethics Lecture.

Rosenthal, M. S. (2018). *Clinical ethics on film*. Springer International.

Sokoloff, L. (1992). The rise and decline of Jewish quotas in medical school admissions. *Bulletin of the New York Academy of Medicine, 68*(4), 497–518. https://www.ncbi.nlm.nih.gov/pmc/art icles/PMC1808007/pdf/bullnyacadmed00005-0055.pdf.

Sorensen, A. A. (1972). Black Americans and the Medical Profession, 1930-1970. *The Journal of Negro Education, 41*(4), 337–342.

Swain, V. (1987). Pioneering research in surgical shock and cardiovascular surgery: Vivien Thomas and his work with Alfred Blalock. *Medical History, 31*(2), 243. https://doi.org/10.1017/S00257 27300046792.

Sorkin I. M. (2007). Obituary of Irving Sorkin. *The Washington Post*, October 18, 2007. https://www.legacy.com/obituaries/washingtonpost/obituary.aspx?n=irving-m-sorkin&pid=96598431.

Schwartz, D. (2015). Yasiin Bey to perform last show ever this week. *Hot New Hip Hop*, October 16, 2016. https://www.hotnewhiphop.com/yasiin-bey-to-perform-last-show-ever-this-week-news.24778.html.

State University. (2019). *Historically black colleges and universities.* https://education.stateunivers ity.com/pages/2046/Historically-Black-Colleges-Universities.html.

Thomas, V. (1985). *Pioneering research in surgical shock and cardiovascular surgery: Vivien Thomas and his work with Alfred Blalock.* University of Pennsylvania Press. [Reprinted with the title: Partners of the Heart: Vivien Thomas and His Work with Alfred Blalock].

Totenberg, N. (2003). The supreme court and 'Brown v. Board of Ed.': The deliberations behind the landmark 1954 ruling. *NPR*, December 8, 2003. https://www.npr.org/templates/story/story.php?storyId=1537409.

Tello, M. (2017). Racism and discrimination in health care: Providers and patients. *Harvard Health Blog*, January 16, 2017. https://www.health.harvard.edu/blog/racism-discrimination-health-care-providers-patients-2017011611015).

Obituary of Vivien Thomas. (1985). Obituary. *Washington Post*, November 28, 1985. https://www.washingtonpost.com/archive/local/1985/11/28/vivien-thomas-researcher-for-hopkins-surgery-dies/6ea67fe4-d92a-4973-9a09-304fbbba48e0/?utm_term=.938df1672100.

Washington, H. (2006). *Medical Apartheid.* New York: Doubleday

Woodson, C. G. (1934). *The Negro Professional Man in the Community.* Washington, D.C.: The Association for the Study of Negro Life and History Inc.

Wood, E. T. (2008). Nashville now and then: A marshall plan for equity. *Nashville Post*, April 11, 2008. https://www.nashvillepost.com/business/education/article/20401895/nashville-now-and-then-a-marshall-plan-for-equity.

Afterword

Healthcare Ethics in a Post-COVID America: How to Use this Book

On March 11, 2020, the World Health Organization (WHO) declared the novel Severe Acute Respiratory Syndrome Coronavirus 2 (SARS-CoV-2), which causes the coronavirus disease 2019 (COVID-19), to be a global pandemic (WHO 2020). The virus had originated in Wuhan, China in the winter of 2019 as an outbreak, and then affected the rest of the world in the spring of 2020. By February, the virus had hit Washington State, while San Francisco declared a state of emergency by that month's end. Already, there were massive disruptions in travel, and several cruise ships were stranded.

When the news emerged of the global pandemic, I was, in fact, still working on this manuscript and completing the chapter on the film *Contagion* (Chap. 6).

As the world watched the virus ravage Italy, which responded with a nationwide lockdown resembling the plague years of medieval Europe, the United States began to see the same emerging pattern, and pre-emptively shut down its institutions and economy in an effort to "flatten the curve" and also preserve its already struggling healthcare system (see Chap. 4 on *Sicko*), which would collapse from the burden of cases. The United States had such a shortage of personal protective equipment (PPE) in early spring 2020 for its healthcare providers, the general public was originally instructed *not* to wear a mask for fear that consumers would stockpile and "panic buy" PPE, making the problem worse. At that time, there were already shortages in the United States of hand sanitizer, cleaning products, and bleach, not to mention basic hospital equipment, and intensive care unit (ICU) resources, such as ventilators. But the worst scenario was yet to come: the U.S. Centers for Disease Control (CDC) had a quality control error in early tests for the virus, which left the United States with no ability to do proper testing and tracing—a gold standard for controlling any outbreak (see. Chap. 6). By mid-March 2020, evidence-based projections (Ferguson et al. 2020) for the United States were a mass-casualty event of over one million dead Americans without extreme measures of mitigation: physical

© Springer Nature Switzerland AG 2020
M. S. Rosenthal, *Healthcare Ethics on Film*,
https://doi.org/10.1007/978-3-030-48818-5

distancing, which amounted to shelter-in-place and shuttering all non-essential businesses. The United States had completely transformed as the "nightmare" pandemic scenario—a virus easily spread through respiratory droplets—unfolded. Those of us who study American healthcare ethics already predicted this was going to be bad.

In March 2020, American bioethicists were concerned with many difficult questions surrounding rationing of scarce resources (see Chaps. 6 and 7) based on the Italian experience with COVID-19. There were three significant pieces of COVID-19 medical news that informed most of our decisions at that juncture: an anonymous healthcare provider wrote a warning to the cavalier Americans, published in *Newsweek* (Anonymous 2020); an interview with an Italian ICU physician, titled "Coronavirus in Italy—Report from the Front Lines" (JAMA 2020); and finally, the Italian Society for Anesthesia Analgesia Resuscitation and Intensive Care (SIAARTI) released its newly drafted triage guidelines (SIAARTI 2020).

The Italian Society guidelines were alarming because they began to ration ICU resources based on age, which many American bioethicists rejected due to overt age discrimination. Indeed, COVID-19 presented unique and new challenges for U.S. hospitals, and many tasked their clinical ethicists with drafting specific COVID-19 rationing protocols that had new considerations with respect to ICU triage and rationing for this particular pandemic. In the context of the U.S., healthcare stakeholders were facing a total collapse of an already inequitable and dysfunctional healthcare system (see Chap. 4). At that time, the healthcare ethics calculation of shut down favored expanding hospital capacity, triaging based on sound ethical frameworks, and saving lives. Shut down was a mitigation of last resort.

The White House Coronavirus Task Force, established in January 2020, began daily briefings in mid-March, with Dr. Anthony Fauci and Dr. Deborah Birx—infectious disease experts from the early days of AIDS (see Chap. 2). On March 31, 2020, the Task Force, in a sobering briefing, projected that between 100,000 and 240,000 Americans would likely die even with strict mitigation efforts of shutdown to flatten the curve; without mitigation, the number would be over 1 million Americans dead (Noack et al. 2020; University of Washington 2020; Ferguson 2020).

By April, the CDC reversed itself on mask-wearing, but made it a soft "suggestion" rather than a requirement. By this point, New York City was mirroring the Italy surge and was near the breaking point. Several New York healthcare providers were journaling about their experiences, moral distress, and essentially combat fatigue. In solidarity, "Flatten the Curve" became the mantra, but many experts, including bioethicists, began to openly debate the risks and harms of economic shutdown, job loss, and other consequences versus the casualties of opening up "blind" without adequate testing and tracing protocols in place. Even in April, testing was still inadequate and there was no sound testing/tracing/quarantine strategy. Meanwhile, mask-wearing and mitigation strategies became a culture war, helping to distinguish the United States as a failed state with respect to setting public health standards to combat the growing pandemic. It was becoming clear to the rest of the world that the United States was amidst a manmade disaster of coronavirus denialism that exacerbated its ability to deal with the pandemic. On April 23, 2020, during a White House Coronavirus Task Force briefing, President Trump suggested the American public might consider drinking or injecting bleach to combat COVID-19 (Rogers et al. 2020). At that juncture, the Task Force's daily briefings ceased.

By May 2020, the United States. began to get a better picture of the demographics associated with its COVID-19 deaths. What was emerging was a familiar and alarming health disparities story in which COVID-19 was ravaging vulnerable populations such as nursing home residents, African Americans, and immigrant worker populations—particularly those who worked in meat packing plants. The United States now had over 1 million cases of coronavirus and by May 8, 2020, 76,000 deaths. Of those who died, almost 60% were African American; another one-third of the deaths were in nursing home residents. But as the health disparities news emerged, many Republican-majority conservative states made the decision to reopen early. By June 2020, President Trump was defying public health guidelines and holding rallies with masses of people in indoor settings without masks, which met the criteria of a biohazard event.

As we near August, more Americans are dead from COVID-19 than who died in World War 1 (over 126,000 as of this writing[1]). Those deaths occurred in a short time frame . We surpassed 9/11 casualties in March; we next surpassed combined casualties from 9/11, and the post-9/11 wars of Afghanistan and Iraq in March. We surpassed Vietnam War casualties in April, and reached 100,000 casualties by Memorial Day. The United States is currently the most infected country with the worst response to the COVID-19 pandemic due to another virus: disinformation and science denialism; sound public health policy is being rejected and refuted by both American leadership and a significant portion of the U.S. population. With the U.S. CDC in ruins (The Lancet 2020), functioning as an attenuated version of its former self, there is no public health agency or expert currently in charge of COVID-19 at a federal level, while individual states are making decisions that are heterogencus. Some states never actually shut down; some opened prematurely without following CDC guidelines. Basic public health principles we learned a century ago have become controversial, such as mask-wearing. Many Americans do not agree on common facts of the virus reported by health experts, medical journals, and the media. As a result, the United States is now considered such a malign pandemic actor, the European Union (EU) is banning Americans from entering EU countries to help control resurgence of outbreaks (Stevis-Gridneff 2020), while Canada has banned Americans from entry as well.

How to Teach this Moment

Whether you're teaching remotely online, or teaching this in a traditional classroom setting, Healthcare Ethics on Film is the primer needed to teach the COVID-19 experience in the United States. In essence, this is the healthcare ethics story of the century, and this book provides you the historical context with which to teach COVID-19 today as part of a healthcare ethics curriculum.

   The COVID-19 pandemic has echoes of the very same health disparities and systemic racism discussed in the Tuskegee study (Chap. 1), a critical primer on what we're seeing in 2020.

   The delayed response to COVID-19 by the Trump Administration, and the failures of its public health officials to manage the early U.S. outbreaks during February 2020 has many echoes of the mismanaged HIV/AIDS crisis (Chap. 2), which, in the words of Randy Shilts' from And the Band Played On, "was allowed to happen". In the same way that policy officials refused to protect the public blood supply because of costs, fully enabling a predictable transfusion-AIDS epidemic, policy officials in 2020 refused to call for a federally mandated national shelter-in-place order, which would have quashed the virus in the absence of a test-trace-quarantine strategy. Instead, a patchwork of diverse mitigation strategies was implemented by individual states with mixed results. Similarly, in the same way the community rejected restrictions on individual liberties in the early days of AIDS (Chap. 2), such as closing the bathhouses which led to superspreading of the virus, or practicing safe sex, many Americans in 2020 balked at the closing of bars and restaurants and wearing masks. There are also echoes of the African American experiences with HIV/AIDS, which disproportionately infected that population by the 1990s due to systemic racism and the legacy of Tuskegee (see Chap. 2).

   Ultimately, in a "deja-vu all over again" scenario, officials in charge of our public health ignored warnings about re-opening the economy too early just as officials ignored warnings in the early 1980s about screening the public blood supply. It was just too inconvenient and expensive to do the right thing. In fact, the same conclusions drawn about HIV/AIDS by the Institute of Medicine in 1995 (see Chap. 2), could be applied to COVID-19 in the United States: "The early history …reflects a failure of individual healthcare bureaucrats to provide moral leadership, which had devastating consequences for public health."

   Another primer for teaching COVID-19 is a discussion about our uniquely unethical and discriminatory American healthcare system, which is the focus of Chap. 4 on *Sicko*. In a system that practices "healthism" and "wealthcare", a pandemic scenario such as COVID-19 could destroy the U.S. healthcare system and uniquely target vulnerable populations with less access to healthcare. Ironically, the Trump Administration even joined several states to try to overturn the *Affordable Care Act* (see Chap. 4) in spring 2020.

   As the COVID-19 pandemic unfolded in the United States, several states also exploited governors' executive orders to cancel elective procedures to expand hospital access, and prioritized blocking all access to abortion services (Habercorn 2020). The Supreme Court heard arguments to block abortion access in Louisiana during this time frame. For these reasons, a primer on reproductive justice and *Roe v. Wade* is important, which I discuss in Chap. 5.

   Indeed, the most relevant film in this collection with respect to teaching COVID-19 is the film *Contagion*, discussed in Chap. 6, a pandemic ethics film that was informed by the SARS epidemic and the preparations for H1N1. *Contagion* is a realistic depiction of a novel virus that originates from bats in China, gets to the United

States via air travel, and completely upends the United States social fabric and healthcare system. As I taught this film in the spring of 2020 to my own students while they were sheltering-in-place, it seemed more like the evening news than a dystopian film about the "not if, but when" nightmare pandemic scenario. Additionally, as rationing scarce resources became a major focus in COVID-19, which included delaying transplant procedures, the rationing frameworks discussed in Chap. 7 on transplant ethics are important primers, too.

COVID-19 also tells the story of professionalism and humanism in healthcare. As healthcare providers became the heroes on the front lines, they also began to suffer from severe moral distress and moral injury, sharing their experiences on social media. Additionally, systemic racism and microaggressions experienced by non-white healthcare providers and trainees on the front lines became a focus, too, during this pandemic. For these reasons, Chaps. 8 and 9 are instructive.

We are in the early days of COVID-19, and I write this on June 30, 2020[1]. The news is not good, and most healthcare ethics experts believe we've lost control of the virus as several states that did not heed public health warnings are now experiencing surges that are on par with New York City and Italy. The United States rivals Brazil in mismanaging this pandemic. We are staring at an unknown future that worked out much better in viral "war games" played during the Obama era than in our current reality. Most Americans will have to go it alone now, and have to calculate their own risks of societal interactions. Many will need to self-isolate in the absence of clear guidelines, or a clear testing and tracing strategy. Vaccine candidates and the ethics of vaccine prioritization and rationing will become the next chapter of this healthcare ethics crisis. But one day, there will be a film about COVID-19 that will tell the story of how the United States lost its way. Perhaps, by teaching this moment with the aid of this book for context, you will be teaching the very student who will help make that film.

M. Sara Rosenthal, University of Kentucky, June 30, 2020.

References

JAMA. (2020). Coronavirus in Italy—Report from the front lines. *JAMA interview.* https://www.youtube.com/watch?v=TKS1pahoPRU.

Anonymous Doctor in Western Europe. (2020). Young and unafraid of the coronavirus pandemic? Good for you. Now stop killing people. *Newsweek*, March 11, 2020. https://www.newsweek.com/young-unafraid-coronavirus-pandemic-good-you-now-stop-killing-people-opinion-1491797.

Ferguson, N., et al. and the Imperial College Covid-19 Response Team. (2020). Report 9: Impact of non-pharmaceutical interventions (NPIs) to reduce COVID-19 mortality and healthcare demand, March 16, 2020. https://www.imperial.ac.uk/media/imperial-college/medicine/sph/ide/gida-fellowships/Imperial-College-COVID19-NPI-modelling-16-03-2020.pdf.

---

[1] When I proofed these pages in late July 2020, 150,000 Americans had died from from COVID-19.

Habercorn, J. (2020). Amid coronavirus outbreak some red states move to restrict abortion, spurring legal fight. *Los Angeles Times*, April 2, 2020. https://www.latimes.com/politics/story/2020-04-02/red-states-block-abortion-coronavirus-outbreak.

Italian Society for Anesthesia Analgesia Resuscitation and Intensive Care (SIAARTI). http://www.siaarti.it/SiteAssets/News/COVID19%20-%20documenti%20SIAARTI/SIAARTI%20-%20Covid-19%20-%20Clinical%20Ethics%20Reccomendations.pdf.

Noack, R., et al. (2020). White House task force projects 100,000 to 240,000 deaths in U.S., even with mitigation efforts. *Washington Post*, April 1, 2020. https://www.washingtonpost.com/world/2020/03/31/coronavirus-latest-news/?fbclid=IwAR23vUfNKjcFMUF2xdLqRapteWdrCOyqm_LO1SrAgPWkrDJ-9Sdsko4hk_E.

Reviving the US CDC. (Editorial). *The Lancet*, *395*, 1521. https://doi.org/10.1016/S0140-6736(20)31140-5.

Rogers, K., et al. (2020). Trump's suggestion that disinfectants could be used to treat coronavirus prompts aggressive pushback. *New York Times*, April 24, 2020. https://www.nytimes.com/2020/04/24/us/politics/trump-inject-disinfectant-bleach-coronavirus.html.

Stevis-Gridneff, M. E. U. (2020). Plans to bar most U.S. travelers when bloc reopens. *New York Times*, June 26, 2020. https://www.nytimes.com/2020/06/26/world/europe/europe-us-travel-ban.html.

University of Washington's Institute for Health Metrics and Evaluation (IHME) Covid-19 Estimation Updates. 2020. http://www.healthdata.org/covid/updates.

World Health Organization. (2020, March 11). *Announcement of global pandemic*. https://www.who.int/dg/speeches/detail/who-director-general-s-opening-remarks-at-the-media-briefing-on-covid-19---11-march-2020.

Index

© Springer Nature Switzerland AG 2020
M. S. Rosenthal, *Healthcare Ethics on Film*,
https://doi.org/10.1007/978-3-030-48818-5

The manufacturer's authorised representative in the EU is Springer
Nature Customer Service Centre GmbH, Europaplatz 3, 69115 Heidelberg,
Germany. If you have any concerns regarding our products, please
contact ProductSafety@springernature.com

Printed and bound by CPI Group (UK) Ltd, Croydon, CR0 4YY

29/04/2026

02099460-0011